Advances in Cement and Concrete Research
ICACCR 2025

4th International Conference on Advances in Cement and Concrete Research (ICACCR 2025), 4-7 of November 2025, Johannesburg, South Africa

Editors

Bolanle D. Ikotun[1], Walied A. Elsaigh[1], and Akeem Ayinde Raheem[2]

[1]University of South Africa, South Africa
[2]Ladoke Akintola University of Technology, Ogbomoso, Nigeria

Peer review statement

All papers published in this volume of "Materials Research Proceedings" have been peer reviewed. The process of peer review was initiated and overseen by the above proceedings editors. All reviews were conducted by expert referees in accordance to Materials Research Forum LLC high standards.

Published under License by **Materials Research Forum LLC**
Millersville, PA 17551, USA

Published as part of the proceedings series
Materials Research Proceedings
Volume 63 (2026)

ISSN 2474-3941 (Print)
ISSN 2474-395X (Online)

ISBN 978-1-64490-404-6 (Print)
ISBN 978-1-64490-405-3 (eBook)

Distributed worldwide by

Materials Research Forum LLC
105 Springdale Lane
Millersville, PA 17551
USA
https://mrforum.com

Manufactured in the United States of America
10 9 8 7 6 5 4 3 2 1

Table of Contents

Preface

ICACCR 2025

The fourth edition of the International Conference on Advances in Cement and Concrete Research (ICACCR 2025) held between November 4-7, 2025, at the University of South Africa, Florida Campus, continued as a series of events marked towards the development of innovative and sustainable construction materials for nation building. The conference themed **"Innovative Construction Materials and Technologies for Sustainable Development"** and accommodated knowledge in the field of supplementary cementitious materials, 4IR technology in concrete, nanotechnology in cement composites, concrete for resilient and durable buildings, water quality in concrete production and concrete materials for sustainable development.

The conference provided a platform for researchers, contractors, academia and industries to deliberate on the challenges and potential way forward as well as new frontier in the production, processing, analysis and testing of cement and concrete materials for sustainable development. In addition, the event captured participants around the world who presented different topics trending in the field of Concrete Technology thus given students an enabling atmosphere to present their research findings. The articles presented in this proceeding had undergone a blind double peer-review process by the international technical scientific committee.

We would like to express our sincere gratitude to all the organising committees and participants for their contributions towards the success of the conference.

Committees

Host

Prof Ednah Onyari-Benecha	Civil and Environmental Engineering and Building Science, UNISA, South Africa

Organising Committee

Conference Chair:	Prof Bolanle Deborah Ikotun	UNISA, South Africa
Conference Co-Chair:	Prof Akeem Ayinde Raheem	Ladoke Akintola University of Technology, Nigeria
Conference secretary:	Dr Rasheed Abdulwahab	UNISA, South Africa
	Dr Walied A. Elsaigh	UNISA, South Africa
	Dr Thomas Chabalala	UNISA, South Africa
	Ms Tebogo Ramorei	UNISA, South Africa
	Mr Veli Nkosi	UNISA, South Africa
	Ms Rikhotso Mixo	UNISA, South Africa
	Ms Zakithi Dube	UNISA, South Africa
	Mr Zimbili Abel Omphetse	UNISA, South Africa
	Dr Opeoluwa Dada	UNISA, South Africa

Dr Madirisha Makungu Marco	UNISA, South Africa
Prof. Johnson Rotimi, Oluremi	LAUTECH, Nigeria
Prof Solomon Olalere Ajamu	LAUTECH, Nigeria
Ms Cleverness Makamu	UNISA, South Africa
Mr Bastiaan Verhoek	UNISA, South Africa
Mr Abiodun Ebenezer Akinwale	UNISA, South Africa

Technical/International Scientific Committee

Dr Walied A. Elsaigh (Chair)	University of South Africa, South Africa
Dr Md Jaynul Abden	Western Sydney University, Australia
Dr Solomon Idowu Adedokun	University of Lagos, Nigeria
Prof Kazeem Kayode Adewole	University of Rwanda, Rwanda
Prof John Babafemi	Stellenbosch University, South Africa
Dr Shankha Pratim Bhattacharya	Indian Institute of Technology Kharagpur, India
Dr Ester Giner Cordero	Archaeology services at the Ministry of Culture. Canton Ticino, Switzerland
Dr Belachew A. Demiss	University of South Africa, South Africa
Dr Chiara D'Erme	University of Applied Sciences and Arts of Southern Switzerland, Switzerland
Dr Thabo Falayi	Namibia University of Science and Technology, Namibia
Prof George Fanourakis	University of Johannesburg, South Africa
Dr Abideen Ganiyu	British University of Bahrain, Bahrain
Prof Francois M. Ilunga	University of South Africa, South Africa

Dr Bahiru Bewket Mitikie	University of South Africa, South Africa
Dr Bukunmi Ogunsanya	Kingston University London, UK
Dr Kolawole Olonade	University of Lagos, Nigeria
Prof. Johnson Rotimi, Oluremi	Ladoke Akintola University of Technology, Nigeria
Dr Blessing Oluwafunmi Orogbade	Olabisi Onabanjo University, Nigeria
Dr Dada Opeoluwa	University of South Africa, South Africa
Dr Damilola Oyejobi	University of Botswana, Botswana
Prof Suvash Chandra Paul	International University of Business Agriculture and Technology (IUBAT), Bangladesh
Dr Shimol Philip	National Institute of Technology Puducherry (NITPY), India
Dr Aaditya P. Sanyal	Indian Institute of Technology (BHU) Varanasi, India
Prof Thandiwe Sithole	University of Johannesburg, South Africa
Prof Simone Stürwald	Eastern Switzerland University of Applied Sciences, Switzerland
Dr Paulo Vaz-Serra	The University of Melbourne, Australia
Dr Moruf O Yusuf	University of Hafr, Eastern Province, Saudi Arabia

Conference Organiser

Ms Thereza Botha	Technoscene, South Africa

Advances in Cement and Concrete Research - ICACCR 2025 | Materials Research Forum LLC
Materials Research Proceedings 63 (2026) 1-11 | https://doi.org/10.21741/9781644904053-1

Strength Performance Evaluation of Wood Ash Concrete Infused with Nanosilica

Abiodun Ebenezer AKINWALE[1,a*], Walied A. Elsaigh[1,b], Akeem Ayinde RAHEEM[2,c]

[1]Department of Civil & Environmental Engineering and Building Science, University of South Africa, Florida Campus, Johannesburg, South Africa

[2]Department of Civil Engineering, Ladoke Akintola University of Technology, Ogbomoso, Oyo State, Nigeria

[a]akinaae@unisa.ac.za, [b]hussiwam@unisa.ac.za, [c]aaraheem@lautech.edu.ng

Keywords: Concrete, Compressive Strength, Wood Ash, Nanosilica

Abstract. Concrete is among the most extensively employed construction materials globally, with an annual production estimated at 27.3 billion tonnes. Approximately 80% of all CO_2 emissions from concrete technology are attributable to the production of Portland cement, the essential binder in concrete, which accounts for 5 - 7% of global CO_2 emissions. Consequently, reducing cement utilization is vital for decreasing carbon dioxide levels. Nonetheless, the strength properties of cement are essential for achieving the desired strength of structures. Therefore, the development of cost-effective additives and alternatives to cement to produce environmentally friendly binding supplements for concrete becomes imperative. In pursuit of this, a 0.5 water/binder (w/b) ratio mix was formulated to evaluate the impact of nanosilica on the compressive strength of wood ash concrete and ascertain the optimal percentage for a partial replacement. The mix design was modified by incorporating wood ash into the cement fraction in proportions of 0, 5, 10, 15, 20, and 25% with nanosilica additions of 0, 0.5, 1.1, and 1.7% based on mixing water percentage. The wood ash was produced by thoroughly burning firewood from wattle trees (Acacia pycnantha). Concrete samples were cast in cube moulds with a w/b ratio of 0.5, and compressive strength testing was conducted on the samples at 3, 7, 14, 28, 56, 90, and 120 days, respectively. Results indicated that, in comparison to the control sample, the wood ash containing nanosilica decreased proportionally, regardless of the percentage of Nanosilica (NS) inclusion. However, the compressive strength of the wood ash (WA5%) group samples, including those without NS, gained 3% of the target strength of 50 MPa within just 14 days. At 28 days, this strength exceeded the target by 6%. This strength increase continued to rise by 6–13% from 28 to 120 days. Samples containing 10% WA with 1.1 and 1.7% NS surpassed the target strength of 50 MPa by 0.6% and 1.3%, respectively, at 28 days. This enhancement was also observed in other samples from 56 to 120 days.

1. Introduction

Concrete is one of the most widely utilized building materials worldwide, with an estimated yearly production of 27.3 billion tonnes [1]. About 80% of all concrete CO_2 emissions are caused by Portland cement production, the necessary binder for concrete, contributing between 5 - 7% of all CO_2 emissions worldwide [2–7]. By 2050, the demand for Portland cement is expected to reach 6,000 million tonnes annually, an increase of almost 200% compared to previous years [8]. These greenhouse gas emissions have caused global climate change, and many nations have responded to this challenge differently [9]. Consequently, reducing cement use is crucial to lowering carbon dioxide levels. However, the importance of cement in the production of composites cannot be overemphasized, as it plays a fundamental role in the development of strength. Therefore, it has become essential to develop cost-effective additives and cement alternatives to produce environmentally friendly binding supplements for concrete [10]. Previous studies have focused on resource efficiency by replacing conventional cement with industrial wastes as supplementary cementitious materials in the development of composites [11,12].

In a study by Farinha et al.[13], a volumetric ratio of 1:4 (binder: aggregates) was selected, and 5, 10, and 15% of wood ash were substituted for cement by volume. The mechanical behaviour tests focusing on compressive strength were conducted on prismatic specimens (40x40x160 mm^3), measuring the strength at 7, 28, 90, 180, 365, and 730 days. The compressive strength of mortar containing 5% WA increased. However, this increase was only noticeable between 28 and 180 days. The reference mortar's compressive strength was higher than that of the other mortars. Furthermore, the compressive strength decreased more at 730 days, ranging from 29% to 45%. Compared to reference mortar, this strength was reduced by adding WA. Similarly, workability and bulk density were unaffected by this substitution, according to the findings of [14,15]. In contrast to the reference samples, the compressive performance was diminished by as much as 45%. Nonetheless, up to 15% of cement can be replaced with wood ash without compromising mortar performance [13,14,15]. In a review by [16], the suggested wood ash percentage should not exceed 15% for optimum concrete performance.

The properties of concrete in both its fresh and hardened stages have led to a recent increase in the use of nanomaterials, such as nanosilica (NS), in manufacturing concrete [17]. Prior studies have demonstrated that the strength and durability of high-strength concrete (HSC) are improved when NS is used as a partial replacement for 0–6% cement. Furthermore, there are very few calcium hydroxide crystals in the dense, compact microstructure of the concrete mixed with NS [18–22]. Additionally, other investigations looked into the possible application of NS in different kinds of concrete. Self-compacting concrete with 2% NS, 0.08 percent nano-carbon, and 40% fly ash (FA) outperformed NS-free admixture, according to Aydin et al. [23]. The fresh and hardened characteristics of concrete at different water-to-binder ratios were improved by substituting NS for 3% of the cement weight, according to Naniz and Mazloom [24].

The quick rise in NS concrete's compressive strength demonstrated that NS serves as both an activating agent for the hydration reaction and a filler to boost the density of the concrete [25]. Adding merely 0.1% NS to ordinary concrete improved its strength by 84%, 93%, and 35% at 3, 7, and 28 days, respectively, for water-cement ratios (w/c) ranging from 0.42 to 0.45, according to experimental results [26]. While the gain was only 6.5 and 0%, respectively, at 0.55 and 0.5 w/b ratios, the concrete with 0.65 w/b ratios significantly increased by 41% in 28 days of compressive strength at 1.5% NS [27]. According to Makarova et al. [26], the concrete's strength was significantly increased when the NS content was raised from 0.05% to 0.5%; however, the strength was unaffected when the NS level was increased to 1%. The pozzolanic reaction at a low dose of 0.3% NS in concrete boosted the early-age compressive strength by decreasing porosity and refined pore structure, according to Du et al. [28].

According to an assessment of the impact of nano-silica in concrete by [29], NS dosages of up to 2–3% could enhance the material's durability and mechanical qualities. Its pore structure refinement, filling effect, and pozzolanic action may be responsible for this. The NS acts as an activator to encourage hydration, and compressive strength rises with NS content. However, the strength can be diminished if the NS dosage exceeds 3%. Because of the agglomeration of NS particles, higher dosages (i.e., 3% or more) may eventually impact the engineering qualities, leading to microcracking, excessive porosity, and decreased compressive strength.

For this reason, the present study focuses on the strength performance evaluation of wood ash-concrete infused with nanosilica. The research aims to investigate the effect of different percentage weight fractions of wood ash mixed with NS on the mechanical property (compressive strength) of wood ash concrete, with the objectives of determining the effect of NS dispersion on the compressive strength of the nanocomposite concrete at the early and long-term strength development.

2. Materials and Methods

2.1 Materials and Properties

Ordinary Portland Cement CEM I 52.5N conforming to SANS 50197-1:2013 [30] was used in this investigation. The mineral fillers used include coarse and fine aggregates with compacting bulk density of 1557 and 1846 kg/m^3, respectively. The coarse aggregates of maximum particle size of 22 mm and fine (crusher sand) of maximum size of 4.75 mm obtained from a local quarry in Johannesburg were used. The particle size analysis of aggregates was conducted in conformity with SANS 3001-AGI:2014 [31]. The wood ash used was sourced from caterers' services situated in Johannesburg. The physical and chemical composition of OPC and the wood ash are provided in Table 1 and 2 respectively. The nanosilica solution (NSS) was used, an aqueous liquid containing colloidal nanosilica in the dispersed state with a 0.3 moles/L concentration. Silicon and oxygen are the only atomic components of the NS used, maintaining a 1:2 atomic ratio consistent with the theoretic composition of SiO_2 (46.74 wt% Si, and 53.26 wt% O).Tap water was used in all mixtures.

The water-to-binder ratio was kept constant at 0.5 for all the mixtures. The fine and coarse aggregates were graded in accordance with SANS 3001-AG1:2014 [31]. The mix ratio of the concrete samples was designed according to C & CI specifications. The cement consumption in the test group concrete (Zawa 5–25) was calculated according to the reduction rate in the concrete preparation process under the same water-binder ratio condition.

Table 1: *Chemical and Physical Composition of the PC*

Property	Test Result (%)	SANS 50197-1 Requirements
1. Oxide composition:		
Alumina, Al_2O_3	4.99	
Silica, SiO_2	19.98	
Ferric Oxide, Fe_2O_3	2.77	
Lime, CaO	64.91	
Sulphuric Anhydride, SO_3	2.10	≤4.0
Magnesia, MgO	1.54	
Sodium, Na_2O	0.08	
Insoluble Residue	0.90	≤5.0
Chloride	0.04	≤0.10
Loss in ignition	2.72	≤5.0
2. Compound composition		
C_3A	6.80	
C_2S	28.21	
C_3S	52.70	
Physical properties of PC		
Property	Test Result (%)	SANS 50197-1 Requirements
Fineness % (Residue on sieve size 0.15 mm)	3	Max. 5
Specific Area (Blaine) m^2/kg	400	No requirement.
Initial setting time (min.)	170	45 minimum
Final setting time (min.)	250	No requirement.
Specific gravity	±3.14	-
Compressive strength (MPa)		
At 2 days	±28	≥20.0
At 28 days	±58	≥52.5
Bulk density, aerated (kg/m^3)	1100 – 1300	
Soundness :		
Le Chatelier Expansion (mm)	1	Max. 10
The manufacturer (PPC) provides Table 2 results.		

***Table 2:** Physical and chemical properties of wood ash*

Properties	**Value**		**Chemical Composition (%)**
		SiO_2	53.57
		Al_2O_3	33.98
		Fe_2O_3	6.12
Conventional parameters		MgO	6.23
Organic material (mg/kg)	<10	CaO	3.24
pH	12.1	MnO	1.66
		SO_3	1.58
Physical properties		Cl	0.07
Density (kg/m^3)	834	Na_2O	0.00
Specific gravity	1.92	K_2O	0.20
Mean size (μm)	0.4	TiO_2	0.20
		P_2O_5	0.30
		SrO_2	0.08
		ZrO_2	0.96
		ZnO	2.66
		(LOI)	**14.2**

The cement and wood ash exhibit both crystalline and amorphous phases owing to the pattern of their spectra as shown in Fig. 1. It could be observed that some points have very sharp humps and others with broader ones. However, edges with sharp humps are more prominent in the case of the cement as compared to the wood ash. Hence, the crystallinity in the cement is more than that present in the wood ash. The lesser crystallinity in the case of wood ash as compared to that of cement is due to the high visibility of broader humps. The peak intensity was observed to occur at 2θ less than 30° in the case of wood ash. This agrees with what was obtained elsewhere [32, 33], and for the cement, the peak intensity was observed to occur in the range of 26-35°. The result agrees with that obtained elsewhere. [34].

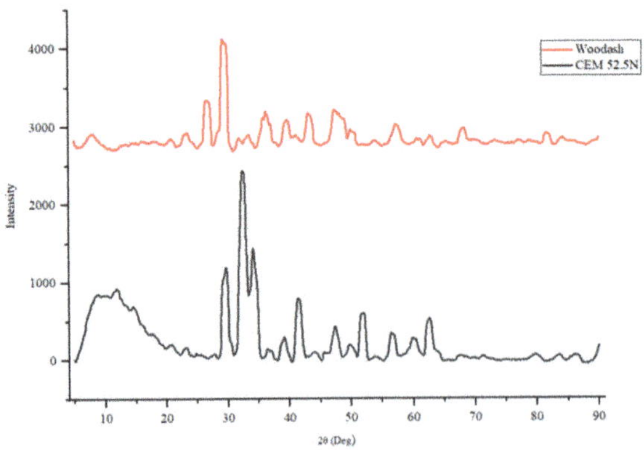

***Fig. 1**. XRD of WA and Cement*

2.2 Research Method

Blended concrete containing cement, wood ash, and NS was prepared. Mixing procedures were carried out in a rotary mixer, according to SANS 5861-3:2006 [35]. Nanosilica (NS) was mixed with water at medium speed. The mixture was allowed to rest for 90 seconds and then mixed continuously for 1 minute at high speed. Details of the mix proportions of the cement concrete are given in Table 3. The mix proportion for the control sample with PC (without nanoparticles) is denoted by CC. At the same time, Zawa 5-25 represents South African wood ash consisting of cement concrete with varying percentages of wood ash between 5 and 25% by weight of cement, respectively. The percentage of nanoparticles varied between 0 and 1.7%, and other parameters, such as fine and coarse aggregates, remain constant as shown in Table 3. The concrete samples were cured in accordance with [35], as shown in Fig. 2.

Table 3: Mix Proportions of M50 Grade Wood Ash Concrete

Mixes	Cement (kg/m^3)	Water (kg/m^3)	Coarse Agg. (kg/m^3)	Fine Agg. (kg/m^3)	NS (kg/m^3)	Wood ash (kg/m^3)	W/B Ratio
CC	385.256	200.333	1112.666	656.3	0	0	0.5
WA05NS0	365.993	200.333	1112.666	656.3	0	14.846	0.5
WA10NS0	346.730	200.333	1112.666	656.3	0	29.692	0.5
WA15NS0	327.468	200.333	1112.666	656.3	0	44.538	0.5
WA20NS0	308.205	200.333	1112.666	656.3	0	59.383	0.5
WA25NS0	288.942	200.333	1112.666	656.3	0	74.229	0.5
WA05NS0.6	365.993	199.131	1112.666	656.3	1.202	14.846	0.5
WA10NS0.6	346.730	199.131	1112.666	656.3	1.202	29.692	0.5
WA15NS0.6	327.468	199.131	1112.666	656.3	1.202	44.538	0.5
WA20NS0.6	308.205	199.131	1112.666	656.3	1.202	59.383	0.5
WA25NS0.6	288.942	199.131	1112.666	656.3	1.202	74.229	0.5
WA05NS1.1	365.993	198.129	1112.666	656.3	2.204	14.846	0.5
WA10NS1.1	346.730	198.129	1112.666	656.3	2.204	29.692	0.5
WA15NS1.1	327.468	198.129	1112.666	656.3	2.204	44.538	0.5
WA20NS1.1	308.205	198.129	1112.666	656.3	2.204	59.383	0.5
WA25NS1.1	288.942	198.129	1112.666	656.3	2.204	74.229	0.5
WA05NS1.7	365.993	196.927	1112.666	656.3	3.406	14.846	0.5
WA10NS1.7	346.730	196.927	1112.666	656.3	3.406	29.692	0.5
WA15NS1.7	327.468	196.927	1112.666	656.3	3.406	44.538	0.5
WA20NS1.7	308.205	196.927	1112.666	656.3	3.406	59.383	0.5
WA25NS1.7	288.942	196.927	1112.666	656.3	3.406	74.229	0.5

CC= Control Concrete with Portland Cement only, ZaWa05 = PC with 5% WA, Zawa10 = PC with 10% WA, Zawa15 = PC with 15% WA, Zawa20 = PC with 20% WA, Zawa25 = PC with 25% WA

Fig. 2: Wood ash +NS concrete samples in a curing tank

For each concrete mixture, 100 x 100 x 100 mm specimens were prepared and tested according to SANS 3001-AG1:2014 [31]. The test specimens were demoulded after 24 hours and then cured in water. A total of 105 concrete cube specimens were cast from each group, including 21 cubes for the control mix, making a total of 441 cubes, covering seven different ages of 3, 7, 14, 28, 56, 90, and 120 days. The compressive properties were tested at 3, 7, 14, 28, 56, 90, and 120 days, respectively, using a Toni Technic hydraulic test frame manufactured by Baustoffprüfsysteme GmbH, Gustav-Meyer-Allee 25, D-13355 Berlin, with a loading capacity of 3000 KN in accordance with [36] as shown in Figure 3. The concrete samples were prepared and tested in accordance with SANS 5836:2006 [39] Standards for test methods of concrete mechanical properties.

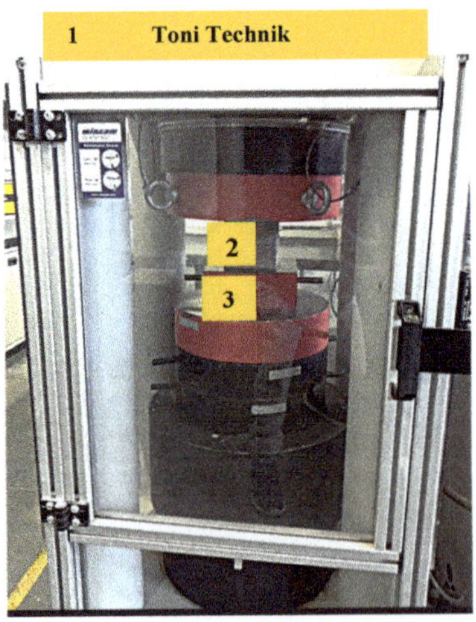

Fig.3: Compressive strength testing: 1-testing machine, 2-spcimen, 3-support.

3. Results and discussion

3.1 Compressive strength

The compressive strength results of the 5% WA replacement group containing various proportions of nanosilica compared to the control sample as shown in Fig. 4a. revealed changes in the compressive strength of the WA concrete samples due to the changing percentage of nanosilica (NS) infused into the mix. The reference mix (CC) achieved compressive strengths of 63.8 and 66.9 MPa at 28 and 120 days, respectively. In addition, including WA as a partial substitute at 5% of PC increases the compressive strength to 42.2 MPa, representing an increase of 9.2% of the control sample at 3 days. Similarly, addition of NS of 0.6 and 1.1% increases the compressive strength to 36.2 and 40.6 MPa, representing an increase of 3.2 and 7.6 % at 3 days, respectively, of the control sample. An increase in the percentage of NS in WA05NS0.6, WA05NS1.1, and WA05NS1.7 mixtures enhanced the compressive strength at 56, 57, and 57 MPa representing a strength gain between 1 and 2% respectively, at 28 days compared to wood ash samples without nanosilica. Similarly, the same mixtures' compressive enhancement stands at 3, 1, and 4% at 120 days.

The results of 10% WA replacement as shown in Fig. 4b displayed an impressive outcome as the compressive strength increases over the testing periods for all the tested samples. Interestingly, samples WA10NS0 and WA10NS1.1 exceeded the control sample by 5% and 3.7% at 3 days, respectively. The results further revealed the effect of NS addition as it improves the compressive strength of tested samples containing NS between 0.6 and 1.7% compared to samples without NS between 14 and 120 days, respectively.

The compressive strength results of the 15% WA replacement group containing various proportions of NS compared to the control sample as shown in Fig. 4c showed that the sample WA15NS1.7 increased in compressive strength by 4.4, 3.5, 2.6, 1.6, and 3.7% compared to sample WA15NS0 (WA without nanosilica) for 7, 14, 28, 56, and 90 days respectively; though this consistent performance shows a linear decrease over the control sample by 12, 16.6, 16.2, 16.7 and 14.8% over the same testing ages.

The performance of 20% WA replacement group containing various proportions of NS compared to the control sample as shown in Fig. 4d indicates addition of NS between 1.1 and 1.7% increases the compressive strength to 40.2 and 40 MPa, compared to WA without NS at 28 days, representing an increase of 3.2 and 3%, respectively. However, samples WA20NS0.6 and WA20NS1.1 increase in compressive strength by 0.9 and 1.9%, respectively, compared to sample WA20NS0 at 120 days.

The increase of WA to 25% play a significant role in strength reduction irrespectively of the amount of Nanosilica inclusion in the 25% wood ash group, as none of the mixtures performance meets the target strength of 50 MPa.

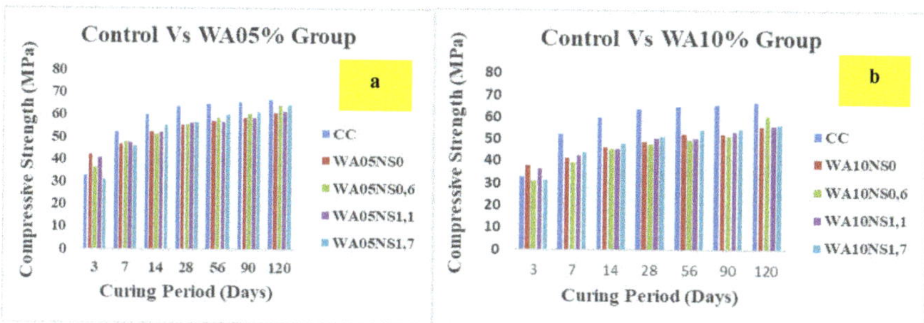

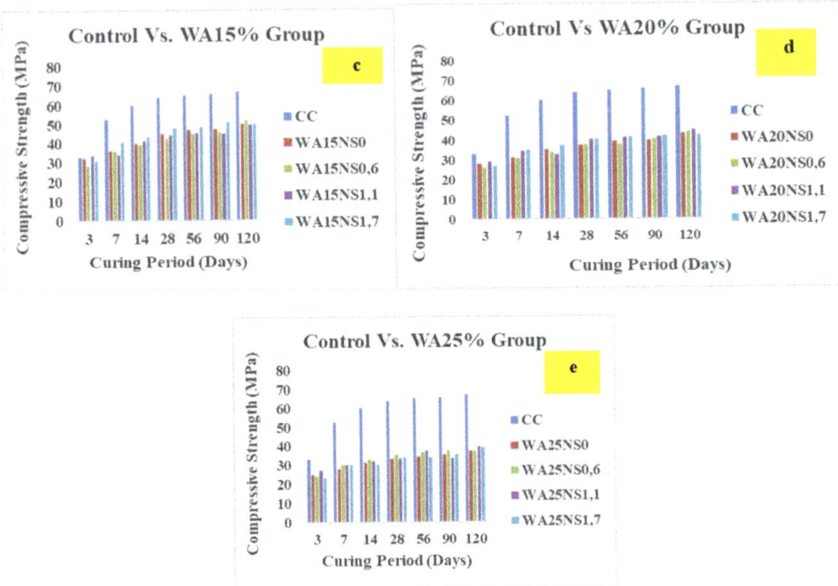

Fig. 4(a-e) Compressive strength of the wood ash containing NS vs. Control Sample

4. Conclusions

According to this study, which investigated the compressive strength of wood ash concrete infused with nanosilica, the following conclusions can be drawn. Compared with the target strength of 50 MPa within the WA5% group, the wood ash containing 1.7 % nanosilica produced the optimum compressive strength of 56.8 and 64.5 MPa, respectively for 28 and 120 days respectively. Similarly, within the WA10% group, wood ash containing 1.7% nanosilica produced the optimum results of 51.3 and 56.8 MPa for 28 and 120 days respectively. The results of these optimum samples are ideal for construction of high-rise buildings and bridges, which require high performance.

1. The replacement of WA at 5% without nanosilica increases the compressive strength to 42.2 MPa, representing an increase of 9.2% of the control sample at 3 days. However, the inclusion of NS of 0.6 and 1.1% increases the compressive strength to 36.2 and 40.6 MPa, representing an increase of 3.2 and 7.6% of the control sample at 3 days, respectively.

2. An increase in the percentage of NS in WA05NS0.6, WA05NS1.1, and WA05NS1.7 samples increased the compressive strength to 55.8, 56.8, and 56.8 MPa, representing a strength gain of 0.6, 1.3, and 1.6% at 28 days, respectively.

3. Samples WA10NS0 and WA10NS1.1 exceeded the control sample by 5, and 3.7% at 3 days, respectively. The results further revealed the effect of NS addition as the compressive strength of tested samples containing NS increases between 0.2 and 2.5% compared to samples without NS between 7 and 120 days, respectively.

4. The results further showed that sample WA15NS1.7 increased in compressive strength by 4.4, 3.5, 2.6, 1.6, and 3.7% compared to sample WA15NS0 (WA without nanosilica) for 7, 14, 28, 56, and 90 days respectively; though this consistent performance shows a linear decrease over the control sample by 12, 16.6, 16.2, 16,7, and 14.8% over the same testing ages.

5. An additional increase of NS between 1.1 and 1.7% increases the compressive strength to 40.2 and 40 MPa, respectively, compared to the WA sample without NS at 28 days, representing an increase of 3.2 and 3%, respectively. However, samples WA20NS0.6 and WA20NS1.1 increased compressive strength by 0.9 and 1.9%, respectively, compared to sample WA20NS0 at 120 days.

5. Recommendation

Future work should focus on how waste wood ash is processed to make it more reactive and effective in concrete strength development. Moreso, fine tuning nano-silica dosages and mixing methods to get better particle distribution is desirable. Long-term durability investigation like water absorption and chloride resistance are essential to understand how these mixes will actually perform over time. Finally, using techniques like scanning electron microscope (SEM), and X-ray diffraction (XRD) to study the concrete's internal structure would help connect the laboratory investigation and the real-world performance and application.

Acknowledgement

The authors would like to acknowledge the University of South Africa's contribution towards the outcome of this research work, authorship, and the conference attendance.

Reference

[1] Abhilash PP, Nayak DK, Sangoju B, Kumar R. (2021) Effect of nano-silica in concrete; a review. Construction and Building Materials. 2021;278: 122347. https://doi.org/10.1016/j.conbuuildmat.2021.122347

[2] Shi C, Jimenez AF, Palomo A.(2011) New cements for the 21st century: The pursuit of an alternative to Portland cement. Cem. Concr. Res. 2011;41(7): 750-763. https://doi.org/10.1016/j.cemconres.2011.03.016

[3] Pacheco-Torgal F, Miraldo S, Ding Y, Labrincha JA. (2013) Nanoparticles for high performance concrete (HPC), Nanotechnol. Eco-Efficient Constr. Mater. Process. Appl. 2013; 38-52. https://doi.org/10.1533/9780857098832.1.38

[4] Khan MI.(2018) Nanosilica/silica fume, Elsevier Ltd. 2018. https://doi.org/10.1016/b978-0-08-102156-9.00014-6

[5] Habert G, Denarie E, Sajna A, Rossi P. (2013) Lowering the global warming impact of bridge rehabilitations by using Ultra High Performance Fibre Reinforced Concretes. Cem. Concr. Compos. 2013; 38: 1-11. https://doi.10.1016/b978-0-08-102156-9.00014-6

[6] Xuan D, Zhan B, Poon CS.(2016) Development of a new generation of eco-friendly concrete blocks by accelerated mineral carbonation. J. Cleaner Prod. 2016;133:1235- 1241. https://doi.org/10.1016/j.jclepro.2016.06.062

[7] Ali MB, Saidur R, Hossain MS. (2011) A review on emission analysis in cement industries. Renew. Sustain. Energy Rev. 2011;15(5):2252-2261. https://doi.org/10.1016/j.rser.2011.02.014.

[8] Pacheco-Torgal F, Jalali S, Labrincha JA, John VM. (2013) Eco-efficient concrete. 2013. https://doi.org/10.1533/9780857098993

[9] Young-Ho K, Yoonsuk P, SaRang B, Soo Young K, Jung-Geun H. (2020) Compressive strength evaluation of Ordinary Portland Cement mortar blended with hydrogen Nano- Bubble water and graphene. Journal of nanoscience and nanotechnology. 2020;20(1): https://doi.org/10.1166/jnn.2020.17287

[10] Hamid Z, Rafiq S. A comparative study on strength of concrete using wood ash as partial replacement of cement. Materials Science and Engineering 2020;955: https://doi.org/10.1088/1757-899X/955/1/102043

[11] Juenger MCG, Snellings R, Bernal SA. (2019) Supplementary cementitious materials: new sources, characterization, and performance insights. Cem. Concr Res. 2019; 122:257-73

[12] Muslim F, Wong HS, Choo TH, Buenfeld NR. (2021) Influences of supplementary cementitious materials on microstructure and transport properties of spacer-concrete interface. Cem. Concr Res. 2021;149:106561

[13] Farinha CB, de Brito J, Veiga R. (2019)Influence of forest biomass bottom ashes on the fresh, water and mechanical behaviour of cement-based mortars. Resour Converv Recycl. 2019;149:750-9.

[14] Siddique R. (2012) Utilization of wood ash in concrete manufacturing. Resour Conserv Recycl. 2012;67:27-33.

[15] Thomas BS, Yang J, Mo KH, Abdalla JA, Hawileh RA, Ariyachandra E. (2021) Biomass ashes from agricultural wastes as supplementary cementitious materials or aggregate replacement in cement/geopolymer concrete: a comprehensive review. J Build Eng. 2021;40:102332.

[16] Jhonatan A, Duitama B, Avellaneda DR. (2022) Pozzolans: A review. Engineering and Applied Science Research 2022;49(4):495-504.

[17] Singh LP, Karade SR, Bhattacharyya SK, Yousuf MM, Ahalawat S. (2013) Beneficial role of nanosilica in cement-based materials – A review. Construct Build Mater 2013;47:1069-77.

[18] Said AM, Zeidan MS, Bassuoni MT, Tian Y. (2012) Properties of concrete incorporating nano-silica. Construct Build Mater 2012;36:838-44.

[19] Ji T. (2005) Preliminary study on the water permeability and microstructure of concrete incorporating nano-SiO2. Cement Concr Res 2005;35(10):1943-7.

[20] Chitra S, Kumar SRRS, Chinnaraju K. (2016) The effect of Colloidal Nano-silica on workability, mechanical and durability properties of High Performance Concrete with Copper slag as partial fine aggregate. Construct Build Mater 2016;113:794-804.

[21] Zareei SA, Ameri F, Bahrami N, Shoaei P, Moosaei HR, Salemi N. (2019) Performance of sustainable high strength concrete with basic oxygen steel-making (BOS) slag and nano-silica. J Build Eng 2019;25:100791.

[22] Hakeem IY, Amin M, Abdelsalam BA, Tayeh BA, Althoey F, Agwa IS. (2022) Effects of nano-silica and micro-steel fiber on the engineering properties of ultra-high performance concrete. Struct Eng Mech 2022;82(3):295-312.

[23] Aydin AC, Nasl VJ, Kotan T. (2018) The synergic influence of nano-silica and carbon nanotube on self-compacting concrete. J Build Eng 2018;20:467-312.

[24] Naniz OA, Mazloom M. (2018) Effects of colloidal nano-silica on fresh and hardened properties of self-compacting lightweight concrete. J Build Eng 2018;20:400-10.

[25] Saloma A, Nasution I, Imran M. (2015) Improvement of concrete durability by nanomaterials. Procedia Eng. 2015;125:608-612. https://doi.org/10.1016/j.proeng.2015.11.078

[26] Makarova NV, Potapov VV, Kozin AV, Chusovitin EA, Amosov AV, Nepomnyashiy AV. (2017) Influence of hydrothermal nanosilica on mechanical properties of plain concrete. Key Eng. Mater. 2017;744:126-120. https://doi.org/10.4028/www.scientific.net/KEM.744.126

[27] Torabian FI, Redaelli E, Lollini F, Li W, Bertolini L. (2016) Effects of nanosilica on compressive strength and durability properties of concrete with different water to binder ratios. Adv. Mater. Sci. Eng. 2016;1-16. https://doi.org/10.1155/2016/8453567

[28] Du H, Du S, Liu X. (2014) Durability performances of concrete with nano-silica. Constr. Build. Mater. 2014;73:705-712. https://doi.org/10.1016/j.conbuildmat.2014.10.014

[29] Abhilash PP, Dheeresh KN, Bhaskar S, Rajesh K. (2021) Effect of nano-silica in concrete: a review. Constr Build Mater. 2021;278:0950-0618. https://doi.org/10.1016/j.conbuildmat.2021.122347

[30] SANS 50197-1:2013. Cement, Part 1, Composition, specifications and conformity criteria for common cements.

[31] SANS 3001-AG1:2014, Civil engineering test methods. Part AG1, Particle size analysis of aggregates by sieving.

[32] Chowdhury S, Maniar A, Suganya O. (2015) Strength development in concrete with wood ash blended cement and use of soft computing models to predict strength parameters. Journal of Advanced Research.2015; 6: 907-913. 2015.

[33] Serafimova E, Mladenov M, Mihailova I, Pelovski Y. (2011) Study on the characteristics of waste wood ash. Journal of the University of Chemical Technology and Metallurgy. 46: 31-34. 2011.

[34] Kothari C, Garg N. (2024) Quantitative phase analysis of anhydrous Portland cement via combined X-ray diffraction and Raman imaging: Synergy and impact of analysis parameters. Cement and Concrete Research. 2024;186: 107662

[35] SANS 5861-3:2006. Concrete tests – Making and curing of test specimens.

[36] SANS 5836:2006 Concrete tests – compressive strength of hardened concrete.

Effect of Activator-to-Binder Ratio on Workability and Strength of Sustainable One-part Geopolymer Mortar

Anuoluwapo Sola KOLADE[1,a,*], Bolanle Debora IKOTUN[1,b] and Damilola Oyewumi OYEJOBI[2,c]

[1]Department of Civil and Environmental Engineering and Building Science, University of South Africa, Science Campus, Florida, Johannesburg, South Africa

[2]Department of Civil Engineering, University of Botswana, Gaborone, Botswana

[a]22838953@mylife.ac.za, [b]ikotubd@unisa.ac.za, [c]oyejobido@ub.ac.bw

Keywords: One-Part Geopolymer, Waste-Based Activator, Activator-to-Binder Ratio, Cement-Free Binder, Sustainable Mortar

Abstract. The widespread reliance on liquid activators in geopolymer systems presents handling, storage and safety challenges that limit industrial scalability. This study addresses these limitations by developing and evaluating one-part geopolymer mortars activated with solid sodium silicate synthesized from waste glass powder and sodium hydroxide. Six mixes with activator-to-binder (AA/B) ratios ranging from 0.50 to 0.75 were prepared using high-volume fly ash and ground granulated blast furnace slag as aluminosilicate precursors. Specimens were cast in 40 × 40 × 40 mm moulds, thermally cured at 80°C for 24 hours and subsequently stored under ambient conditions for up to 28 days. Results showed that increasing the AA/B ratio reduced flow from 120% to 70% and decreased compressive and flexural strengths from 16.24 MPa and 5.02 MPa to 8.77 MPa and 3.55 MPa, respectively, due to excessive alkalinity disrupting gel cohesion and matrix densification. Regression models developed from the experimental data effectively predicted the relationship between the AA/B ratio and both fresh and hardened properties (R^2 > 0.90), identifying an optimum range of 0.51 – 0.55 that achieved balanced flowability (105 – 115%) and mechanical strength (14 – 16 MPa compressive and 4 – 5 MPa flexural). Although the models were not experimentally validated, they provide a robust basis for optimizing mix design in future studies. The findings, therefore, confirm the reactivity and viability of waste-derived sodium silicate as a sustainable solid activator, offering mechanistic and predictive insights that advance the design of low-carbon, practical one-part geopolymer mortars for precast and low-to-medium performance repair applications.

Introduction

Geopolymers are sustainable, cement-free binders synthesized through the alkali activation of aluminosilicate precursors. Based on their activation approach, they are classified as two-part or one-part systems. Two-part geopolymers rely on liquid alkaline activators such as sodium silicate and sodium hydroxide (NaOH), which are mixed separately with solid precursors like fly ash (FA) or metakaolin [1,2]. Although effective, these systems pose handling and safety challenges due to the corrosive nature of the liquid activators and the need for accurate on-site mixing, limiting large-scale application [3].

One-part geopolymers address these challenges by incorporating the activator in solid form. In this approach, the precursor and solid activator are pre-blended and only water is added during mixing. This dry formulation improves storage stability, reduces corrosion risks, enhances worker safety and simplifies field handling, which make one-part systems more practical and scalable for real-world applications [3,4].

Various aluminosilicate sources, including FA, ground granulated blast furnace slag (GGBFS), metakaolin and silica fume, have been studied in one-part formulations. Among these, FA–GGBFS blends have gained attention for their synergistic reactivity, where GGBFS provides calcium ions that promote early strength, while FA improves workability and long-term polymerization [2,5]. Their widespread availability as industrial by-products in South Africa further underscores their suitability for sustainable binder production.

However, the development of one-part systems remains constrained by the dependence on commercial sodium silicate, which is energy-intensive and costly to produce. Recent research has explored waste-derived silicates, such as those synthesized from waste glass powder (WGP), as greener alternatives. Yet, limited studies have examined their reactivity, performance and optimal mix parameters in one-part systems, especially regarding the activator-to-binder (AA/B) ratio, a key factor controlling dissolution, gel formation, workability and strength [4–7].

Accordingly, this study investigates the influence of the AA/B ratio on the fresh and mechanical properties of one-part fly ash–slag geopolymer mortars activated by sodium silicate synthesized from waste glass and NaOH. Regression models were developed to predict property–parameter relationships and identify an optimum AA/B range that achieves balanced flowability and strength. The study thereby contributes to a deeper mechanistic and predictive understanding of waste-derived activators in one-part systems, supporting the design of low-carbon, locally sourced and practical geopolymer mortars for precast non-structural and low-to-medium performance repair applications.

Materials and Methods

Materials

The materials used in this study (Fig. 1) include fly ash (FA) and ground granulated blast furnace slag (GGBFS) as the primary aluminosilicate binders, with their oxide composition presented in Table 1. The density and D_{10}, D_{50} and D_{90} particle size distributions of FA are 2.36 g/cm^3, 0.24 μm, 4.13 μm, 41.51 μm respectively, while that of GGBFS are 2.92 g/cm^3, 0.22 μm, 38.46 μm, 59.40 μm, respectively. FA and GGBFS were chosen because they are abundant industrial by-products in South Africa with consistent chemical composition and high reactivity, unlike many natural or agricultural pozzolans that often exhibit variable properties.

The alkaline activator (sodium silicate) was produced from WGP and NaOH pellets. The WGP had a maximum particle size below 150 μm and amorphous content of approximately 95%, while the NaOH had purity exceeding 97%. This approach was adopted to promote waste valorization and reduce dependence on commercial sodium silicate that is energy-intensive and costly to produce [8]. Also, well-graded standard sand with 2.57 fineness modulus and 0.08% dust content was used, conforming to EN 196-1 [9].

Table 1: Chemical composition by weight of FA, GGBFS, WGP and Na_2SiO_3

Materials	SiO_2	Al_2O_3	Fe_2O_3	CaO	TiO_2	MgO	K_2O	P_2O_5	Na_2O	SO_3	Others	LOI
FA	53.83	33.18	4.16	3.63	1.51	0.93	0.68	0.44	0.11	0.08	1.45	0.80
GGBFS	34.69	17.37	0.72	35.08	0.78	7.71	1.01	0.01	0.26	1.13	1.24	–
WGP	70.08	2.16	1.13	10.50	0.27	1.06	0.69	0.23	12.41	0.38	1.09	0.86
Na_2SiO_3	50.08	0.97	0.07	0.46	0.05	0.13	0.06	0.02	48.11	0.01	0.04	0.06

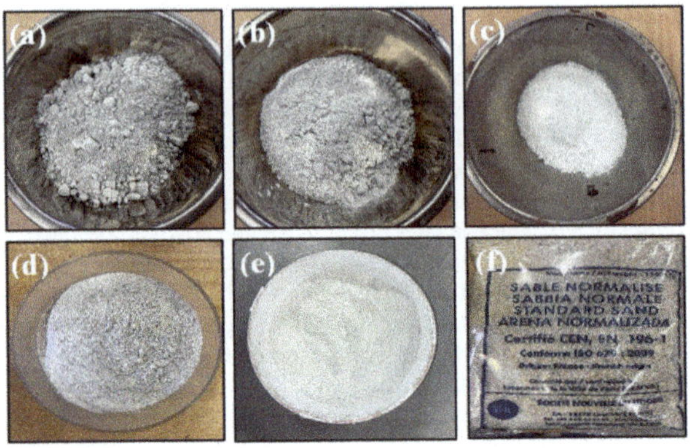

Fig. 1: Images of materials used (a) FA (b) GGBFS (c) ground NaOH (d) WGP (e) sodium silicate (f) sand

Sodium Silicate Preparation and Characterization

Sodium silicate (Na_2SiO_3) was synthesized thermochemically via alkali fusion at 500°C for 3 hours, as shown in Fig. 2. Following Eq. 1, ground NaOH pellets and WGP were dry-mixed in a 2:1 mass ratio (656 g NaOH : 345 g WGP) to yield approximately 1 kg of Na_2SiO_3, based on the 70.08 % SiO_2 content of WGP and the purity of NaOH. The fusion parameter was adopted from preliminary trials by an earlier study [8], which explored temperatures between 500 – 650°C and durations of 2 – 3 hours, and identified 500°C for 3 hours as producing the most favourable mortar strength.

$$2NaOH_{(s)} + SiO_{2\,(s)} \xrightarrow{\Delta} Na_2SiO_{3\,(s)} + H_2O_{(g)} \quad (1)$$

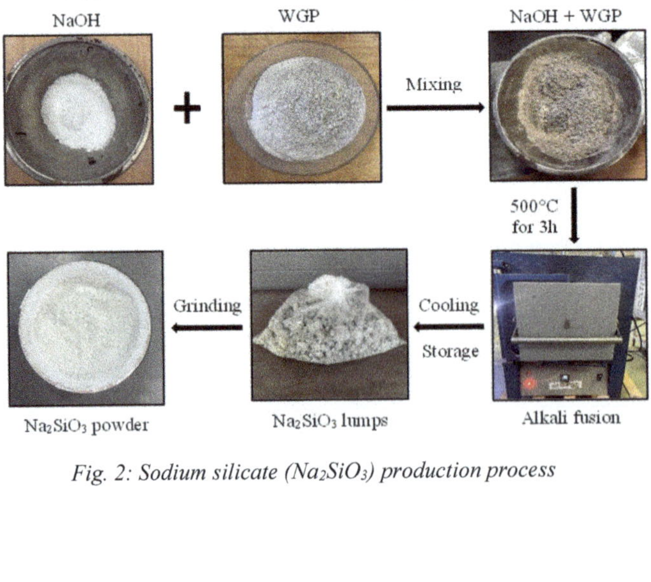

Fig. 2: Sodium silicate (Na_2SiO_3) production process

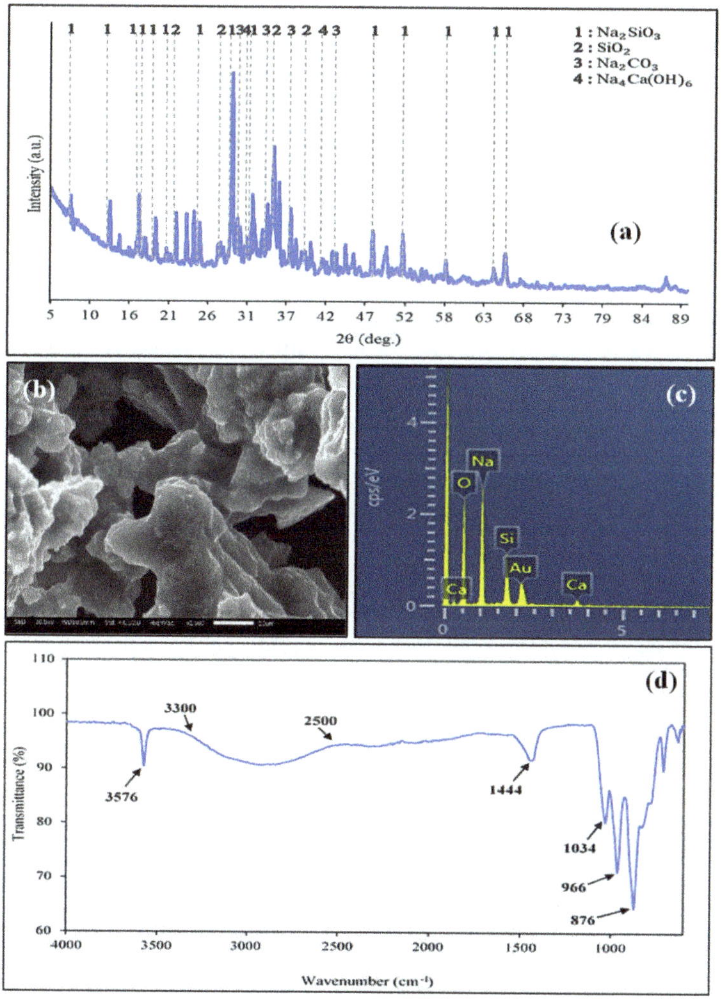

Fig. 3: Synthesized sodium silicate's (a) XRD (b) SEM (c) EDS (d) FTIR analysis results

The oxide composition of the synthesized greenish-white granular sodium silicate powder with a silica modulus (SiO_2/Na_2O) of approximately 1.0 is presented in Table 1. The results of the material's X-ray diffraction (XRD), scanning electron microscopy (SEM), energy-dispersive X-ray spectroscopy (EDS) and Fourier-transform infrared spectroscopy (FTIR) analyses are shown in Fig. 3. The XRD pattern, recorded over a 2θ range of 5 – 90°, exhibited a nearly flat baseline with sharp peaks between 18 – 35°, indicating partial crystallinity. The main diffraction peaks correspond to sodium metasilicate (Na_2SiO_3), while additional reflections between 29 – 42° are associated with sodium carbonate (Na_2CO_3), likely formed through atmospheric carbonation during cooling. Minor peaks between 31 – 42° were linked to $Na_4Ca(OH)_6$, attributed to residual

CaO in the WGP, and broad humps within 22 – 39° suggest the presence of amorphous SiO_2. The crystallinity of the sodium silicate was estimated semi-quantitatively at approximately 84.8%. SEM micrographs also revealed a rough, irregular morphology with crater-like surface features, while EDS spectra confirmed Na, Si, O and Ca as the principal elements. The FTIR spectrum further supports these findings, showing a dominant broad band between 1100 – 800 cm^{-1} corresponding to asymmetric Si–O–T (T = Na^+) stretching, indicative of partial depolymerization of the glass network and formation of silicate oligomers. Secondary peaks at 1034, 966 and 876 cm^{-1} represent Si–O–Na vibrations arising from non-bridging oxygens, while minor features at 3576 cm^{-1} and 1444 cm^{-1} are attributed to O–H stretching and slight carbonation, respectively. Collectively, these complementary analyses confirm the effective breakdown of waste glass and formation of reactive sodium silicate species.

Geopolymer Mortar Preparation and Testing

Six different AA/B ratios and equivalent FA and GGBFS quantities based on a 450 g reference cement presented in Table 2 were used to prepare the geopolymer mortars. The constituent materials were dry-mixed to ensure uniform distribution of the activator and water was then added along with thorough mixing following SANS 5861-1 [10]. Workability was measured according to ASTM C1437 [11] to determine the spread of fresh mortar. Afterwards, the mortar was cast into moulds and subjected to thermal curing at 80°C for 24 hours, followed by ambient curing at 23 ± 2°C. The compressive and flexural strengths of the geopolymer mortars (Fig. 4) were evaluated at 7 and 28 days per SANS 50196-1 [12].

Table 2: Mortar mix design per 3 specimens

Mixes	Sand [g]	Precursor ratio	Fly ash [g]	GGBFS [g]	AA/B ratio	Sodium silicate [g]	Water-to-binder ratio	Water [g]
Mix 1-0.50	1350	7:1	296	53	0.50	174.50	0.5	225
Mix 2-0.55	1350	7:1	296	53	0.55	191.95	0.5	225
Mix 3-0.60	1350	7:1	296	53	0.60	209.40	0.5	225
Mix 4-0.65	1350	7:1	296	53	0.65	226.85	0.5	225
Mix 5-0.70	1350	7:1	296	53	0.70	244.30	0.5	225
Mix 6-0.75	1350	7:1	296	53	0.75	261.75	0.5	225

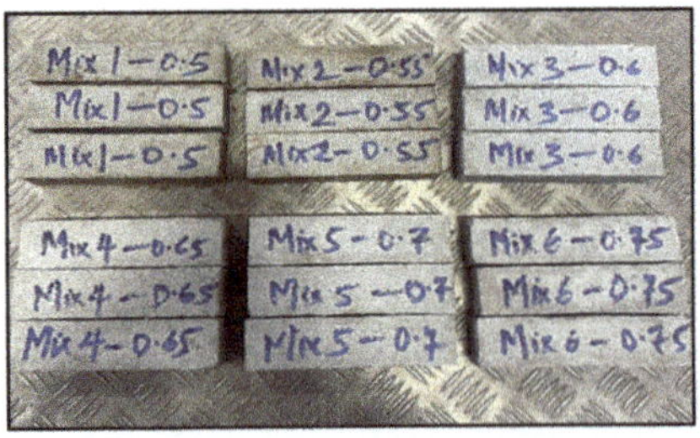

Fig. 4: Hardened mortar specimens

Results and Discussion

Effect of AA/B Ratio on Workability

The flow test results (Fig. 5) indicate a notable reduction in workability as the AA/B ratio increases from Mix 1-0.50 (120%) to Mix 6-0.75 (70%). The decreasing trend suggests that higher activator content restricts particle mobility, increases inter-particle friction and accelerates polymerization [13,14].

Effect of AA/B Ratio on Compressive and Flexural Strengths

The 28-day compressive strength (Fig. 5) showed a consistent reduction with increasing AA/B ratio, decreasing from 16.24 MPa at AA/B = 0.50 (Mix 1) to 8.77 MPa at AA/B = 0.75 (Mix 6), indicating that higher activator content adversely affected mortar strength. This reduction can be attributed to excessive alkalinity, which disrupted polymeric network formation, weakened gel bonding and promoted microstructural defects [4,15]. Similarly, the flexural strength decreased from 5.02 MPa to 3.55 MPa over the same range, reflecting reduced matrix integrity and increased brittleness [16]. The high activator content also hindered the formation of N–A–S–H and C–A–S–H gels, resulting in higher porosity and microcracking within the matrix. Consequently, both compressive and flexural strengths decreased as gel cohesion and interparticle bonding weakened [8]. Hence, increasing the AA/B ratio beyond the optimum range destabilizes the composite's microstructure and compromises its performance.

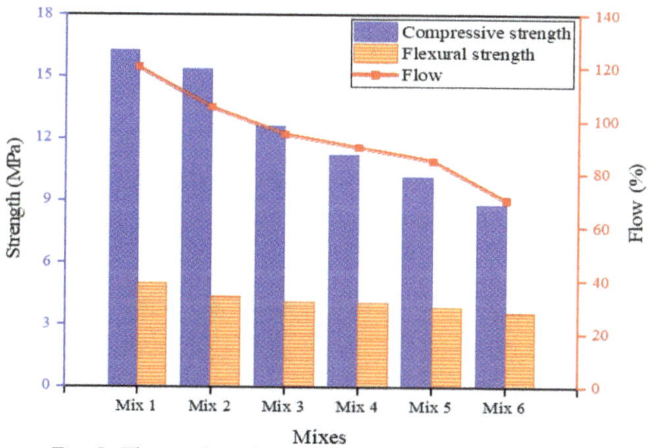

Fig. 5: Flow and mechanical strength of mortar specimens

Optimum Mix Selection

Based on the experimental results, the optimal AA/B ratio falls within 0.50 – 0.55, which achieves a practical balance between mechanical strength and workability requirements. Within this range, the mixes exhibited sufficient flow (105 – 115%) to ensure adequate compaction and ease of placement, while maintaining compressive and flexural strengths (14 – 16 MPa and 4 – 5 MPa, respectively), consistent with the flow specification in ASTM C1437 [10]. Beyond this range, excessive activator content reduced workability and strength due to microstructural instability, whereas lower ratios yielded inadequate reactivity and insufficient gel formation.

To further validate the experimental observations, linear regression models were developed to predict compressive strength (CS), flexural strength (FS) and flow (F) as a function of the AA/B ratio (x), as shown in Eqs. 2–4. These models exhibited high predictive accuracy, with R^2 values of 0.978, 0.910 and 0.964, respectively, demonstrating that 97.8% of the variation in compressive strength, 91.0% in flexural strength and 96.4% in flow can be reliably explained by the AA/B ratio. This high degree of correlation confirms that the regression equations provide an effective tool for forecasting mix performance and optimizing activator dosage in one-part geopolymer formulations. Fig. 6 illustrates the strong agreement between predicted and experimental values, while Table 3 presents the predicted properties for AA/B ratios between 0.50 and 0.55.

$$CS = 31.810 - 31.091x \tag{2}$$

$$FS = 7.371 - 5.143x \tag{3}$$

$$F = 206.667 - 180x \tag{4}$$

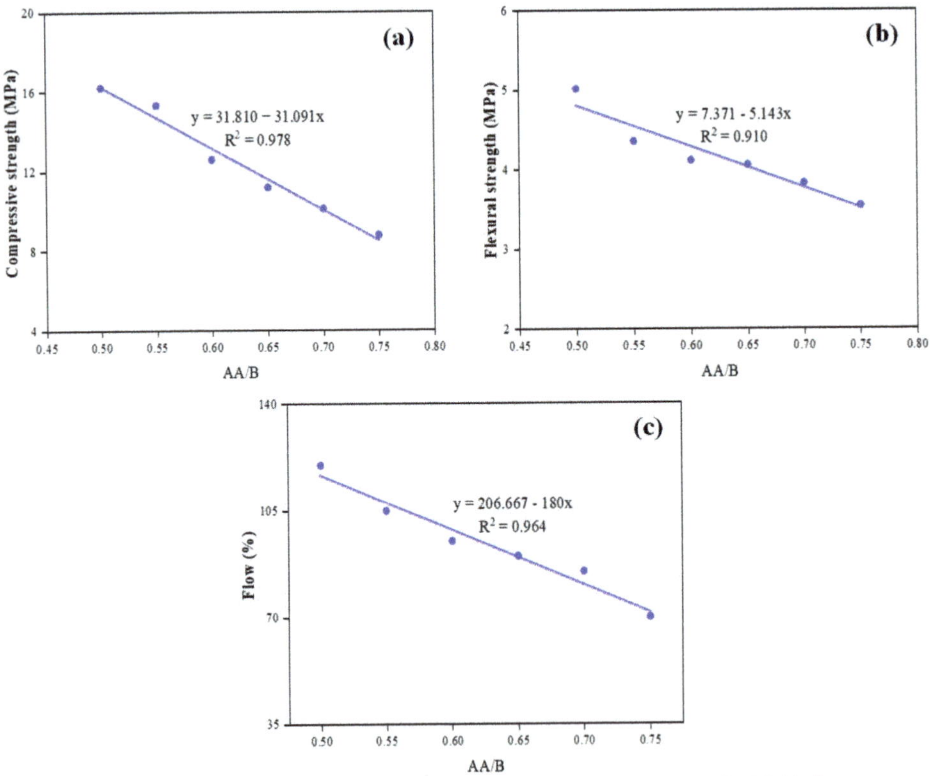

Fig. 6: Regression lines showing relationship between AA/B and (a) CS (b) FS (c) flow

Table 3: Predicted flow, compressive strength and flexural strength

AA/B	Experimental			Predicted		
	CS [MPa]	FS [MPa]	F [%]	CS [MPa]	FS [MPa]	F [%]
0.50	16.24	5.02	120	16.27	4.80	117
0.51	–	–	–	15.96	4.75	115
0.52	–	–	–	15.65	4.70	113
0.53	–	–	–	15.34	4.65	111
0.54	–	–	–	15.03	4.59	109
0.55	15.34	4.36	105	14.71	4.54	108

Conclusion

At the end of the research, the following inferences could be drawn:

i. The combined use of FA and GGBFS exhibited strong pozzolanic synergy, where the calcium-rich GGBFS accelerated early geopolymerization and strength gain, while the silica and alumina-rich FA enhanced workability and long-term reaction stability.

ii. The sodium silicate synthesized from WGP and NaOH showed high reactivity and effective dissolution potential, which confirm its capability to activate aluminosilicate precursors and serve as a sustainable, low-cost alternative to energy-intensive commercial activators.

iii. Optimal performance was achieved at an AA/B ratio of 0.51 – 0.55, which yielded balanced properties, including compressive strength of 14 – 16 MPa, flexural strength of 4 – 5 MPa and flow of 105 – 115%, thus defining a practical compositional window for one-part geopolymer mortars.

iv. The regression models developed provide a predictive framework for optimizing mix design parameters to enable strength and workability estimation without extensive experimental trials.

v. The study validates the feasibility of low-carbon, waste-valorized one-part geopolymer mortars for precast non-structural and low-to-medium strength repair applications (flow of 100 – 120% and strength of 10 – 20 MPa in accordance with ASTM C1437 [11] and EN 1504-3 [17]), supporting circular economy goals and sustainable construction.

Acknowledgment

The authors gratefully acknowledge the University of South Africa's support in providing the resources, equipment and facilities required for this research.

References

[1] N.B. Singh, B. Middendorf, Geopolymers as an alternative to Portland cement: An overview, Constr. Build. Mater. 237 (2020) 117455. https://doi.org/10.1016/j.conbuildmat.2019.117455

[2] A.S. Kolade, B.D. Ikotun, D.O. Oyejobi, A review of the chemistry, waste utilization, mix design and performance evaluation of geopolymer concrete, Iran. J. Sci. Technol. - Trans. Civ. Eng. (2025). https://doi.org/10.1007/s40996-025-01801-w

[3] K. Allali, N. Bella, I.A. Bella, O. Khodjet El Fehem, M.V. Borrachero, J. Payá, J. Monzó, A review of geopolymer cement, from two-part geopolymer to one-part geopolymer cement and its geotechnical applications, Emirates J. Eng. Res. 29(4) (2024). https://scholarworks.uaeu.ac.ae/ejer/vol29/iss4/2

[4] N. Ye, J. Yang, S. Liang, Y. Hu, J. Hu, B. Xiao, Q. Huang, Synthesis and strength optimization of one-part geopolymer based on red mud, Constr. Build. Mater. 111 (2016) 317-325. https://doi.org/10.1016/j.conbuildmat.2016.02.099

[5] A.S. Kolade, B.D. Ikotun, D.O. Oyejobi, Composition and performance driven mix design methodology for geopolymer mortars, Discov. Civ. Eng. 2(162) (2025). https://doi.org/10.1007/s44290-025-00327-4

[6] Z. Xu, J. Yue, G. Pang, R. Li, P. Zhang, S. Xu, Influence of the activator concentration and solid/liquid ratio on the strength and shrinkage characteristics of alkali-activated slag geopolymer pastes, Adv. Civ. Eng. 2021(1) (2021) 6631316. https://doi.org/10.1155/2021/6631316

[7] P. Prochon, Z. Zhao, L. Courard, T. Piotrowski, F. Michel, A. Garbacz, Influence of activators on mechanical properties of modified fly ash based geopolymer mortars, Materials 13(5) (2020) 1033. https://doi.org/10.3390/ma13051033

[8] A.S. Kolade, B.D. Ikotun, D.O. Oyejobi, G.E. Aderinto, M.M. Makungu, Performance evaluation of geopolymer mortars prepared with waste glass powder-derived sodium silicate solution, Iran. J. Sci. Technol. - Trans. Civ. Eng. (2025). https://doi.org/10.1007/s40996-025-01952-w

[9] EN 196-1, Methods of Testing Cement – Part 1: Determination of Strength, European Standards, 2016.

[10] SANS 5861-1, Concrete Tests: Mixing Fresh Concrete in the Laboratory, South African Bureau of Standards, Pretoria, 2006.

[11] ASTM C1437, Standard Test Method for Flow of Hydraulic Cement Mortar, ASTM International, West Conshohocken, PA, 2015.

[12] SANS 50196-1, Methods of Testing Cement. Part 1: Determination of strength, South African Bureau of Standards, Pretoria, 2006.

[13] H.A. Ghanim, U.J. Alengaram, N.M. Bunnori, M.S.I. Ibrahim, Innovative in-house sodium silicate derived from coal bottom ash and its impact on geopolymer mortar, J. Build. Eng. 99 (2025) 111428. https://doi.org/10.1016/j.jobe.2024.111428

[14] T-T. Pham, N-L. Nguyen, T-T. Nguyen, T-T. Nguyen, T-H. Pham, Effects of superplasticizer and water–binder ratio on mechanical properties of one-part alkali-activated geopolymer concrete, Build. 13(7) (2023) 1835. https://doi.org/10.3390/buildings13071835

[15] Y. Lv, C. Wang, W. Han, X. Li, H. Peng, Study of the mechanical properties and microstructure of alkali-activated fly ash–slag composite cementitious materials, Polym. 15 (2023) 1903. https://doi.org/10.3390/polym15081903

[16] A.M. Tahwia, D.S. Aldulaimi, M. Abdellatief, O. Youssf, Physical, mechanical and durability properties of eco-friendly engineered geopolymer composites, Infras. 9 (2024) 191. https://doi.org/10.3390/infrastructures9110191

[17] EN 1504-3, Products and Systems for the Protection and Repair of Concrete Structures: Definitions, Requirements, Quality Control and Evaluation of Conformity – Part 3: Structural and Non-structural Repair, European Standards, 2006.

Development of Quarry Dust and Fly Ash-based One-Part Geopolymer: Potential Stabilizer for Sustainable Pavement Construction

Gbenga Emmanuel ADERINTO[1,a *], Jacob Olumuyiwa IKOTUN[1], Valentine Yato KATTE[2], Bolanle Deborah IKOTUN[3], Makungu Marco MADIRISHA[3], Rasheed ABDULWAHAB[3]

[1]Department of Civil Engineering, Durban University of Technology, Pietermaritzburg, P.O. Box 3201, South Africa

[2]Department of Civil and Mining Engineering, University of Namibia, Ongwediva P.O. Box 3634, Namibia

[3]Department of Civil and Environmental Engineering and Building Science, University of South Africa, Johannesburg P.O. Box 1710, South Africa

[a]22176502@dut4life.ac.za

Keywords: Geopolymer, Sustainable Pavement, Quarry Dust, Fly Ash, Alkalination, Circular Economy, Alternative Binders

Abstract. The rising demand for sustainable and cost-effective pavement materials has spurred research into alternative binders to replace Ordinary Portland cement (OPC) and lime, both of which are associated with high carbon emissions and resource depletion. This study developed a one-part geopolymer composite using fly ash (FA) and quarry dust (QD) as aluminosilicate precursors, activated with sodium silicate (SS) synthesized from waste glass and sodium hydroxide. X-ray fluorescence (XRF) analysis confirmed the suitability of the raw materials, with high SiO_2 contents (73.92% in QD and 54.49% in FA) and Al_2O_3 contents (13.13% in QD and 33.74% in FA), while the synthesized sodium silicate, characterized by X-ray diffraction (XRD) shows distinct Na_2SiO_3 peaks of approximately 61% amorphous content, which are vital for geopolymerization. The geopolymer mixes were prepared with an average FA:QD: SS mass ratio of approximately 6:3:1 and an activator-to-precursor (AL/P) ratio between 0.35 and 0.38. The developed mixes demonstrated good workability, with flowability ranging from 110% to 126%, facilitating easy application. Compressive strength tests show an increase with curing age, reaching 4.2–6.7 MPa at 7 days, 19.9–24.4 MPa at 14 days, and 25.5–30.7 MPa at 28 days, all exceeding the minimum strength requirements for stabilized pavement base layers. Similarly, flexural strength improved from 3.2 MPa at 7 days to 5.6 MPa at 28 days, indicating improved load-bearing capacity and long-term performance under traffic conditions. The synthesized sodium silicate exhibited high reactivity that significantly enhanced geopolymerization. These findings demonstrate that the FA–QD geopolymer binder activated with synthesized sodium silicate exhibited excellent mechanical strength and good workability, confirming its suitability as a sustainable pavement stabilizer and advancing circular economy goals through the valorization of multiple industrial wastes.

Introduction

Pavements are one of the most important infrastructural systems for economic development. However, their design is often fraught with concerns, such as the inadequate bearing capacity of the supporting layers, such as subgrade, subbase, and base course, to support the intended wheel loads imposed on them. Pavement construction relies heavily on natural materials, with 90% of aggregates coming from virgin sources or quarries [1-2]. Most developing nations face a shortage of high-quality materials, leading to two costly options: finding new quarries or using conventional stabilizers like Ordinary Portland cement (OPC) and lime, which are often associated with severe environmental consequences. Therefore, alternative materials and stabilization techniques in pavement construction are highly sought after.

However, the mining and processing industry produces significant quantities of waste every year [3]. Disposing such a large volume of waste is not only expensive but also causes a variety of environmental and ecological problems, including land occupation, windblown dust, contamination of surface and subsurface water, and tailing dam collapse [4-5]. Therefore, using industrial waste as a construction material for roads could be a feasible solution to the problems mentioned above. To use these wastes as pavement material, either alone or in combination with natural materials, will require some stabilization [6]. One of the studies that addressed this problem might be the one by Qian et al. [7], who investigated quarry dust that had been stabilized with 5% cement as a road subbase layer. The stabilized specimens reached an Unconfined Compressive Strength (UCS) of 4.37 and 7.17 MPa after 7 and 28 days of wet curing, respectively, which satisfied the strength requirement of the stabilized road material. Most of the existing research looks at using OPC or lime as stabilizers. Thus, the demand for more sustainable waste stabilization procedures with minimal environmental impact is a current challenge.

Geopolymerization is a novel, eco-friendly method for producing construction materials by turning industrial and natural waste such as fly ash, mine tailings, and slags into strong, durable polymers using alkaline hardener solutions such as sodium hydroxide (NaOH) and sodium silicate (Na_2SiO_3) [4]. Unlike traditional methods that use a lot of energy and release harmful emissions, this process helps reduce pollution, waste, and the use of limited natural resources, making it a more sustainable option for the future.

Quarry dust is a byproduct of the mining industry obtained during the quarrying process. Quarry dust exhibits high shear strength, which is highly beneficial for its use as a geotechnical material [8]. It has good permeability, and its variations in water content do not affect its engineering properties [8]. To eliminate the harmful effects of waste materials, they must be disposed of properly and safely; otherwise, they can be blended with other construction materials that can be used for various construction purposes, such as subgrade, foundations, and embankments

Fly ash is the by-product of burning coal, which is fine particles separated in flue gases from the burning of coal [9]. Fly ash consists of spherical vitreous particles with the basic components being silica, alumina, and lime. Approximately 80% of the total ash generated during coal combustion is released as fly ash, while the rest forms bottom ash [9]. Using industrial by-products instead of natural materials is becoming a practical solution, especially as we aim to reduce environmental impact. While these by-products may not always match the quality of natural resources, they offer good performance at a lower cost. This makes them a compelling alternative in construction and infrastructure projects. [6-7], [10]. Therefore, this study developed a one-part geopolymer binder composite using quarry dust (QD) and fly ash (FA) with synthesized sodium silicate for pavement construction. It is noteworthy to state that this research is in line with the circular economy agenda.

Materials and Methods

Materials. The materials and chemical reagents used, and their sources, are listed in Table 1 below.

Table 1: *Details of materials used in this study*

Raw Material/Chemical	Source/ Location
Coal Ash (FA)	Lethabo Power Station, Sasolburg, South Africa.
Quarry dust (QD)	Jukskei Quarry Company Limited, Midrand, South Africa.
Waste Glass powder (WGP)	African Pegmatite (Pty) Ltd, Meyerton, South Africa.
Sodium Hydroxide (NaOH)	MINEMA Chemicals, Laser Park, Johannesburg, South Africa
Water (H_2O)	Civil Engineering Laboratory

Research Methods.

Characterization of Materials. The materials used for this research, as illustrated in Fig. 1, are fly ash, quarry dust, and waste glass powder (WGP), designated as (a), (b), and (c), respectively. The study materials are assessed chemically using XRF and SEM-EDX; physically using specific gravity and particle size analysis and mechanically using flexural and compressive strength tests to determine their usage as a binder in pavement construction material. The particle size of the quarry dust used in the study was particles passing 0.3 mm (fine sizes). It is important to note that geopolymer materials currently lack standardized mix design procedures and specifications. Therefore, this study adopted conventional paste and mortar preparation methods, with the final mix proportions determined through laboratory trial mixes.

Fig. 1: *Representative sample of (**a**) Quarry dust (**b**) Fly ash (**c**) Waste glass powder*

Sodium Metasilicate (SS) Preparation. The sodium metasilicate (Na_2SiO_3) used in this study was produced from waste glass powder (WGP) by adopting the procedure of Mori, Vinai, and Soutsos reported [11], with some modifications as shown in Fig. 2. The thermochemical method was used to prepare SS by dry mixing NaOH pellets with WGP in a ratio of 2:1 and heated in a furnace for 3 hours at a temperature of 500 °C. It is worth mentioning that the stoichiometric calculation, as shown in Equation 1, considered the percentage of SiO_2 in WGP obtained from the XRF analysis. The obtained SS was crushed into powder and characterized by X-ray diffraction (XRD).

$$2NaOH + SiO_2 \xrightarrow{heat} Na_2SiO_3 + H_2O \quad (1)$$

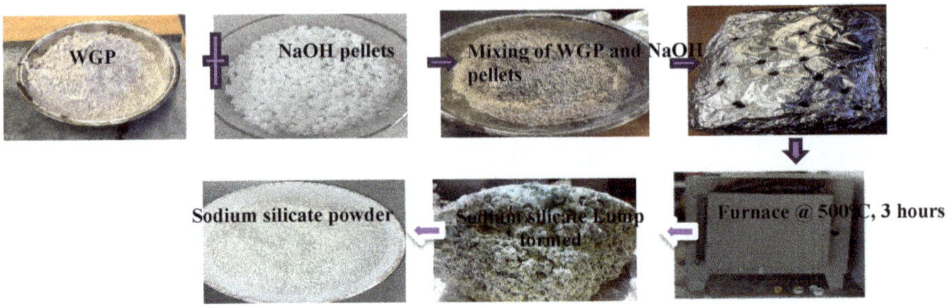

Fig. 2: Preparation process of sodium silicate

Mix Design. Table 2 below shows the mix ratio of the geopolymer specimen.

Table 2: *Mix design compositions of geopolymer binder paste*

Mixes	FA(g)	QD(g)	SS(g)	H2O(g)	AL/B	FA:QD:SS:H_2O
Run-1	1162	684	202,5	460	0,35	6: 3: 1: 2
Run-2	1133	702	205	470	0,37	6: 3: 1: 2
Run-3	1104	684	202,5	460	0,37	5.5: 3.5: 1: 2
Run-4	1133	666	205	470	0,38	5.5: 3: 1: 2

Testing Methods. The sample preparation and tests conducted are in accordance with South African National Standards (SANS). The geopolymer mixes were prepared according to the SANS 5861-1:2006 standard [12] for mixing and casting. The mixed composite was subjected to compressive strength and flexural strength tests.

Flowability test. The workability of the mortar specimens was evaluated using the flow table test according to ASTM C1437 (2015) [13]. The test was performed twice for each mix, and the flow value reported is the average of four measurements calculated using Eq. 2.

$$\text{Flow (\%)} = \frac{\text{Spread diameter (mm) - Internal base diameter of mould (mm)}}{\text{Internal base diameter of mould (mm)}} \times 100\% \tag{2}$$

Compressive strength test. Compressive strength tests were also performed on the geopolymer specimens at 7, 14, and 28 days of ambient curing, using a 300 kN capacity compression testing machine with a constant load rate of 2.4 kN/s. For each mix, four (6) specimens sized 40 x 40 x 80 mm were tested per curing day, and their compressive strengths were recorded after fracture to compute the mean strength. The test was conducted on the hardened geopolymer specimen in accordance with SANS 50196-1:2015 [14].

Flexural strength test. The flexural test was performed on the hardened geopolymer composite to determine its flexural strength in accordance with SANS 50196-1:2015[14]. Flexural strength tests were conducted on the hardened geopolymer samples at 7, 14, and 28 days of curing, using a bending testing machine with a capacity of 10 kN and a load rate of 0.050 kN/s. Three (3) specimens measuring 40 x 40 x 160 mm were assessed for each curing day at ambient conditions of ±23°C, and their flexural strengths were recorded after failure to calculate the average flexural strength for the mix.

Results and Discussion

Chemical Properties. Table 3 below shows the XRF results of the study materials. The SEM-EDX is shown in Fig. 4 below.

Table 3: *Chemical composition of precursor material*

Oxides (%)	SiO_2	Al_2O_3	Fe_2O_3	CaO	TiO_2	MgO	K_2O	P2O5	Na_2O	SO_3	BaO	MnO
FA	54,49	33,74	3,08	3,59	1,69	1,01	0,67	0,39	0,13	0,18	0,1	-
QD	73,92	13,13	1,45	1.13	0,21	0,45	4,54	0.07	3,37	0	0,07	-

Geotechnical Properties. Table 4 below shows the geotechnical properties of the sample materials.

Table 4: *Geotechnical properties of precursor material*

Properties	Test code/specification	Observed value	
Grain Size distribution	SANS 3001- AG [16]	Quarry dust (QD)	Fly ash(FA)
• Gravel (%)		7,67	0
• Sand (%)		76,4	13,03
• Fines (%)		15,87	86,86
• Fineness Modulus		3,7	0,40
• Coefficient of Uniformity		2	0
Bulk Density (kg/m^3)	SANS 5845:2006 [16]	1668	1062
Specific gravity SG (g)		2,605	2,1
Colour		Grey	Pale grey

XRD analysis of the prepared sodium metasilicate. The XRD analysis was used to determine the phase morphology of the synthesized SS. The XRD analysis of the synthesized SS was conducted over a diffraction angle (2θ) range of 4° to 90°, as shown in the Figure. 3. The diffractogram exhibits a near-flat baseline with distinct medium-to-high intensity peaks at 2θ values of 16.56°, 24.75°, 29.09°, 34.59°, 36.99°, 37.72°, 47.91°, 51.83° and 65.49°, corresponding to Na_2SiO_3. These results are similar to the findings by Vinai and Soutsos [11], Wang et al. [17], and Li et al. [18]. In addition, the crystallinity and amorphousness of the Glass Powder Sodium Silicate (GPSS) were estimated semi-quantitatively to be 39.13% and 60.87% via the method proposed by Tiainen et al. [19]. These findings confirm the successful alkali fusion reaction between WGP and NaOH, establishing Na_2SiO_3 as the primary mineral phase contributing to the strength and stability of the geopolymer mortar matrix. However, the substantial amorphous phase observed at a fusion temperature of 500°C for over 3 hours suggests an incomplete SiO_2 reaction due to unreacted WGP. If WGP had fully reacted at a higher fusion temperature or prolonged duration, the crystallinity would have increased, potentially reducing GPSS reactivity and affecting its effectiveness in geopolymerization [11].

Materials Research Forum LLC
https://doi.org/10.21741/9781644904053-3

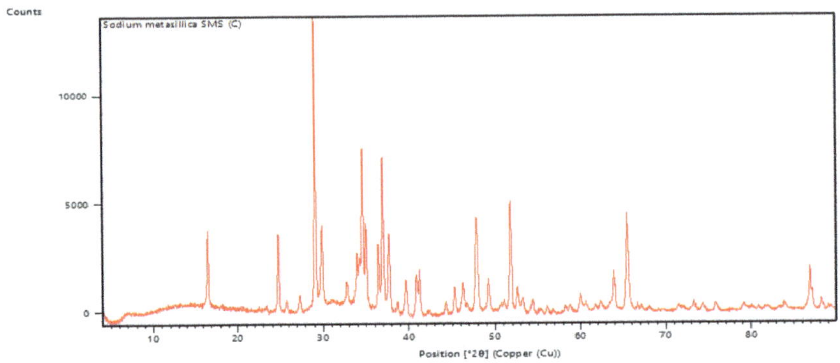

Fig. 3: *XRD diffractograph of SS*

Mechanical Properties.

Flowability. Flowability for geopolymerized specimens ranged from 110% to 126% in all the Runs. Runs 1, 2, 3, and 4 have the workability of 112%, 110%, 121% and 126%, respectively. Statistically, the geopolymerized specimen showed significantly higher workability compared to conventional binders such as cement and lime, been reported in literature [11, 17]. The workability difference can be attributed to the distinct properties of the activator. The lower SiO_2/Na_2O ratio in sodium silicate resulted in shorter silicate chains, accelerating geopolymer gelation [20]. The sodium silicate's higher density is indicative of increased dissolved solids and oligomers, further affecting mix rheology and water dispersion [22, 23]. Similarly, trace impurities in GPSS, such as $Na_4Ca(OH)_6$, may also have accelerated reaction kinetics, reducing flow [21], [23]. Furthermore, the fly ash-to-quarry dust ratio also influenced workability, with higher fly ash content increasing flow due to its finer, spherical, and porous particles [24], while quarry dust's angular particle shape and increased water demand limit flow. This explains the gradual reduction in flow with increased quarry dust.

Compressive Strength. Fig.4 presents the compressive strength results of the geopolymerized specimens. The compressive strength of the specimens shows an increase with increasing FA and hardener content across all mixes, ranging from 4.2 – 30.68 MPa. The activation of FA by the synthesized sodium silicate facilitated the geopolymerization process, resulting in stronger bonds and the formation of a well-integrated geopolymer network structure, which improved the resistance to compressive stresses in the geopolymer composite [25]. It was observed that the early strength of the hardened geopolymerized binder specimen was low. This was because there was an initial dissolution of the monomers of the precursors before the cross-linking to form oligomers with a stronger bond, which was seen in the later curing stage. The synthesized sodium silicate also contributed to the late strength due to the source of the base material, which is the glass powder, which is attributed to late strength in concrete production [22]. QD was observed to primarily act as a filler rather than an active participant in the geopolymerization process. This was seen in the mixes with increased QD. Similarly, the unreacted particles in the mixes with lower fly ash content likely contributed to increased porosity and weaker interparticle bonding.

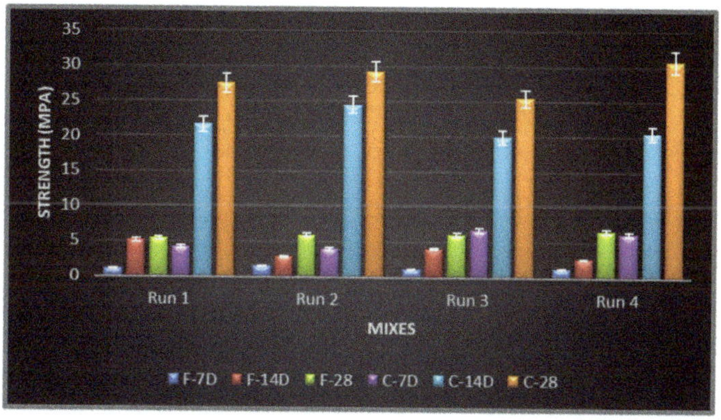

Fig. 4: Compressive (C) and Flexural (F) of geopolymer mixes at 7, 14, and 28 days

Flexural Strength. A similar trend was also observed in the flexural strength of the QD-FA-based geopolymers, as shown in Fig.4. The flexural strength of the specimens shows an increase ranging from 1.05- 6.59 MPa. This can be attributed to the activation of FA by synthesized sodium silicate, facilitating the geopolymerization process, resulting in stronger bonds and the formation of a dense geopolymer network structure for improved resistance to bending stresses in the geopolymer composite [26]. QD-FA particle sizes can also contribute to the increased mechanical strength due to the fineness of the dust, which can fill voids in the matrix and their geopolymeric reactions, leading to a synergistic effect in the flexural strength. QD primarily acted as a filler rather than an active participant in the geopolymerization process. Thus, flexural strength reduced with increasing QD content in the mixes. This strength reduction is due to the inert nature of QD, which limited the formation of a continuous geopolymer network, as noted by [27].

Conclusion and Recommendation

This study investigates the physiochemical and mechanical properties of quarry dust and fly ash-based geopolymer binder for use as a stabilizer in pavement construction material. Based on the findings and discussions presented in the preceding sections, the following conclusions and recommendations are derived:

1. The physicochemical characteristics of the sample quarry dust and fly ash materials suggest they are suitable for use as a precursor for geopolymer binder in pavement applications.
2. The results revealed that the combined use of quarry dust and fly ash as a geopolymer precursor exhibited excellent mechanical properties due to efficient micro-filling ability and geopolymeric reaction.
3. All the geopolymer mixes can be used as a binder in the sub-layer (base and subbase) in pavement construction, especially in the construction of light traffic roads.
4. The developed geopolymer matrix can be used as a surface layer with further investigation
5. Utilizing quarry dust and fly ash encourages recycling of industrial by-products and reduces waste, contributing to more sustainable construction practices.
6. Optimizing the mix design and more testing, such as scanning electron microscopy (SEM) and Fourier transform infrared spectroscopy (FTIR), are required for deeper insight into the reaction mechanisms, gel composition, and microstructural evolution of the geopolymer mixes to ensure that the product meets required performance standards for pavement.

References

[1] B. Aytekin, and A. Mardani-Aghabaglou, Sustainable materials: A review of recycled concrete aggregate utilization as pavement material, Transportation research record 2676, no. 3 (2022): 468-491. https://doi.org/10.1177/03611981211052026

[2] A.T. Papagiannakis, and E.A. Masad, Pavement design and materials, John Wiley & Sons, 2024.

[3] G.M. Mudd, and D. V. Boger, The ever-growing case for paste and thickened tailings—towards more sustainable mine waste management, Journal of Australian Institute of Mining and Metallurgy, 2 (2013): 56-59.

[4] J.O. Ikotun, G.E. Aderinto, M.M. Madirisha and V.Y. Katte, Geopolymer cement in pavement applications: bridging sustainability and performance, Sustainability 16, no. 13 (2024): 5417. https://doi.org/10.3390/su16135417

[5] B.A. Méndez-Ortiz, A. Carrillo-Chávez, and M.G. Monroy-Fernández, Acid rock drainage and metal leaching from mine waste material (tailings) of a Pb-Zn-Ag skarn deposit: environmental assessment through static and kinetic laboratory tests, Revista mexicana de ciencias geológicas 24, no. 2 (2007): 161-169.

[6] L. Manjarrez, and L. Zhang, Utilization of copper mine tailings as road base construction material through geopolymerization, Journal of Materials in Civil Engineering 30, no. 9 (2018): 04018201. https://doi.org/10.1061/(ASCE)MT.1943-5533.0002397

[7] G. Qian, T. Huang, and S. Bai, Use of cement-stabilized granite mill tailings as pavement subbase, Journal of Materials in Civil Engineering 23, no. 11 (2011): 1575-1578.https://doi.org/10.1061/(ASCE)MT.1943-5533.0000276

[8] T.G. Soosan, A. Sridharan, B.T. Jose, and B. M. Abraham, Utilization of quarry dust to improve the geotechnical properties of soils in highway construction, Geotechnical Testing Journal 28, no. 4 (2005): 391-400. https://doi.org/10.1520/GTJ11768

[9] M.W. Heyns, Fly ash as an alternative stabiliser for road pavement materials: a case study in South Africa, PhD dissertation, Bloemfontein: Central University of Technology, Free State, 2016. http://hdl.handle.net/11462/1387

[10] E. Tunçer, Sustainable Rubberized Pervious Concrete, Master's thesis, Middle East Technical University (Turkey), 2023.

[11] R. Vinai, and M. Soutsos, Production of sodium silicate powder from waste glass cullet for alkali activation of alternative binders, Cement and Concrete Research 116 (2019): 45-56. https://doi.org/10.1016/j.cemconres.2018.11.008

[12] SANS 5861-1 (2006) Cement test methods, Part 1: Mixing of standard mortar and preparation of test specimens for strength tests in the laboratory, South African Bureau of Standards, Pretoria

[13] ASTM C1437-15 (2015) Standard test method for flow of hydraulic cement mortar, ASTM International, West Conshohocken, PA

[14] SANS 50196-1 (2006) Methods of testing cement. Part 1: determination of strength, South African Bureau of Standards, Pretoria

[15] SANS 3001-AG1:2014: South African National Standard, Civil engineering test methods Part AG1 Particle size analysis of aggregate by sieving (2014)

[16] SANS 5845:2006: South African National Standard, Civil engineering test methods, Bulk densities, and voids content of aggregates (2006).

[17] S. Wang, P. Hao, S. Li, A. Zhang, Y. Guan, and L. Zhang, Synthesis of glycerol carbonate from glycerol and dimethyl carbonate catalyzed by calcined silicates, Applied catalysis a: general 542 (2017): 174-181. https://doi.org/10.1016/j.apcata.2017.05.021

[18] B. Li, H. Li, X. Zhang, P. Fan, L. Liu, B. Li, W. Dong, and B. Zhao, Calcined sodium silicate as an efficient and benign heterogeneous catalyst for the transesterification of natural lecithin to L-α-glycerophosphocholine, Green Processing and Synthesis 8, no. 1 (2019): 78-84. https://doi.org/10.1515/gps-2017-0190

[19] M. Tiainen, J. Daavitsainen, and R. S. Laitinen, The role of amorphous material in ash on the agglomeration problems in FB boilers, A powder XRD and SEM-EDS study, Energy & fuels 16, no. 4 (2002): 871-877. https://doi.org/10.1021/ef010269j

[20] M. Criado, A. Fernández-Jiménez, A. Palomo, I. Sobrados, and J. Sanz, Effect of the SiO2/Na2O ratio on the alkali activation of fly ash. Part II: 29Si MAS-NMR Survey, Microporous and Mesoporous Materials, 109(1-3), (2008.): pp.525-534. https://doi.org/10.1016/j.micromeso.2007.05.062

[21] H.A.A.E. Ghanim, U.J. Alengaram, N.M. Bunnori, and M.S.I. Ibrahim, Innovative in-house sodium silicate derived from coal bottom ash and its impact on geopolymer mortar, Journal of Building Engineering 99 (2025): 111428. https://doi.org/10.1016/j.jobe.2024.111428

[22] A.S. Kolade, B. Ikotun, D.O Oyejobi, G.E Aderinto, and M.M. Madirisha, Performance Evaluation of Geopolymer Mortars Prepared with Waste Glass Powder-Derived Sodium Silicate Solution, Iranian Journal of Science and Technology, Transactions of Civil Engineering (2025): 1-20. https://doi.org/10.1007/s40996-025-01952-w

[23] S. Dadsetan, H. Siad, M. Lachemi, O. Mahmoodi, and M. Sahmaran, Sodium glass liquid from glass waste as a user-friendly hardener in structural geopolymer systems, Cement and Concrete Composites 130 (2022): 104525. https://doi.org/10.1016/j.cemconcomp.2022.104525

[24] Y. Hu, Z. Tang, W. Li, Y. Li, and V.W.Y. Tam, Physical-mechanical properties of fly ash/GGBFS geopolymer composites with recycled aggregates, Construction and Building Materials 226 (2019): 139-151. https://doi.org/10.1016/j.conbuildmat.2019.07.211

[25] M. Zhang, T. El-Korchi, G. Zhang, J. Liang, and Mi. Tao, Synthesis factors affecting mechanical properties, microstructure, and chemical composition of red mud–fly ash based geopolymers, Fuel 134 (2014): 315-325. https://doi.org/10.1016/j.fuel.2014.05.058

[26] S. Rios, C. Ramos, A. Viana da Fonseca, N. Cruz, and C. Rodrigues, Mechanical and durability properties of a soil stabilised with an alkali-activated cement, European Journal of Environmental and Civil Engineering 23, no. 2 (2019): 245-267. https://doi.org/10.1080/19648189.2016.1275987

[27] E. Cohen, A. Peled, and G. Bar-Nes, Dolomite-based quarry-dust as a substitute for fly-ash geopolymers and cement pastes, Journal of Cleaner Production 235 (2019): 910-919. https://doi.org/10.1016/j.jclepro.2019.06.261

Modelling of Sodium Hydroxide Treated Rubber Crumb – Mortar's Compressive Strength Using Multilayer Perceptron Approach

Abiodun Joseph KILANI[1,a*], Bolanle Deborah IKOTUN[1,b] and Rasheed ABDULWAHAB[1,c]

[1]Department of Civil & Environmental Engineering and Building Science, College of Science, Engineering and Technology, University of South Africa, Florida Science Campus, Roodepoort, Johannesburg, South Africa

[a]20994680@mylife.unisa.ac.za; [b]ikotubd@unisa.ac.za; [c]abdulr@unisa.ac.za

Keywords: Artificial Neural Network, Rubber Crumb - Mortar, Compressive Strength, Rubber Crumb, Model, Multilayer Perceptron

Abstract. The structural integrity of mortar is dependent on its performance in service. The ability of mortars to withstand compressive stress solely depends on its compressive strength. Incorporation of rubber crumb (RC) in mortar is one of the approaches used in environmental waste management campaigns. The literature revealed the flaw associated with the use of untreated RC in mortar, and this requires urgent solutions. In this study, the compressive strength of RC – mortar was determined in accordance with SANS 50196 -3:2006. The RC – mortars were produced from the mixture of 1.5 – 5.0% of 0.075 - 2.0 mm RC (by the weight of cement) which has been treated with 7% of sodium hydroxide (NaOH) solution, 1:3 of cement and sand, and 0.5 water – cement ratio. The 40 x 40 x160 mm mortar cubes were used to determine the compressive strengths of the RC - mortar on Universal Testing Machine (UTM) at 7 and 28 days of standard water curing. The experimental values of mortar's compressive strength were modelled using multilayer perceptron (MP) of artificial neural network (ANN) to predict the compressive strength of the RC - Mortar. Among the ANN model predicted values, the optimum compressive strength of RC-mortar was observed to be 61.98 MPa at 1.5% of RC. The model predicted optimum compressive strength (61.98 MPa) is in correlation with that of experimented value (61.91 MPa) at excellent performance metrics of sum of squares error (SSE), mean absolute percentage error (MAPE), root mean square error (RMSE) and coefficient of determination (R^2) values of 0.0010, - 0.0125, 0.0166 and 0.988 respectively. In conclusion, MP of ANN model provided matchable prediction of RC-mortar's property, contributing to the development of environmental management and advance the production of sustainable composite material in the global construction industries.

Introduction

The importance of mortar to construction industries cannot be overemphasized. As a matter of fact, mortars perform significant roles in buildings and construction sectors. On many occasions, cement mortar contributes to the enhancement of buildings' structural integrity through strong bonding within the bricks and blocks, efficiency in repairing and patching of concrete surface, preventing building external parts from erosion and weathering [1]. Although the performance of mortar in structural buildings is excellent, its usage in construction industries is not without deterioration which call for quick remedy for better sustainability of mortar. As investigated by [2], the rate at which mortar develops complex matrix bond in hydration is low. This is caused by high quantity of water contributing to less expansive capacity of mortar in service. As a result, the mortar develops low strength to withstand cracks in elastic modulation. Mortar deterioration has caused a lot of damage to the construction sector [3]. For instance, the deterioration in mortar has resulted in wearing of bonds between the mortar and blocks, stones, and bricks of structures, leading to weakening of the whole mortar-brick structures, most especially in building structures.

This has resulted in several collapses of structures in the global communities with costly repairs and calls for immediate solutions [4]. According to [5], the hardened mortars were found to decrease in flexural strength as their curing ages in water were increasing. This is caused by its microporous properties, and it encourages deterioration and cracking of mortar. To address the weak properties of mortar, the lightweight synthetic aggregates (LSA) were incorporated in mortar to enhance its structural properties [8]. The results showed that mortar's compressive and flexural strengths decreased with increasing LSA content. This trend was observed in both high and normal – strength – mortars. Correcting these deteriorations associated with mortar's strength reduction, incorporation of sustainable construction material will be the best alternative for mortar's strengths enhancement.

In recent times, inappropriate disposal of scrap tyres has turned to environmental problems in the global communities. This makes the inhabitants of those communities to be liable to environmental risk such as microplastic pollution, fire hazards, water and soil pollution [7]. In the global communities, up to 1 billion scrap - tyres are generated in a year. Globally, about 4 billion scrap tyres were disposed of into landfills in a year, turning to environmental problems [8]. The incorporation of processed scraps tyres (in a form of rubber crumb) in mortar is one of the best methods of enhancing the sustainability of mortar and its structural properties, and control environmental pollution emanating from wrong disposal of waste tyres. Although, the incorporation of rubber crumb in mortar has been found decreasing the mechanical properties of mortar from the literature due to its poor treatment [9], however, this study works on addressing the issue by incorporating a sodium hydroxide treated rubber crumb in mortar to enhance its performance, and also make a prediction of related values using multilayer perceptron approach of artificial neural network.

Materials and Methods

Materials

The materials used for this investigation are water, Ordinary Portland Cement (OPC) of CEM I 52.5N, standard sand, and rubber crumb (RC). The rubber crumb (RC) was obtained from the shredding of the scraps of resilience tyres - type called offroad tyres (OT) (Figure 1).

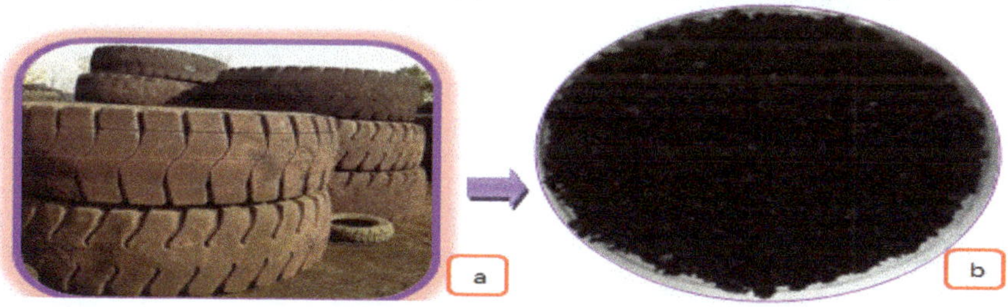

Figure 1 *Scraps of OT and its extracted Rubber Crumbs: (a) Offroad Scrap Tyres (b) Extracted rubber crumbs from scraps of OT*

Methods

Method of RC-mortar production

RC-mortar was produced from the mixing of 1350g of standard sand with 450g of CEM I 52.5N, 6.8 – 22.5g of RC, and 225g of water using aggregates mix proportion of 1:3 and 0.5 water – to - cement ratio as shown in Figure 2. The 0.075 - 2.0 mm of RC was added to the mortar by the weight of CEM I 52.5N in varying contents of 0, 1.5, 2.5, 3.5 and 5.0%, respectively after being treated with 7% of Sodium hydroxide (NaOH) solution by the weight of CEM I 52.5N for better result. The early age properties such as consistency, initial and final setting times, and compressive strength of mortar were determined in accordance with SANS 50196 – 3: 2006 [10], and EN 196 – 3: 2005 [11]

***Figure 1** Method of Producing RC - mortar*

Method of RC – mortar Modeling with Artificial Neural Network model

The artificial neural network (ANN) model uses 40% of its input variables such as percentage of RC, cement, mass of RC, water and standard sand for training and 60% of the variable for testing the model prediction of 28 days compressive strength of RC-mortar as shown in Figure 3. The network architecture of the model (as shown in Figure 3) presented the model prediction in three layers, that are, input, hidden, and output layers. The three of the samples trained are hidden at the hidden layers with only compressive strength at output layer according to Figure 3. The model performance was evaluated by using the sum of squares error (SSE), mean absolute percentage error (MAPE), mean square error (RMSE), Mean Absolute Deviation (MAD) and coefficient of determination (R^2) of the model's predicted values as shown in equation 1- 4.

$$MAPE = \left(\frac{1}{N}\right) x \left[\sum_{i=1}^{N} \left[\frac{(O_i - P_i)}{O_i}\right] x\ 100\% \right] \quad (1)$$

$$RMSE = \sqrt{\Sigma}(P_i - O_i) \quad (2)$$

$$MAD = \sum_{i=1}^{N} \frac{(O_i - P_i)}{N} \quad (3)$$

$$R^2 = \left[\frac{\Sigma_i^n (O_i - O)(P_i - P)}{\Sigma_i^n (O_i - O)^2\ x\ \Sigma_i^n (P_i - P)^2}\right]^2 \quad (4)$$

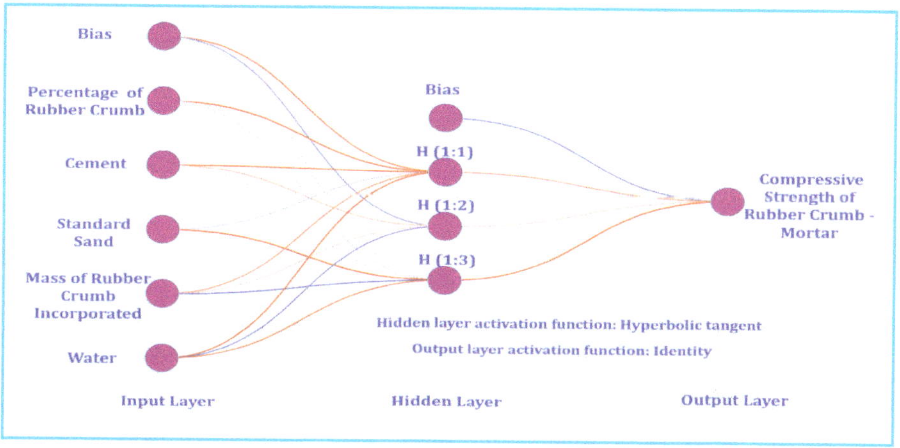

Figure 3 *Network Architecture of Predicting the Optimum Compressive Strength of RC-Mortar*

Results and Discussions

Result of RC – Cement Paste's Consistency and Setting Times

The incorporation of 1.5 – 5.0% of RC in mortar increased its consistency by 0.5 - 1.5%, making it more suitable for RC- mortar production with the addition of less water. Also, RC reduced the initial setting time of the mortar's paste by 31 minutes, and its final setting time was delayed by 16 minutes (Figure 4). Therefore, RC have a good consistency and setting times with cement paste, satisfying the SANS 50196 – 3: 2006 [10] standard.

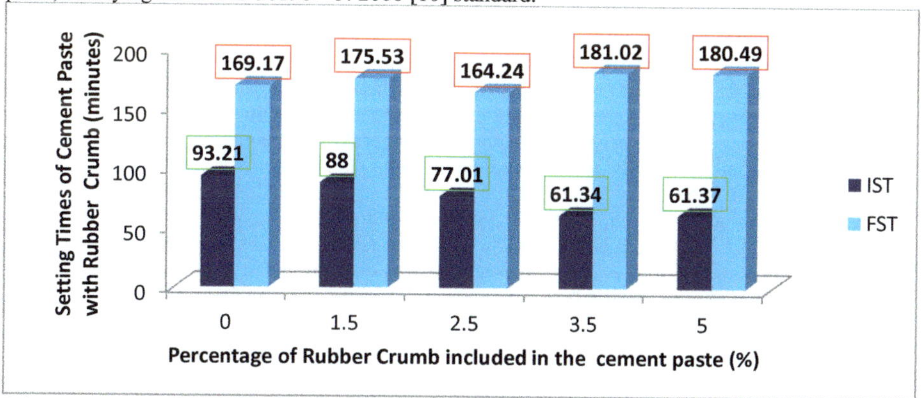

Where IST and FST are the initial and final setting times

Figure 4 *Setting Times of Crumbed Rubber – Cement Paste.*

Result of RC – Mortar's Compressive Strength

The compressive strength of RC-mortar experiment is presented as shown in Figure 5. The optimum compressive strengths of RC- mortar, 51.00 MPa and 61.78 MPa, were observed at incorporation of 1.5% of RC at 7 and 28 days which are by 3.3% and 0.2% more than that of mortar with 0% of RC (control) respectively. Thus, treatment of RC with sodium hydroxide influenced its performance in increasing the mortar's compressive strength. This result agrees with the findings of [12] and [13] which showed that mortar's compressive strength increases with incorporation of RC.

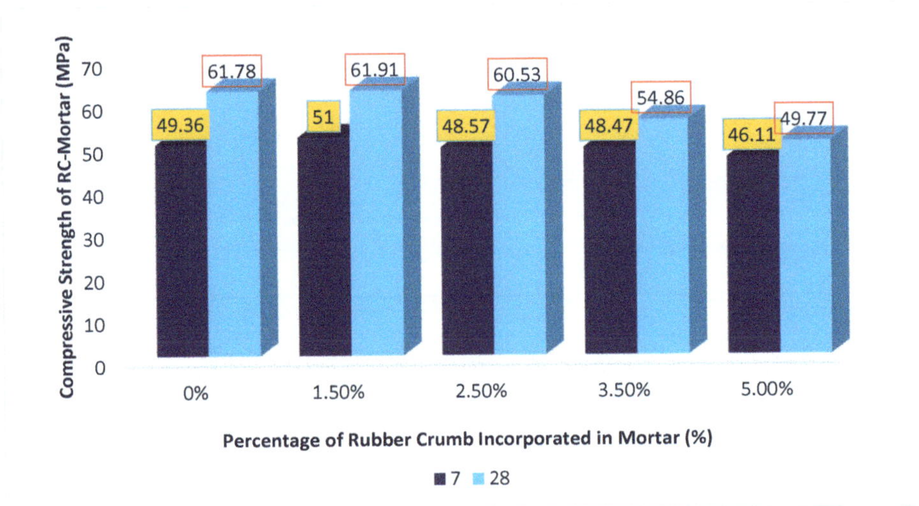

***Figure 5** The Compressive Strength of RC-Mortar Experimented*

Result of RC – Mortar's Compressive Strength MP-ANN Model, and its compression with that of 28 day experimented Values

According to the 28-day results of MP-ANN model predictions as shown in Figure 6, the optimum compressive strength of RC-mortar was predicted by the model (61.96 MPa) at 1.5% of RC is in correlation with that of experimented optimum compressive strength value of 61.96 MPa. Comparing the all MP-ANN model predicted values of RC-mortar's compressive strength produced with 0 – 5.0% of RC (61.85, 61.98, 60.53, 54.69 and 49.98 MPa) with that of experimented values of 61.78, 61.91, 60.53, 54.86 and 49.77 MPa with percentage strengths difference of 0.32, 0.11, 0, 0.31 and 0.42% , it can be deduced that, MP - ANN model made excellent prediction of RC-mortar's compressive strength. Also, the MP – model contributed to the development in RC usage for better performance of mortar.

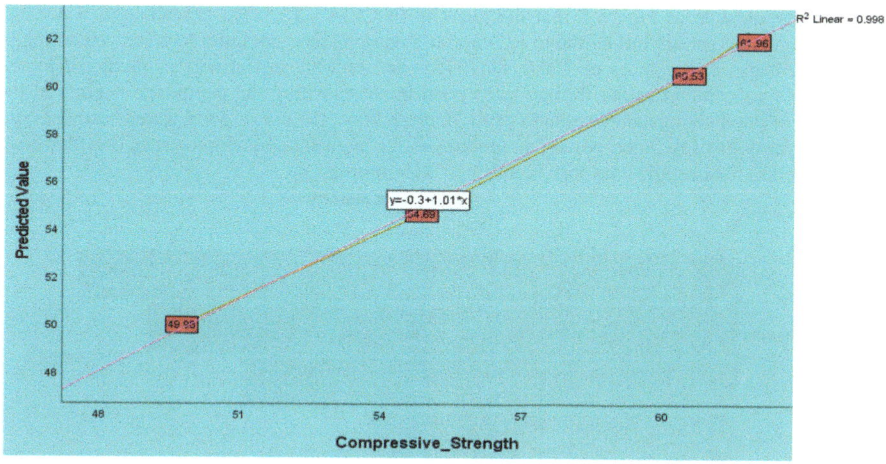

***Figure 6** MLP-ANN Model's Predicted Compressive Strength of Rubber Crumb - Cement Mortar (in MPa)*

Residual, Performance Metric and Normalize Importance of MP-ANN Model's Prediction

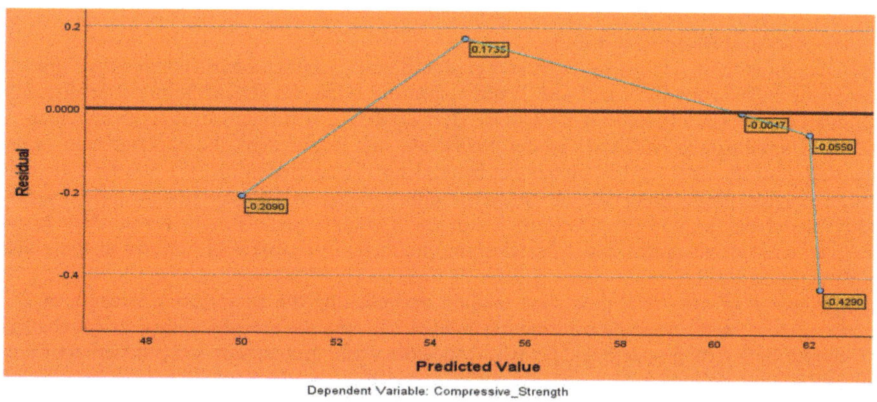

***Figure 7** Residual Chart of MP-ANN model Prediction of RC-mortar's Compressive Strength*

As shown in Figure 7, it was only at 3.5% of RC that compressive strength of RC-mortar experimented is more than that of predicted value with residual value of 0.1735. All the other negative residual values indicated that MP-ANN models have higher prediction of RC-mortar's compressive strength, indicating good result of strength prediction. In addition, the performance metric values of - 0.0125, 0.0166, - 0.036, 0.998 and 0.001 for MAPE, RMSE, MAD, R^2 and SSE indicated that MP-ANN model made the excellent prediction of RC-mortar compressive strength.

It is very clear from Figure 8 that the quantity of CEM I 52.5, RC-percentages, water and standard sand content are influential to RC-mortar's compressive strength increment according to their percentages importances of 100.0, 89.7, 75.3 and 56.1%, respectively as predicted by the model. This contributed to the formation of good bonding within the composite leading to the formation of good compressive strength of RC-mortar. Also, the use of good water - cement ratio and the percentage of RC incorporated in mortar should be given a high importance in both design and production to normalize the sustainability of RC - mortar.

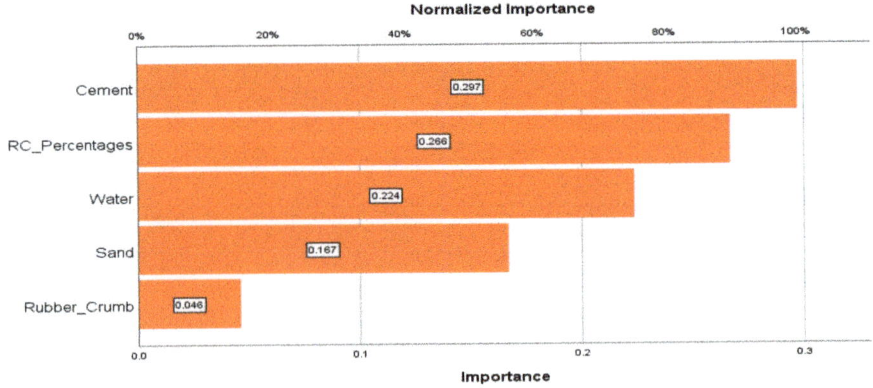

Figure 8 *Normalized Importance of RC-Mortar Variables*

Conclusions and Recommendations

The incorporation of 1.5% of RC in mortar produced the optimum mortar's compressive strength increment of 3.3% and 0.2% at 7 and 28 days of curing. Also, the MP-ANN model predicted an optimum compressive strength at 1.5% of RC during 28 days of curing with strength increment of 0.21%. Therefore, the MP-ANN model predictions are validated by the experimented values obtained, thus, the MP – ANN model can be used for the prediction of RC – mortar's strength before its actual production in the current construction practice. Also, it can be concluded that MP-ANN model has a good prediction of RC-mortar's compressive strength contributing to the environmental management and advancing the production of sustainable mortar in the global construction industries.

Application of MP-ANN prediction method in predicting the compressive strength of RC-mortar is recommended for the construction industries to have the preview of expected strength of RC-mortar before its production. This will reduce the waste of materials and production of low-strength RC-mortar.

References

[1] Zeeshan M. (2024). Application of Mortar in Construction, https://alsyedconstruction.com/application-of-mortar-in-construction/ (Accessed 14th April, 2025).

[2] Klangvijit W. and Sookramoon K. (2018). Study of the Mix Cement Properties of Mortar Cement Used in Masonry and Plaster from the Waste Biscuit Firing of Ceramic, IJRET: International Journal of Research in Engineering and Technology, Volume: 04 Special Issue: 01, Available on http://www.ijret.org

[3] Flyriver (2025). Mortar Joint Deterioration: Causes, Consequences, and Cures. Retrieved from https://www.flyriver.com/g/mortar-joint-deterioration

[4] Bethel (2025). Understanding Mortar Deterioration: Causes & Prevention Tips, https://bethelcustombrick.com/blog/understanding-mortar-deterioration-causes-prevention-tips.html, Accessed on 16th April, 2025

[5] Tararushkin E. and Shchelokova T. (2020). Daily strength testing of the Portland cement mortars. IOP Conference Series Materials Science and Engineering 862(2):022031, DOI: 10.1088/1757-899X/862/2/022031

[6] Swan C. and Bonora A. (2017). Compressive, Flexural and Tensile Strengths of various Mortar mixes Containing Synthetic Lightweight Aggregate, 2017 World Coal Ash (WOCA) Conference in Lexington, KY-May 9 – 11, 2017; http://www.flyash.info/

[7] Asda – Tyres (2025). Environmental Impact of Vehicle Tyres: Recycling Statistics by Country,https://www.asdatyres.co.uk/blog/environmental-impact-of-vehicle-tyres-recycling-statistics-by-country/ (Accessed 16th April, 2025)

[8] Hailstone (2022). What Happens When Tires Reach the End of The Road? https://www.forbes.com/sites/jamiehailstone/2022/03/18/what-happens-when-tires-reach-the-end-of-the-road/?sh=5b17e8ae6fed (Accessed 16th April 2025)

[9] Claudio de Souza Kazmierczak, Samuel Dutra Schneider, Orlando Aguilera, Cristhiana Carine Albert, Maurício Mancio (2020), Rendering mortars with crumb rubber: Mechanical strength, thermal and fire properties and durability behaviour, Construction and Building Materials, Volume 253, 2020,119002, ISSN 0950-0618, https://doi.org/10.1016/j.conbuildmat.2020.119002.

[10] SANS 50196 – 3:2006 Edition 2: Method of Testing cement, https://www.afrisam.co.za/wp-content/uploads/2021/10/2010277-AfriSam-Technical-Ref-Guide-update-2021.pdf

[11] EN 196-3:2005. Methods of testing cement - Part 3: Determination of setting times and soundness. Retrieved from https://standards.iteh.ai/catalog/standards/cen/483fd029-2859-43a2-b3ec-eb3bf21ae307/en-196-3-2005#.

[12] Zang M., Ding J., Pang L. and Ding K. (2023). Influence of mechanical properties of modified mortar on In-plane and Out-of-plane seismic performance of RC frame structure with infill walls, Structures, Vol. 50, pp. 1686 -1698, https://doi.org/10.1016/j.istruc.2023.02.118

[13] Kilani, A.J., Ikotun, B.D. & Abdulwahab, R. (2025). Production of Sustainable Mortar with Sodium Hydroxide Treated Rubber Crumb. Iranian Journal of Science and Technology Transaction of Civil Engineering. https://doi.org/10.1007/s40996-025-01768-8

Advances in Cement and Concrete Research - ICACCR 2025 Materials Research Forum LLC
Materials Research Proceedings 63 (2026) 38-46 https://doi.org/10.21741/9781644904053-5

Assessment of the Properties of Concrete Produced with Blended Diutan Gum and Ground Granulated Iron and Steel Slag as Partial Replacement for Portland Limestone Cement in Concrete

Oluwaleke Adekunle OLOWU[1,a *], Akeem Ayinde RAHEEM[2,b], Hezekiah Olatunbosun OMOLAYO[1,c], Adesina Emmanuel ALADELOBA[1,d], Abiodun Yesiru AKINSANYA[3,e]

[1]Department of Building, Yaba College of Technology, Yaba, Lagos, Nigeria

[2]Department of Civil Engineering, Ladode Akintola University of Technology, Ogbomoso, Oyo State, Nigeria

[3]Department of Building, Lagos State University of Science and Technology, Ikorodu, Lagos, Nigeria

[a]oluwaleke.olowu@yabatech.edu.ng, [b]aaraheem@lautech.edu.ng, [c]bosun.omolayo@yabatech.edu.ng, [d]adesina.aladeloba@yabatech.edu.ng, [e]engrabiodunyesiru@lasutech.edu.ng

Keywords: Diutan Gum, Ground Granulated Iron and Steel Slag, Portland Limestone Cement, X-Ray Fluorescence, Compressive and Flexural Strengths

Abstract. The study assesses the properties of concrete produced with the combination of Diutan gum and Ground Granulated Iron and Steel Slag (GGISS) as replacement to Portland Limestone Cement (PLC) in Concrete with the intention to achieve green and sustainable concrete. X-ray fluorescent (XRF) and Fineness were used to characterize both PLC and Blended Diutan gum and GGISS; tests conducted on aggregate include gradation test, aggregate crushing value (ACV) and aggregate impact tests. Water absorption, compressive and flexural strength tests were conducted on the hardened concrete. The workability and bleeding resistance of the fresh concrete produced attest to reducing agent of Diutan gum; the fineness test showed that PLC is finer than blended Diutan gum and GGISS. The aggregate (coarse and fine) is well graded and of adequate strength, the major chemical compositions of PLC are calcium oxide, 69.79% and silicon oxide, 17.07%. The composition for the blended Diutan gum and GGISS is Titanium dioxide, 34.49% and calcium oxide, 27.57%. The amount of water absorbed decreases as the percentage replacement of cement with blended Diutan gum and GGISS increases; while both compressive and flexural strengths decrease significantly as the amount of blended Diutan gum and GGISS increases. In conclusion the concrete produced with blended Diutan gum and GGISS is eco-friendly as it reduces the amount of PLC used and can be used as self-compacting concrete which can saved cost of labour and reduce wear and tear of formwork. However, when the blend of Diutan gum and GGISS are used between 10 to 20 replacement level it improves the compressive strength of concrete.

Introduction

Concrete is composed basically of aggregates, and paste which comprised of cement and water [1] and sometimes admixture to regulate the properties of concrete. Cement as a concrete ingredient exhibits properties of setting and hardened when mixed with water to form paste. Cement is complex and variable in composition and constitution. It is an important constituent of concrete. The composition of PLC greatly affects the strength of the concrete; PLC is a conglomeration of components, which includes clay, silica, iron oxide and aluminium. Cement is divided into two classes namely: hydraulic and non-hydraulic. Hydraulic cement has the ability to set and harden under water; it includes Portland cement, Aluminium cement and Pozzolanic cement. Non-hydraulic cement does not have the ability to set and harden under water but requires air to harden; a very good example of non-hydraulic cement is lime cement [2].

The emissions of carbon dioxide during the production of cement couple with its universal usage makes cement to be categorized as non-green building material and non-sustainable. Limestone, a major component in the production of cement is nature given and it is depleted on a daily basis universally which makes cement production non sustainable.

A blend of Diutan gum and Ground Granulated Iron and Steel Slag (GGISS) was used as partial replacement to Portland limestone cement in this study. GGISS is an industrial waste product which means using industrial waste to replace substance that has economic value, that is, waste to wealth; while Diutan gum is a polysaccharide, a chemical admixture that have significant impact on fresh concrete properties like workability which include stability, mobility and compatibility with concrete; using this substance produced to reduce pollution and harm GGISS pose to the environment and impact specific characteristics such as water reduction, self-compaction, segregation resistant properties to the concrete produced.

Sustainable and green building is the ability to continue building with no adverse effects on the environment hence; to achieve sustainable and green building, effort must be made to provide alternative binding agent in concrete to PLC whose production and usage generate gases responsible for the depletion of ozone layer causing global warming, the major cause of general rise in sea level, environmental degradation, and pollution. PLC uses between 5 to 15% of limestone to reduce the carbon footprint of cement production unlike the Ordinary Portland Cement which use 5 % as processing addition; hence, PLC is better green material than Ordinary Portland Cement. Both the PLC and OPC contains limestone as one of the primary ingredients; but the rate at which PLC re-absorb carbon back into the system of production is better than that of OPC.

Research towards reduction of these green gasses to the atmosphere include diversification from the use of PLC to the use of alternatives like Rice Husk Ash, combination of Calcined clay and limestone among many others; capturing of carbon dioxide in the atmosphere for other industrial usage, incorporation of blended substance to replace PLC in concrete production in order to reduce the production of these green gases is what this study is based upon. The main objective of the study is developing a sustainable cement and concrete material with a view to reduce the carbon foot-print and producing eco-friendly construction material.

Materials and Methods

Materials used for the study include aggregate (granite and sharp sand), Portland limestone cement (PLC) of grade 43 complying with the Nigerian Industrial Standard [3] as binder and water. The granite aggregate conformed to the British Standard BS EN 13055 - 2 [4], it was free of silt, clay, dirt and deleterious materials; it was obtained from Ratcon Limited quarry site along Lagos-Ibadan expressway. The sharp sand was obtained from the Ogun River at Owode, Ogun State, Nigeria. The cement was sourced from Lafarge Cement Company, Ewekoro, Ogun State, Nigeria. Diutan gum was provided from a local vendor at Ojota, Lagos, Nigeria; it was mixed with GGISS in ratio 1:1 before use to cast concrete cubes and beams for the study.

Concrete materials were batched by weight, treatment and control concrete cubes of size 150 mm and beam size of 100 × 100 × 500mm were cast. A total of 108 concrete cubes and beams were cast for compressive strength and flexural strength tests respectively. While 18 number concrete cubes were cast for water absorption test. The coarse aggregate was sieved to maximum sizes of 37.5 mm before use to cast concrete. Concrete Mix Ratios (MR) adopted was 1:1½:3, while Water Cement Ratio (W/C) adopted was 0.60 and % Replacement (RR) of blended Diutan gum and GGISS to PLC adopted were 10, 20, 30, 50 and 60 %. The specimens were de-mould after 24 hours and put into the water tank for curing for 7, 14, 21, 28, 35 and 42 days, maintaining curing temperature at $25^0C \pm 5^0C$ as per [5]. The XRF analysis was carried out to determine the percentage oxide composition in the samples of the PLC and blended Diutan Gum and GGISS. The procedure adopted was outlined in [6]; while the Fineness test was carried out in accordance with the requirement of [7]. The test was carried out at a temperature of 27±2 °C. Other physical properties on the granite investigated include gradation, whose test procedure adopted is as outlined in [8]; aggregate crushing value (ACV), procedure for the test is as contained in [9, 1], aggregate impact value (AIC) test was conducted as prescribed by [11]; Compressive strength test was conducted in accordance with [12]. Fresh concrete properties were assessed using slump and compacting factor tests. The compressive strength test machine used to test the cubes specimens was in accordance with [13]; while the flexural strength test was conducted in accordance with [14]. The third point loading arrangement of the machine was set-up; the actuator gradually released the load steadily and without shock at 0.06 ± 0.04 N/ (mm^2 s). The rate of loading was maintained without change until failures occurred. The flexural strength was calculated as shown in equation 1.

$$F_{cf} = \frac{F \times L}{d_1 \times d_2^2} \quad (1)$$

Where:
F_{cf} = Flexural strength
L = Effective span length between supports
F = Maximum load at fracture
d_1 = Width of beam at the fracture plane
d_2 = Depth of the beam at the fracture plane

Results and Discussion

Composition of PLC and Blended Diutan gum and GGISS. The results of XRF Analysis on samples of blended Diutan gum and GGISS is presented in Tables It can be observed from Table 1 that the predominant oxides in the blended sample of Diutan gum and GGISS are: Titanium dioxide, (TiO_2), Calcium oxide (CaO), Magnesium oxide (MgO), and Silicon oxide (SiO_2) in the following proportions 34.491, 27.573, 5.69, and 2.561% respectively. The table showed that major composition of PLC is Calcium oxide (CaO), Silicon oxide (SiO_2), Aluminium oxide (Al_2O_3), and Sulphur trioxide (SO_3) and in the following proportion 69.786, 17.065, 4.721, and 3.7424% respectively. The higher tensile strength of concrete produced with PLC compared to the blend of Diutan gum and GGISS steamed from the higher metallic oxides in the PLC.

Olonade, K. A., Kadiri, M. B. and Aderemi, P. O. (2015) [15] reported that steel slag, a form of GGISS is composed of SiO_2, Ferric oxide, Fe_2O_3, CaO, and K_2O; [16] stated that the major constituents of steel slag to include di calcium silicate (C_2S), tri calcium silicate (C_3S) and tetra calcium aluminates ferrite (C_4AF). These compounds in cement participated in the hydration process and enhanced the strength of concrete. The presence of similar elemental oxides in both blends of Diutan gum and GGISS and PLC are the major arguments used to justify that the blend of Diutan gum and GGISS will produce a mixture that will be able to replace Portland Limestone Cement in concrete production.

Fineness of Portland Limestone Cement and Blended Diutan gum and GGISS. Table 2 showed the results of a fineness test conducted on samples of Portland Limestone Cement and Blended Diutan gum and Ground granulated Iron and Steel Slag. From Table 2 it is evident that Portland Limestone Cement contains more of Silt grain size (96.56 %) than Blended Diutan Gum and GGISS (73.09 %). The table also showed that Blended Diutan gum and GGISS contains higher amounts of granite, coarse sand, and fine sand, very fine sand than Portland limestone cement. Hence, cement used is finer than the blend of Diutan gum and GGISS and has a greater binding power than blended Diutan gum and GGISS.

Gradation of Aggregate. Sieve analyses conducted on coarse granite aggregate and sharp sand are presented in Figures 1 & 2. The coefficients of uniformity (C_U) and coefficient of concavity (C_C) for granite aggregate are 4.64 and 1.76 for maximum aggregate size of 37.5mm. These values were greater or equal to 4.00 for C_U and less than 3 for C_C [17] hence the granite aggregate is well graded and the resulting concrete produced are dense and of high strength.

Table 1: XRF Analysis of Blended Diutan gum and GGISS and PLC Sample

S/N	PARAMETERS		
		PLSC (%)	Blended Diutan gum and GGISS
	Colour	Browns	Dark Brown solid (%)
1.	SiO_2	17.065	2.561
2.	MgO	2.19	5.69
3.	TiO_2	0.1	34.491
4.	Al_2O_3	4.721	1.407
5.	CaO	69.786	27.573
6.	SO_3	3.742	0.319
7.	Fe_2O_3	1.121	1.106

Table 2: Particle Size (Fineness) Analysis Result

S/N	SAMPLE ID	Gravel	Coarse Sand	Fine Sand	Very Fine Sand	Silt
				(%)		
1.	Specimen A	9.00	4.54	6.15	7.22	73.09
2.	Specimen B	0.46	2.68	0.10	0.20	96.56

Where:
Specimen A = Blended Diutan gum + GGISS
Specimen B = Portland Limestone Cement (PLC)

Deduction from the particle size distribution of sharp sand used showed in figure 2 that the coefficient of uniformity for sharp sand was 13.3 which was greater than 6, hence the aggregate is well graded and the resulting concrete produced is dense and of adequate strength. However, the strength needs to be verified with mechanical testing.

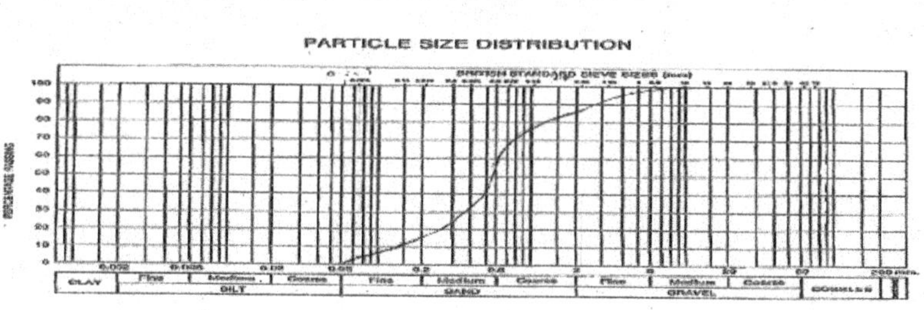

Figure 1: Particle size Distribution for Granite aggregate

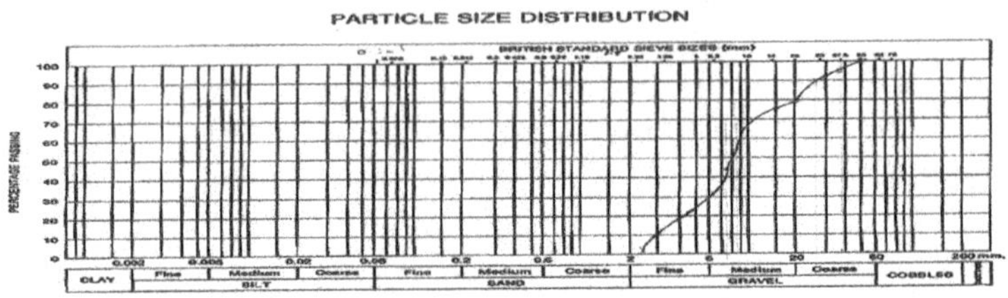

Figure 2: Particle size Distribution for Sharp sand

Aggregate Crushing Value (ACV) Test. The result of the aggregate crushing value test on granite aggregate is presented in Table 3. From the table, granite aggregate has ACV values of 8.58. This value is less than 10 which met the requirements of [10, 11]. Hence, the aggregate is classified as exceptionally strong and can withstand crushing force under load. The aggregate is durable and can be used as aggregate in concrete production.

Aggregate Impact Value (AIV) Test. The result of aggregate impact value test conducted on granite aggregate is presented in Table 4. From the table, granite aggregate AIV value is 20 which is less than 30 as specified by [11]. This revealed that the aggregate can be used for heavy duty concrete flooring, concrete pavement floor and any other concrete works as specified in [12].

Table 3: Aggregate Crushing Value for Granite Aggregate

PARAMETER	GRANITE AGGREGATE		
	1st Trial	2nd Trial	3rd Trial
W_1(kg)	3.25	3.25	3.25
W_2(kg)	0.28	0.27	0.29
ACV	8.50	8.31	8.92
Av. ACV		8.58	

Where:
W_1 = Total weight of aggregate
W_2 = Weight of aggregate passing 2.36 sieve size;
ACV = Aggregate crushing value.

Table 4: Aggregate Impact Value (A.I.V) for Granite Aggregates

PARAMETER	GRANITE		
	1st Trial	**2nd Trial**	**3rd Trial**
M_1	234	235	233
M_2	47	47	49
AIV	20	20	21
Av. AIV		20	

Where:
M_1 = Oven dried Sample (g)
M_2 = Oven dried Sample Passing 2.36mm Sieve
Aggregate Impact Value (AIV) = $\frac{M_2}{M_1} \times 100$

Slump and Compacting factor Test Results. The result of the slump and compacting factor (CF) tests conducted on fresh concrete is as shown in Table 5. From the table it was observed that the slump values of concrete produced with mix ratio 1:1½:3 for maximum aggregate sizes of 37.5 are 180, 120, 100, 60, 40 and 35mm respectively for 0, 10, 20, 30, 50 and 60% replacement level of Portland limestone cement with Blended Diutan Gum and GGISS; while the compacting factor values are 0.98, 099, 0.99, 0.96, 0.94 and 0.97 respectively.

The higher values of CF and Slump values observed at 10 and 20 % replacement levels and relative reduction of the same values observed at 30, 50 and 60% replacement level indicated that the concrete becomes less workable as the blended Diutan Gum and GGISS increase; providing resistance to bleeding and make it more cohesive and stable resulting into a self-compacting concrete.

Water Absorption Test Result. The result of the water absorption test conducted at 28 days curing on samples of concrete cubes cast without and with blended Diutan gum and GGISS referred to as control and experiment concrete cubes is presented in Table 6. From the table, it was observed that control concrete cubes absorbed more water than the experiment concrete cubes containing 10 to 20% replacement levels, but the water absorbed by experiment concrete cubes containing 30 to 50 % increases more than the control concrete cubes before finally reducing to the lowest at 60% replacement level. It was also observed that the amount of water absorbed by the experiment concrete cubes increases up to 30% replacement level before reduction of water absorbed at 50 to 60% replacement level.

Compressive Strength Test Result. The result of the compressive strength test is as presented in Figure 3. From the results, compressive strength increases with curing time across all replacement levels. However, as the percentage of replacement with GGISS and Diutan Gum increases, the compressive strength consistently decreases, especially at 30% and above. Reduction in concrete strength experienced was because there is reduction of active clinker necessary for the hydration of concrete with replacement of Diutan Gum - a natural polymer. All the test results for the control cubes and experiment cubes containing 10 to 20 % replacement levels met the 20N/mm^2 requirement at 28 days curing [18]. These values are greater than the requirement of 20 N/mm^2 for mix ratio 1:1.5:3 which the study adopted.

Table 5: Slump and Compacting Factor Test Results of Control and Experiment Concrete Cubes

Duitan Gum + GGISS	0%	10%	20%	30%	50%	60%
W1	3.2	3.2	3.2	3.2	3.2	3.2
W2	14.8	14.9	14.7	14.7	14.5	14.55
W3	15	14.95	14.75	15.1	15.2	14.8
W4	11.6	11.7	11.5	11.4	11.3	11.35
W5	11.8	11.75	11.55	11.7	12	11.6
CFV	0.98	0.99	0.99	0.96	0.94	0.97
SV	180	120	100	60	40	35

Where:
W1 = Weight of Cylinder
W2 = Weight of Cylinder + Weight of Partially Compacted Concrete
W3 =Weight of Cylinder + Weight of Fully Compacted Concrete
W4 = Weight of Partially Compacted Concrete
W5 = Weight of Fully Compacted Concrete
C.F.V = Compacting Factor Value = $\mathrm{C.F.V} = \frac{W1}{W2}$

S.V = Slump Value = Difference in fall of concrete from the top of cone to the lowest concrete point.

Table 6: Water Absorption of Control and Experiment Concrete cubes

Days of curing	Water Cement Ratio	Percentage Content of Blended Diutan Gum and GGISS IN (MM)					
		0	10	20	30	50	60
28	0.60	2.55	2.47	2.39	3.31	3.23	2.15

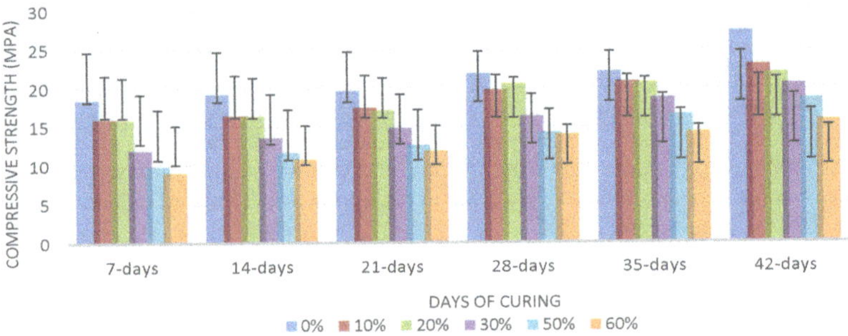

Figure 3: Compressive Strength Results of Control and Experiment Concrete cubes

Flexural Strength Test Results. Figure 4 showed that the flexural strengths of control concrete beams are higher than the flexural strengths of experimental concrete beams. The flexural strength of the control concrete beam at 28-day curing is 3.28 N/mm^{2}while that of the experimental concrete beams were 3.13, 3.07, 3, 2.45, 2.22 and 2.10 N/mm^2 at 10, 20, 30, 50 and 60 % replacement levels respectively. It was also observed that the flexural strength of experimental concrete beams reduces as the replacement level of blended Diutan gum and GGIS increases. The results are because GGISS has lower binding capacity compared to cement, and the polymeric nature of Diutan Gum introduces more shrinkage and affects the interfacial transition zone (ITZ), which weakens the tensile stress resistance.

Statistical Interpretation of Results on Compressive Strength Test. From Table 7 the adjusted R-square is 0.812, which means that approximately 81.2% of the variability in the dependent variable is accounted for by the independent variables included in the model. A higher adjusted R-square indicates a better fit of the model to the data. Between (0 -20) %, moderate improvement in compressive strength is observed, this suggested that at low dosage blend of Diutan gum and GGISS enhance particle bonding and improves internal cohesion; beyond 30% compressive strength is reduce due to excessive organic interference in microstructural packing; it is also observed that higher doses of blended Diutan gum and GGISS to replace PLC does imply higher compressive strength of concrete.

The errors bars (standard deviation) of 1.54 is small, this indicates good consistency and reproducibility across all samples. The variation is controlled, so the trend is reliable.

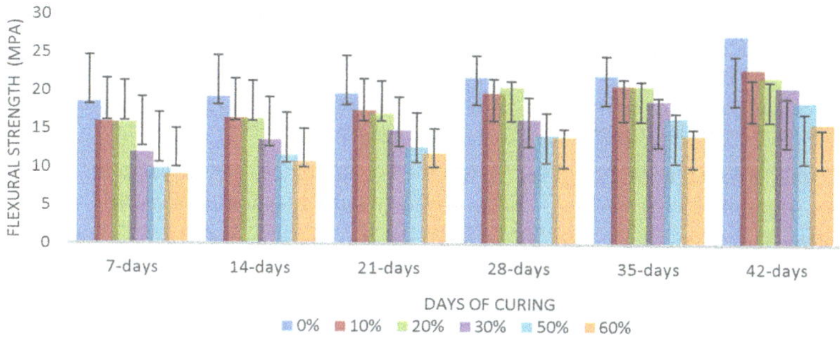

Figure 4: Flexural Strength of Control and Experimental Beams.

Table 7: Model Summary

Model	R	R Square	Adjusted R Square	Std. Error of the Estimate
1	.922[a]	.849	.812	1.54146

a. Predictors: (Constant), Blended Diutan Gum

Conclusion

Concrete produced with blended Diutan gum and GGISS is eco-friendly as it reduces the amount of PLC used and can be used as self-compacting concrete which can save cost of labour and reduce wear and tear of formwork. However, the optimal range of replacement of blended Diutan gum and GGISS to PLC should be maintained between 10 - 20% for satisfactory performance of concrete strengths and workability.

Reference

[1] Gamigboye, G. O., Edet, A. N., Egwuatu, C., Olowu, O. A. and Odewumi, T. (2015). Assessment of Compressive Strength of Concrete Produced from Different Brands of Portland Cement, Civil and Environmental Research, 7, (8): 33 – 38.

[2] Neville, A. M. (2014). Properties of Concrete. (5th Ed.). Pitman Books Ltd, London.

[3] NIS 444 - 1 (2003). Composition, Specifications and Conformity Criteria for Common Cements, Nigerian Industrial Standard, Standard Organisation of Nigeria, Lagos, Nigeria.

[4] British Standard Institution European Norm (2004), Natural lightweight aggregate for concrete, mortar and grout, BS EN 13055-2, British Standard Institution, London.

[5] British Standard Institution European Norm (2000), Concrete- Part 1- Specification, Performance, Production and Conformity, BS EN 206-1, British Standard Institution, London.

[6] British Standard Institution European Norm (2021), Method for determination X-ray Fluorescence BS EN ISO 22940 - 1: 2021, British Standard Institution, London.

[7] British Standard Institution European Norm (2018) Method for determination of Fineness Test, BS EN 196 - 6: 2018, British Standard Institution, London.

[8] British Standard Institution European Norm (2020), Methods for Determination of Particle Size Distribution. BS EN 933 – 2:2020, British Standard Institution, London.

[9] British Standard Institution European Norm (1998). Methods for determination of aggregate crushing value (ACV), BS EN 1097-2, British Standard Institution, London.

[10] British Standard Institution (1990) Testing aggregate, Method for determination of aggregate crushing value (ACV), BS 812-110, British Standard Institution, London.

[11] British Standard Institution (1990) Methods for determination of aggregate impact value (AIV), BS EN 812 - 112, British Standard Institution, London.

[12] British Standard Institution European Norm (2002) Method for determination of Compressive Strength, BS EN 12390- 3: 2002, British Standard Institution, London.

[13] British Standard Institution European Norm (2002) Specification for Compressive Strength Test Machine, pr EN 12390- 4: 2002, British Standard Institution, London.

[14] British Standard Institution European Norm (2000). Method for determination of Flexural Strength, BS EN 12390: Part 5, British Standard Institution, London.

[15] Olonade, K. A., Kadiri, M. B. and Aderemi, P. O. (2015). Performance of Steel Slag as Fine Aggregate in Structural Concrete. *Nigerian Journal of Technology*. 34 (3): 452 - 458. https://doi.org/10.4314/njt.v34i3.4

[16] Yongchang, G., Jionhe, X., Jianbai, Z. and Kexian, Z. (2019). Utilization of unprocessed steel slag as fine aggregate in normal and high strength ccncrete, Construction and Building Materials Journal, 204 (1): 41-49. https://doi.org/10.1016/j.conbuildmat.2019.01.178

[17] Micheal, S. M. and John, P. Z. (2006). Materials for Civil and Construction Engineers, Pearson Education Inc. (2nd Ed.). Upper Saddle River, New Jersey.

[18] British Standard Institution (2023). Specification for requirement of Compressive Strength Test BS 8500 – 1 – (2023), British Standard Institution, London

Evaluation of Compressive Strength of Fibre Reinforced Concrete using Randomizable Filtered Classifier

Abideen Adekunle GANIYU

Department of Civil Engineering, British University of Bahrain, Saar 527, Bahrain

a.ganiyu@bub.bh

Keywords: Fibre-reinforced Concrete, Compressive Strength, Machine Learning, Randomizable Filtered Classifier

Abstract. Fibre reinforced concrete (FRC) is a matrix material that adds discrete fibres to mixed concrete with the aim of increasing its mechanical properties. Conventional techniques for the evaluation of compressive strength of concrete entail time-consuming, laborious laboratory tests. Machine Learning (ML) models have the capabilities to utilise data on mix proportions to train a model that can accurately estimate strength outcomes for new mixes. This research employs a codeless ML tool to predict the compressive strengths of FRC using 173 datapoints with 7 input features of binder, water, fine aggregate, coarse aggregate, volume fraction of steel, aspect ratio and age. An 80:20 data split was adopted for training and testing sets, while pre-processing of data, visualisation, hyper-parameter optimisation, and feature selection were all done in the model development stage. Randomizable Filtered Classifier (RFC) algorithm yielded the optimal model, with a coefficient of determination, R^2 of 0.9886 and root-mean-square-error of 14.4529 as values of the performance measures. This implies that the model's predictions are closely aligned with the true values, underscores the strong correlation between the input features and the compressive strength, and asserts the model's validity. The study leverages data-driven soft computing methods to provide timely, cost-effective, accurate and reliable processes for the design of compressive strength of fibre-reinforced concrete for practical usage.

Introduction

The inclusion of discrete fibres into normal concrete to form Fibre reinforced concrete (FRC) is a significant advancement in concrete technology. In comparison with normal concrete, FRC exhibits enhanced mechanical and durability properties[1, 2], increased toughness and ductility[3], reduced micro-cracks resulting from shrinkage and other volume changes[4], reduced permeability [5], improved resistance to acid, alkali, and sulphate attacks [4], enhanced longevity in harsh environments [6], and reduced maintenance costs over the lifespan of a structure [7]. FRC is suitable for a wide range of structural applications, and can be tailored for specific applications by varying the type and amount of fibres used [8]. Compressive strength is a fundamental property for the structural design of concrete; it helps in understanding the stress-strain behavior which is crucial for predicting the structural behavior of concrete members. The compressive strength test of FRC is a critical evaluation of the material's ability to withstand axial loads [9]. FRC superior performance over conventional concrete is associated with increased peak stress, yield strain, ultimate strain, toughness index, and fracture toughness [10, 11] induced by fibre inclusion, these properties make FRC suitable for high performance structural applications[12, 13].

Fibres in FRC could be natural or synthetic, and are distributed throughout the concrete matrix, proper distribution can prevent fire balling, which can negatively impact fibre's effectiveness [14]. Steel, glass, carbon, basalt, polypropylene, polyester, coir, sisal, jute, hemp, flax, kenaf, bamboo, and hybrid fibres have been used in FRC matrix [15-18]. Long, short, crimped, hooked, hooked-end, twisted, milled, mil-cut, copper-coated, wave-shaped are examples of steel fibres used in FRC [19-21]. Steel fibres are commonly associated with the enhancement of tensile and flexural strength of FRC, however, their impact on compressive strength increments have been reported from moderate (13.69%) [22] to as high as marking a 233.1% (355.0 MPa) increase over the control mix [23]. The volume fraction of fibres in FRC is crucial to its compressive strength, increasing fibre content generally improves compressive strength, but there is an optimal range[24]. The composition of the concrete mix, including the cement content, aggregate type, water-cement ratio, use and type of admixtures, superplasticizers, etc. also influences the compressive strength[25].

Machine learning (ML) is a branch of artificial intelligence which mimics human-like learning processes in a structured, mathematical framework. It enables computer models to learn patterns from historical data and make informed decisions about new, unseen data based on patterns it learned [26]. ML has emerged as a powerful tool for predicting the compressive strength of concrete by offering a more efficient and accurate alternative to traditional experimental methods [27]. This approach leverages data-driven models to capture complex interactions between various concrete components, enabling precise predictions that are crucial for structural safety and material optimization [28].

Pal et. al. [29] assessed thirteen different machine learning models for the prediction of compressive strength of fibre-reinforced rubberized concrete using 905 experimental data points collated from diverse datasets. It deploys grid search and 10-fold cross-validation for hyperparameter tuning to optimize each model and provides an in-depth feature importance analysis to tackle the complex relationship between diverse concrete constituents. The results reveal that rubber content and the water/cement ratio are the major contributors to compressive strength, and CatBoost algorithm, with R^2 value of 0.865, and lowest error values emerged the best model. This analysis informs engineers which mix components to prioritize or modify when designing rubberized recycled aggregate FRC, and the viability of CatBoost algorithm as a reliable data-driven model for the compressive strength prediction.

An innovative framework for the prediction of compressive strength of recycled aggregate FRC combining ML with optimization algorithms was presented by Duan [30]. XGBoost algorithm was coupled with Tunicate Swarm Algorithm (TSA) to yield an outstanding result of R^2 value of 0.9897 and low error measures. Through extensive simulations, the research explicates the effects of steel fibre types on compressive strength, highlights the optimal fibre dosage, and offers insights into the trade-offs between material performance and fibre content. The research delves into the microstructural analysis of the FRC pore structures and explains the relationship between the internal structural characteristics and its compressive behavior. This link between microscopic observations and macroscopic strength provides a more holistic understanding of concrete performance.

Li et. al. [31] develops ML models to predict the 28-day compressive strength of steel FRC exploring both individual algorithms and ensemble techniques. The SVR AdaBoost model demonstrated the highest accuracy with an R^2 value of 0.96, and lowest error values of 4.4 MPa & 8 MPa for mean absolute error (MAE), and root mean square error (RMSE), respectively. The sensitivity analysis revealed that cement, silica fume, water, and superplasticizer have the most significant impact on the strength predictions, and this highlights the FRC constituents that must be closely controlled in practice to achieve reliable strength predictions. The research provides strong evidence that ML techniques could replace time-consuming traditional methods of estimating compressive strength of FRC.

WEKA is an open-source ML software with a wide array of algorithms for analyzing both qualitative and quantitative datasets [32]. The Random Filtered Classifier algorithm addresses both feature selection and classification simultaneously, it uses a randomization technique to filter out less relevant features, thereby focusing on the most informative ones for building the classifier, the approach is particularly effective in reducing the initial search space, which can significantly enhance the computational efficiency of the classification process, and lead to more interpretable models [33].

The application of ML to compressive strength of FRC determination saves time, materials and costs. Using a codeless tool like WEKA for the prediction enables concrete materials experts without coding background to carry out the prediction. This research explores multiple algorithms in WEKA to develop and validate predictive models for the estimation of compressive strength of FRC. All the important modelling steps were undertaken before the optimal model was established. The correlation analysis of the input attributes with the output attribute were determined, likewise is the sensitivity analysis of the input attributes with the developed model.

Methodology

The dataset was obtained from GitHub, a public data repository. It was gathered from 12 experimental research on concrete collated by Hoang [34]. It contains 173 datapoints and 9 features namely, binder, water, fine aggregate, coarse aggregate, volume fraction of steel fibre, aspect ratio, age, splitting tensile strength, and compressive strength. The attributes are mainly the constituent materials in FRC with details about the steel fibre i.e. volume fraction, aspect ratio, and testing age of concrete. The only exception is the splitting tensile strength, which is a mechanical property of FRC. The statistical analysis of the features is presented in Table 1.

Table 1: Statistical Summary of the dataset

Variables	Unit	Min	Mean	Max	Standard deviation	Skewness
Binder	kg/m^3	230.00	403.59	651.00	105.92	0.24
Water	kg/m^3	133.00	205.61	762.00	120.90	3.98
Fine aggregate	kg/m^3	400.00	699.72	1049.20	122.69	-0.17
Coarse aggregate	kg/m^3	293.30	976.41	1236.00	234.99	-1.59
Volume fraction of steel fibre	%	0.20	0.99	5.00	0.54	2.55
Aspect Ratio	-	32.42	64.18	82.67	14.52	-0.45
Age	Day	7.00	27.68	91.00	11.74	3.26
Compressive strength	MPa	7.75	54.99	145.84	29.16	0.43
Splitting tensile strength	MPa	0.79	4.52	11.50	1.79	0.45

The dataset appears clean, with no missing values. The first preprocessing step was to remove the split tensile strength values from the dataset; this is essential because the value of compressive strength is independent of it both theoretically and in practice. This step ensured that the model learns patterns specific to the required output without interference from unrelated features. Next, the correlation between the remaining seven input features and the output feature was determine using the "*CorrelationAttributeEval*" feature of WEKA which was paired with a Ranker search method to list features in order of correlation strength.

For the model development, the dataset was randomly split into 80:20 ratio for training and testing, respectively. The random split is to ensure unbiased evaluation and generalization of the model's performance. In essence, 138 datapoints were used for the training, and 35 datapoints were used for the testing. N-folds of 5-fold and 10-fold, and Random splits of 66/34; 70/30, 75/25, 80/20; and 85/15 were simulated for available algorithms under lazy, meta, functions, rules, and trees classifiers. The effects of data standardization and normalization were later investigated on leading algorithms.

Further, hyperparameters were tuned for the leading algorithms to improve predictive accuracy, optimize learning efficiency, and enhance the robustness of the leading models. The evaluative metrics for the constructed model encompass the correlation coefficient R, which assesses the statistical relationship between actual and predicted values, the coefficient of determination R^2, which elucidates the extent of variability accounted for by the fitted model, the mean absolute error (MAE) which quantifies the average magnitude of the errors in a set of predictions, and the root mean squared error (RMSE), which signifies the sample standard deviation of the discrepancies between observed and predicted values. Mathematical representations of *RMSE, MAE* and R^2 are presented in equations (1-3).

$$RMSE = \sqrt{\frac{1}{n}\sum_{i=1}^{n}(y_i - \hat{y}_i)^2} \quad (1)$$

$$MAE = \frac{1}{n}\sum_{i=1}^{n}|y_i - \hat{y}_i| \quad (2)$$

$$R^2 = 1 - \frac{\sum_{i=1}^{n}(y_i - \hat{y}_i)^2}{\sum_{i=1}^{n}(y_i - \bar{y}_i)^2} \quad (3)$$

where y_i, $\hat{y}_i$, $\bar{y}_i$ and n are the experimental output, model output, mean of the experimental output, and number of samples, respectively. The lower the values of *RMSE* and *MAE*, the better the model, the closer the value of R and R^2 value to 1, the better the model. Sensitivity analysis was also done using the optimal model to identify how each input feature influenced the model's prediction.

Results and Discussions

Figure 1 presents the outcome of correlation analysis done; it is observed that the order of linear dependence of the attributes on compressive strength is binder, coarse aggregate, age, water, aspect ratio, and volume fraction. Water and volume fraction have negative correlations, i.e. their increase leads to a decrease in compressive strength, while increase in other features leads to an increase in compressive strength as they are positively correlated.

	Binder	Water	Fine aggregate	Coarse aggregate	Volume fraction	Aspect Ratio	Age
Water	-0.424						
Fine aggregate	0.145	-0.578					
Coarse aggregate	0.152	-0.212	0.235				
Volume fraction	-0.089	0.261	-0.249	0.007			
Aspect Ratio	-0.117	0.213	-0.019	-0.405	-0.036		
Age	0.235	-0.080	0.242	0.200	-0.065	-0.011	
Compressive strength	0.686	-0.291	0.249	0.397	-0.031	0.129	0.341

Figure 1: Pearson correlation between features

The most promising results of the simulations were obtained for the 80/20 split; the top five algorithms are presented in Table 2. Randomizable Filtered Classifier (RFC) led the pack with R = 0.9861, and RMSE = 5.3382. Hence, it was applied for the testing stage where it yields R = 0.9943 corresponding to R^2 of 0.9886, this implies 98.86 % of compressive strength is accurately predictable from the input variables using the RFC model, and this underscores its' accuracy. The error values are MAE = 10.5868, and RMSE = 14.4529 which are higher than the values obtained from the training stage but are highly reasonable. The hyperparameters of the RFC – filter, filter options, base classifier and seed - were tuned, and the default values yield the best results. Figure 2 shows the relationship between the actual values and the predicted values of the compressive strength for the testing dataset.

Table 2: Performance Evaluation of Different Algorithm at Training

Name	R	R^2	MAE	RMSE
Meta- Randomizable Filtered Classifier	0.9861	0.9724	3.7144	5.3382
Meta- Regression by Discretization	0.9767	0.9539	5.5017	7.0973
Functions - Multilayers Perceptron	0.967	0.9351	11.1129	13.6195
Rules - M5Rules	0.9564	0.9147	6.8387	9.3252
Lazy - KStar	0.9463	0.8955	6.6291	10.3352

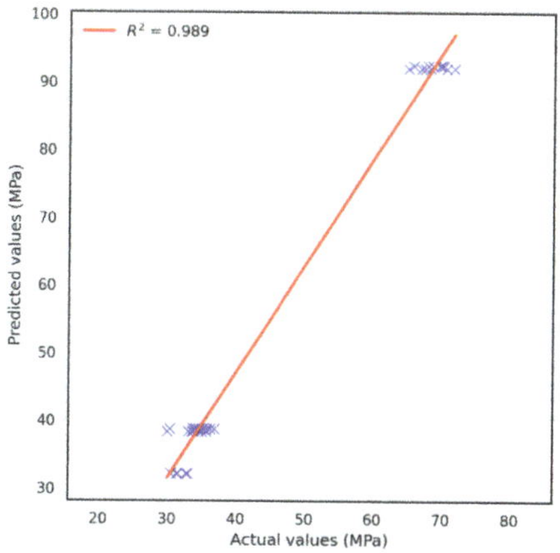

Figure 2: Actual vs. Predicted values of Compressive strength of RFC model

The result of the RFC model along with the outcome of the sensitivity analysis carried out on the features are presented in Table 3. According to the table, binder is the most important feature influencing the developed model, and the accuracy of the model will drop to 9.35% if it is removed from the input features. The next significant features are the volume fraction of fibre, water, and coarse aggregate, which will reduce the accuracy of the developed model to 25%, 30% & 36%, respectively when removed from the input features before modelling. Table 4 presents the R^2 values of RFC model developed in this research in comparison with other published research on ML prediction of FRC using WEKA software. The RFC model deployed in this research has the highest value of R^2, thereby yielding the highest accuracy.

Table 3: Sensitivity Analysis for Randomizable Filtered Classifier Algorithm

Name	R	R^2	MAE	RMSE
Randomizable Filtered Classifier, RFC (All features)	0.9943	0.9886	10.5868	14.4529
RFC (Excluding Binder)	0.3058	0.0935	17.3727	24.0887
RFC (Excluding Water)	0.5496	0.3021	17.2384	21.074
RFC (Excluding Fine Aggregate)	0.7146	0.5107	13.1953	15.7042
RFC (Excluding Coarse Aggregate)	0.6036	0.3643	16.0004	19.4675
RFC (Excluding Volume fraction of fibre)	0.5012	0.2512	19.0693	22.4879
RFC (Excluding Aspect Ratio)	0.8603	0.7401	20.2381	24.0631
RFC (Excluding Age)	0.8596	0.7389	19.7281	22.8833

Table 4: Comparison of Performance with Previous Models

Reference	Algorithm	R^2
[35]	Support Vector Machine	0.6801
[36]	K-Nearest Neighbour	0.984
[37]	Multilayer Perceptron	0.9107
This study	Randomizable Filtered Classifier	0.9886

Conclusions

The research utilised codeless software to develop an accurate model for the prediction of compressive strength of FRC. The dataset has 173 instances and seven input features which are important in the determination of compressive strength. A comprehensive approach was adopted for the modelling which produced the optimal RFC model with a R^2 of 0.9886. Binder is the most consequential input features in the developed model. The study demonstrates that data-driven soft computing techniques offer a reliable, cost-effective, and efficient approach for predicting the compressive strength of FRC, supporting their practical application in structural design.

Data Availability Statement

The dataset utilised in this study is available at https://github.com/NHDDTUEDU/DE_XGBoost_SFRC_fspt

References

[1] D.A. Joshi, G. Mounika, S. Sai Teja, V. Sarayu, K. Narendar, N.B. Ibrahim, D.M. Sangeetha, Novel Mechanical Strength Prediction Models of Fibre Reinforced Concrete Using Statistical Analysis, E3S Web of Conferences, 559 (2024). https://doi.org/10.1051/e3sconf/202455904019

[2] Z. Wang, H. Li, X. Zhang, Y. Chang, Y. Wang, L. Wu, H. Fan, The Effects of Steel Fiber Types and Volume Fraction on the Physical and Mechanical Properties of Concrete, Coatings, 13 (2023). https://doi.org/10.3390/coatings13060978

[3] G. Plizzari, S. Mindess, Fiber-reinforced concrete, in: Developments in the Formulation and Reinforcement of Concrete, 2019, pp. 257-287. https://doi.org/10.1016/B978-0-08-102616-8.00011-3

[4] F.M.D.S. More, S.S. Subramanian, Impact of Fibres on the Mechanical and Durable Behaviour of Fibre-Reinforced Concrete, Buildings, 12 (2022). https://doi.org/10.3390/buildings12091436

[5] T.E.T. Buttignol, J.L.A.O. Sousa, T.N. Bittencourt, Ultra High-Performance Fiber-Reinforced Concrete (UHPFRC): a review of material properties and design procedures, Revista IBRACON de Estruturas e Materiais, 10 (2017) 957-971. https://doi.org/10.1590/s1983-41952017000400011

[6] F. Kassimi, K.H. Khayat, Mechanical properties of fiber-reinforced concrete with adapted rheology, Cement and Concrete Composites, 118 (2021). https://doi.org/10.1016/j.cemconcomp.2021.103958

[7] N. Nassif, M. Talha Junaid, K. Hamad, Z. Al-Sadoon, S. Altoubat, M. Maalej, Performance-Based Prediction of Shear and Flexural Strengths in Fiber-Reinforced Concrete Beams via Machine Learning, Structural Engineering International, 34 (2024) 651-656. https://doi.org/10.1080/10168664.2024.2310520

[8] R. Wang, J. Zhang, Y. Lu, J. Huang, Towards Designing Durable Sculptural Elements: Ensemble Learning in Predicting Compressive Strength of Fiber-Reinforced Nano-Silica Modified Concrete, Buildings, 14 (2024). https://doi.org/10.3390/buildings14020396

[9] M. Amar, Comparative use of different AI methods for the prediction of concrete compressive strength, Cleaner Materials, 15 (2025). https://doi.org/10.1016/j.clema.2025.100299

[10] H. Alguhi, D. Tomlinson, Experimental and analytical study of steel and chopped glass fibre reinforced concrete under compression, Construction and Building Materials, 418 (2024). https://doi.org/10.1016/j.conbuildmat.2024.135421

[11] H. An, Y. Song, L. Liu, X. Meng, Experimental Study of the Compressive Strengths of Basalt Fiber-Reinforced Concrete after Various High-Temperature Treatments and Cooling in Open Air and Water, Applied Sciences, 11 (2021). https://doi.org/10.3390/app11188729

[12] M. Pourbaba, R. Chakraborty, M. Pourbaba, A. Belarbi, J. Yeon, A New Insight into the Design Compressive Strength of Ultra-High Performance Concrete, Buildings, 13 (2023). https://doi.org/10.3390/buildings13122909

[13] A.U. Abubakar, Statistical Evaluation of Compressive Strength in High Performance Concrete (HPC) with Steel Fiber Addition, Sustainable Structures and Materials, An International Journal, 2 (2019) 14-31.

[14] S. Kim, S. Han, C. Park, K.K. Yun, Compressive Behavior Characteristics of High-Performance Slurry-Infiltrated Fiber-Reinforced Cementitious Composites (SIFRCCs) under Uniaxial Compressive Stress, Materials (Basel), 13 (2020). https://doi.org/10.3390/ma13010159

[15] M. Abdulkareem, F. Ayeronfe, T.M. Jassam, A.H. AlAteah, K.A.A. Al-Sodani, M.M.H. Al-Tholaia, H. Yam, A. Ganiyu, S.C. Alih, Compressive and Flexural Strengths of Bio-Recycled Concrete Incorporated with Kenaf Fibre, Journal of Natural Fibers, 21 (2024). https://doi.org/10.1080/15440478.2023.2296913

[16] M. Abdulkareem, A. Ganiyu, A.H. AlAteah, K.A.A. Al-Sodani, M.M. Al-Tholaia, F. Ayeronfe, Static and Dynamic Resistance of Bio-concrete Incorporated with Kenaf Fiber, Journal of Natural Fibers, 22 (2025) 2462976. https://doi.org/10.1080/15440478.2025.2462976

[17] J. Li, J. Chang, H. Qiao, Research on optimal design of mechanical properties of hybrid fiber-reinforced concrete, Structural Concrete, 23 (2022) 1625-1641. https://doi.org/10.1002/suco.202100635

[18] H.H.Z. Khalel, M. Khan, A. Starr, N. Sadawi, O.A. Mohamed, A. Khalil, M. Esaker, Parametric study for optimizing fiber-reinforced concrete properties, Structural Concrete, 26 (2024) 88-110. https://doi.org/10.1002/suco.202300509

[19] Y.M. Abbas, M.I. Khan, Robust Machine Learning Framework for Modeling the Compressive Strength of SFRC: Database Compilation, Predictive Analysis, and Empirical Verification, Materials (Basel), 16 (2023). https://doi.org/10.20944/preprints202310.1573.v1

[20] W. Ben Chaabene, Machine Learning Prediction of Shear Capacity of Steel Fiber Reinforced Concrete, (2020).

[21] A.M. Salim, A.M. Salim, V. Gribniak, A. Rimkus, K. Holschemacher, U. Quapp, A. Singh, Mechanical Properties of Different Types of Fiber-Reinforced Concrete: Benefits and Limitations, Proceedings of International Structural Engineering and Construction, 11 (2024). https://doi.org/10.14455/ISEC.2024.11(2).MAT-09

[22] A.P. Adewuyi, O. Animbom, Performance-Based Evaluation of Steel Fibre Reinforced Normal- and High-Strength Concretes Using Statistical Analysis of Experimental Database, Journal of Civil Engineering and Urbanism, 14 (2024) 238-246. https://doi.org/10.54203/jceu.2024.25

[23] W. Ni, X. Cui, J. Yuan, W. Sun, C. Cui, Y. Wu, J. Feng, The influence of fiber, aggregate and cementitious materials on the mechanical properties of ultra-high content steel fiber reinforced reactive powder concrete, Construction and Building Materials, 431 (2024) 136530. https://doi.org/10.1016/j.conbuildmat.2024.136530

[24] M. Iqbal Khan, Y.M. Abbas, Significance of fiber characteristics on the mechanical properties of steel fiber-reinforced high-strength concrete at different water-cement ratios, Construction and Building Materials, 408 (2023). https://doi.org/10.1016/j.conbuildmat.2023.133742

[25] L.R. Mailyan, A.N. Beskopylny, B. Meskhi, S.A. Stel'makh, E.M. Shcherban, O. Ananova, Optimization of Composition and Technological Factors for the Lightweight Fiber-Reinforced Concrete Production on a Combined Aggregate with an Increased Coefficient of Structural Quality, Applied Sciences, 11 (2021). https://doi.org/10.3390/app11167284

[26] A.A. Ganiyu, F.R. Alalawi, Efficacy of instance-based learning algorithms for the prediction of ultimate bearing capacity of piles, in: 4th Asia-Pacific Conference on Physical Modelling in Geotechnics, 2024.

[27] I.B. Mustapha, Z. Abdulkareem, M. Abdulkareem, A. Ganiyu, Predictive Modeling of Physical and Mechanical Properties of Pervious Concrete using XGBoost, Neural Computing and Applications, (2024). https://doi.org/10.1007/s00521-024-09553-w

[28] I.B. Mustapha, M. Abdulkareem, T.M. Jassam, A.H. AlAteah, K.A.A. Al-Sodani, M.M.H. Al-Tholaia, H. Nabus, S.C. Alih, Z. Abdulkareem, A. Ganiyu, Comparative Analysis of Gradient-Boosting Ensembles for Estimation of Compressive Strength of Quaternary Blend Concrete, International Journal of Concrete Structures and Materials, 18 (2024). https://doi.org/10.1186/s40069-023-00653-w

[29] A. Pal, K.S. Ahmed, F.M.Z. Hossain, M.S. Alam, Machine learning models for predicting compressive strength of fiber-reinforced concrete containing waste rubber and recycled aggregate, Journal of Cleaner Production, 423 (2023). https://doi.org/10.1016/j.jclepro.2023.138673

[30] S. Duan, Compressive strength prediction of fiber-reinforced recycled aggregate concrete based on optimization algorithms, Frontiers in Built Environment, 10 (2024). https://doi.org/10.3389/fbuil.2024.1509714

[31] Y. Li, Q. Zhang, P. Kamiński, A.F. Deifalla, M. Sufian, A. Dyczko, N.B. Kahla, M. Atig, Compressive strength of steel fiber-reinforced concrete employing supervised machine learning techniques, Materials, 15 (2022) 4209. https://doi.org/10.3390/ma15124209

[32] M. Nidhya, P.K. Shah, Acquiring Knowledge by Performing Classification and Clustering of Datasets using WEKA: Intelligence through MLP, RF, DT, and RepTree, in: 2023 International Conference on Communication, Security and Artificial Intelligence (ICCSAI), IEEE, 2023, pp. 833-837. https://doi.org/10.1109/ICCSAI59793.2023.10421077

[33] V. Mhetre, M. Nagar, Classification based data mining algorithms to predict slow, average and fast learners in educational system using WEKA, in: 2017 International Conference on Computing Methodologies and Communication (ICCMC), IEEE, 2017, pp. 475-479. https://doi.org/10.1109/ICCMC.2017.8282735

[34] N.-D. Hoang, Predicting tensile strength of steel fiber-reinforced concrete based on a novel differential evolution-optimized extreme gradient boosting machine, Neural Computing and Applications, (2024). https://doi.org/10.1007/s00521-024-10458-x

[35] A.U. Abubakar, K.A. Kau, M. Hassan, M.S. Tabra, Prediction of Compressive Strength in High Strength Concrete with Steel Fiber Addition using Support Vector Machine Algorithm, Sustainable Structures and Materials, An International Journal, 3 (2020) 25-36.

[36] A.U. Abubakar, M.S. Tabra, Prediction of Compressive Strength in High Performance Concrete with Hooked-End Steel Fiber using K-Nearest Neighbor Algorithm, International Journal of Integrated Engineering, 11 (2019). https://doi.org/10.30880/ijie.2019.11.01.016

[37] A.U. Abubakar, M.S. Tabra, Predictive Capabilities of Multilayer Perceptron (MLP) in WEKA Algorithm for High Strength Concrete with Steel Fiber Addition, Journal of Cement Based Composites, 1 (2020) 13-18. https://doi.org/10.36937/cebacom.2020.002.003

Impacts of Cement, Sawdust and Eggshell Powder on the Properties of Laterite Bricks

Wasiu Olabamiji AJAGBE[1,a], Mariam IDRIS[1,b], John BRAIMAH[2,c], Abideen Adekunle GANIYU[3,d*]

[1]University of Ibadan, Ibadan, Oyo State, Nigeria

[2]Bells University of Technology, Ota, Ogun State, Nigeria

[3]British University of Bahrain, Saar 527, Bahrain

[a]ajagbewas@gmail.com, [b]idrismariam25@gmail.com, [c]jibraimah@bellsuniversity.edu.ng, [d]a.ganiyu@bub.bh

Keywords: Sawdust, Eggshell Powder, Laterite Bricks, Cement Stabilisation, Sustainable Building Materials

Abstract. Bricks are fundamental construction units, valued for their durability, thermal properties, and versatility in load-bearing and non-load-bearing structures. With the growing concerns over the environmental footprint of conventional building materials, a promising approach is the utilisation of agricultural and industrial by-products in lieu of conventional materials to reduce environmental impact while meeting structural and durability needs. This study investigates the impacts of incorporating sawdust (SWD) and eggshell powder (ESP) in the production of cement-stabilised laterite bricks. The index properties of laterite sample were determined, different composite mixes of SWD, ESP, cement and laterite in different proportions were evaluated for the compressive strength at 7, 14 and 28 days, water absorption, and thermal conductivity tests were also conducted. The results show that an increase in SWD contents resulted in lower structural strength and higher water absorption, and the inclusion of ESP enhances both the strength and thermal insulation. ESP also enhances the durability of bricks by reducing the water absorption ability. By examining these alternative raw materials for brick production, the research contributes to the development of eco-friendly construction practices which aligned with global sustainability and resource efficiency goals.

Introduction

The demand for sustainable, cost-effective, and environmentally friendly construction materials has intensified due to rising costs of conventional materials, environmental concerns, and the need for affordable housing. This is driven by the construction industry's significant environmental impact, resource consumption, and the growing demand for eco-friendly solutions. Sustainable materials offer a pathway to mitigate these impacts by reducing carbon emissions, conserving resources, and promoting healthier living environments. These materials are not only crucial to environmental sustainability but also offer economic and social benefits, making them an essential component of modern construction practices[1].

Laterite, a soil rich in iron and aluminium, is prevalent in tropical regions and can be utilised in various forms for construction purposes. Laterite is widely regarded as an eco-friendly construction material due to its low toxicity, recyclability, and minimal processing requirements [2]. Cement-stabilised laterite bricks offer a viable alternative to traditional materials, reducing embodied energy and carbon emissions associated with fired clay bricks. Utilising laterite in brick production can mitigate environmental issues associated with the over-extraction of traditional raw materials like sand and clay [3].

To improve the performance and sustainability of lateritic bricks, research has explored the incorporation of supplementary cementitious materials (SCMs) and agricultural waste. SCMs are natural or artificial substances that contain silica, alumina, and calcium which when finely divided, react chemically with calcium hydroxide in the presence of water at room temperature to form cementitious properties [4]. In other words, they are materials which in themselves do not possess cementitious properties but when combined with calcium hydroxide (from lime or Portland cement) can form cementitious compounds [5, 6].

Eggshell powder (ESP), a calcium-rich by-product of food processing, has shown potential as a partial cement replacement due to its high calcium carbonate content, which is similar to limestone used in cement production [7, 8]. The utilisation of ESP significantly reduces CO_2 emissions associated with cement manufacturing, and addresses waste management issues by repurposing eggshells, which are otherwise disposed of in landfills, contributing to environmental pollution. ESP also enhances the mechanical properties of concrete including compressive, tensile and flexural strengths [9-11]. Chemical analysis confirms that ESP has pozzolanic properties, containing high percentages of silica (SiO_2), iron oxide (Fe_2O_3), alumina (Al_2O_3), and calcium oxide (CaO) [12-14]. Lack of industrial-scale supply of ESP is a likely barrier to its use in concrete production.

Likewise, sawdust (SWD), a lignocellulosic waste from timber processing, has been investigated as a soil modifier and lightweight aggregate in construction materials primarily due to its potential to reduce environmental impact and construction costs. Generated in large quantities from sawing, milling, and sanding operations, sawdust is often discarded, leading to environmental hazards such as air and water pollution [15]. Incorporating sawdust into construction materials such as concrete and mortar offers a dual advantage: reducing construction costs and mitigating the environmental impact of sawmill waste. Sawdust has low thermal conductivity which helps to improve thermal insulation, its lightweight nature yields reduced density, making it suitable for non-structural applications [16-18].

Both ESP and SWD offer environmental and economic benefits by reducing reliance on cement and promoting waste reuse. Despite their potential, limited studies have investigated the combined impact of ESP and SWD on the engineering properties of cement-stabilised laterite bricks. This study addresses that gap by evaluating the effects of partial cement replacement with ESP and partial laterite replacement with SWD on the physical, mechanical, and thermal performance of unfired laterite bricks. The research aims to contribute to sustainable construction practices by developing low-cost, energy-efficient building materials from locally available resources.

Methodology

Laterite was sourced from Ajibode, University of Ibadan extension (7.469495N, 3.904925E), while Sawdust was sourced from the Bodija sawmill (7.433889N, 3.915556E), both in Ibadan, Nigeria. Commercially available Portland lime cement (Grade 32.5N) was used, eggshells were sourced from local snack vendors and blended with a commercial blender, and water from the material laboratory of the University of Ibadan was used. Laboratory tests on undisturbed samples of the soil collected from the field were carried out at the University of Ibadan Geotechnical Laboratory. Both the sieve analysis and the hydrometer test were done to determine the particle size distribution of the laterite. The Atterberg limit tests conducted include both the liquid limit and the plastic limit, the plastic index was determined from the outcome of both tests. Specific gravity and moisture content tests were also conducted; the test procedures are in accordance with ASTM-D4318-17 and ASTM-D-2216-19 [19, 20].

Wooden moulds were used in the preparation of 200 x 100 x 65mm laterite bricks according to the ASTM-C62-23 [21]. The formwork was lubricated with oil for easy removal when demoulding. The samples were mixed manually with the aid of shovel and hand trowel, and water was gradually added while mixing until homogeneity was achieved. Mixed sample was put in the mould in three layers, each layer was filled following 25 blows using tamping rod, 16 ± 0.5 mm diameter and 600 ± 2 mm long. The sample was kept under the shade to avoid direct rainfall. Curing by water sprinkling started after 24 hours when the bricks were demoulded and continued until the day of testing. A total of 80 bricks were made, 3 samples each for compressive strength tests on 7, 21 and 28 days, 3 and 2 samples each for water absorption, and thermal conductivity tests, respectively, on the 28th day. Table 1 presents the material mix proportions and the number of tests carried out. Both the compressive strength test and water absorption tests were done according to ASTM-C67/C67M-24 [22], while the thermal conductivity test was carried out according to ASTM-D5334-22 [23].

Table1: Material mix proportions and test designations

	Mix Proportion (%)				Number of tests done		
Mix Name	Laterite	Sawdust	Cement	ESP	Compressive Strength	Water Absorption	Thermal Conductivity
A	95	-	5	-	3 (3)	3	2
B	90	5	5	-	3 (3)	3	2
C	85	10	5	-	3 (3)	3	2
D	95	-	3	2	3 (3)	3	-
E	95	-	2	3	3 (3)	3	-
F	90	5	3	2	3 (3)	3	2

Results and Discussion

The initial moisture content of the lateritic soil was 16.73%, the specific gravity was 2.65, the particle size distribution analysis result shows 8.17% clay, 13.73% silt and 78.10% sand, the liquid limit is 27.92, plastic limit is 17.97, and plasticity index is 9.95. Basically, it is classified as silty sand (SM) of low plasticity, according to USCS soil classification [24]. Figure 1 presents the compressive strength results of the bricks of different mix proportions at 7, 14 and 28 days.

Mix A with Laterite (95%) and 5% cement consistently has the highest compressive strength at all ages. The replacement of Laterite with 5% SWD (mix B) led to a significant drop in the compressive strength. Further reduction in laterite and increase in SWD, i.e. 85% laterite and 10% SWD (mix C) led to a further reduction in the compressive strength, with sample C consistently having the lowest compressive strength through the three test days (7, 14, & 28 days). For mixes A, B, and C, no ESP was added, and the cement content is constant at 5%, hence, the variation in compressive strength is directly attributed to the quantity of SWD contents. SWD is organic and lightweight, with more pores and voids, all these would create weak zones, and reduce the density, hence the load bearing capacity of the bricks. SWD has limited bonding with laterite and interrupts the interlocking and compaction of laterite particles. This limitation weakens the internal structure and lowers the structural integrity of the samples leading to the reduction in compressive strength of the bricks, from 7.51 MPa to 1.11 MPa at 28 days. The results align with earlier research on the impact of SWD on concrete [17, 18].

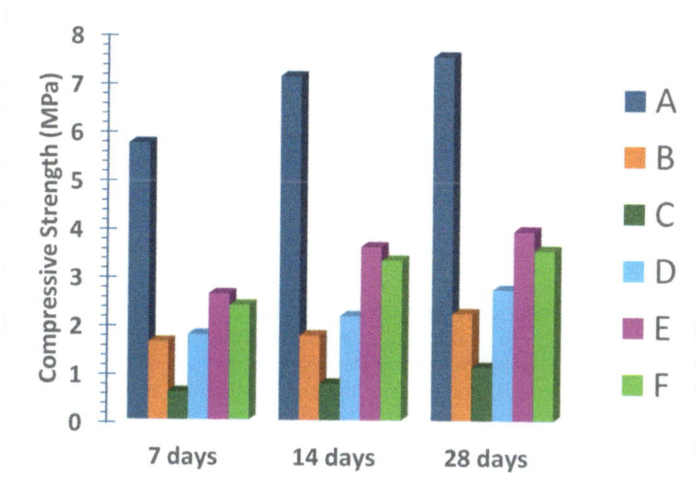

Figure 1: Compressive strength of SWD and ESP bricks at 7, 14 and 28 days

For mixes D and E, with no SWD but partial replacements of cement with ESP, the compressive strength rose again, due to the pozzolanic properties of ESP. The fine particle size of ESP helps to fill voids in the soil matrix thereby reducing porosity, improving the particle packing, and enhancing density, hence increasing the compressive strength [25]. The compressive strength of mix E (3.91 MPa) is higher than that of mix D (2.7 MPa) because the ESP content is slightly higher. The reduction in the compressive strength of mix F (3.52 MPa) is due to the re-introduction of SDW into the mix, and the consequent reduction of the laterite contents. It is pertinent to mention that mixes A, D, E, and F satisfy the minimum requirement of 2.5 MPa of compressive strength of brick according to the Nigerian Building Code [26]. Basically, mix E (3.91 MPa) has the best structural performance after the control mix, A (7.51 MPa). All values are at 28 days test.

Table 2 presents the outcome of water absorption test on the laterite bricks soaked by full immersion for 24 hours after 28 days. Comparing mixes, A, B and C, the higher the SWD contents, the higher the water absorption. Mix C with the highest SWD contents (10%) has the highest absorption, while Samples F and B with 5% SWD have subsequent high absorption rate. It is noted that this is due to the porous and hydrophilic nature of sawdust [27, 28]. Mix with ESP contents, D and E also have higher absorption when compared to mix A which has none of the SCMs. It is relevant to mention that none of the bricks dissolved after removal from a 24-hours total immersion in water. For better performance, low water absorption implies greater durability, less subjectivity to efflorescence, cracking, weathering and erosion, and stronger structural performance. In addition, low water absorption prevents moisture penetration, mold growth, and decay of internal finishes such as plaster and painting [29, 30]. In summary, Mixes D and E containing ESP without SWD have the best absorption performance after the control mix A (without both SWD and ESP).

Table 2: Water absorption computations

	Number of tests done		
Mix Name	Dry weight at 28 days (Kg)	Saturated Weight after 24 hours (Kg)	Water Absorption (%)
A	2.434	2.720	11.75
B	1.888	2.304	22.03
C	1.360	1.896	39.41
D	2.306	2.606	13.00
E	2.186	2.510	14.82
F	1.938	2.378	22.76

The result of thermal conductivity test on the laterite bricks is presented in Table 3. The initial temperature taken is referred to as 0 minutes, and subsequent temperature was taken after two minutes according to ASTM-C67M-24 [23]. The result shows that laterite brick samples partly replaced with 5% SWD, Mix B, have higher thermal conductivity than those with 10% SWD composition, Mix C. This is because sawdust is highly porous, adding more sawdust increases air content and voids within the brick resulting in lower thermal conductivity and better insulation. Mix F, with 2% ESP contents has the lowest thermal conductivity, implying the best insulator from the tested mixes. For performance purposes, brick mix F with best insulation helps keep indoor temperatures stable, reducing reliance on heating or cooling systems thereby reducing electricity/fuel costs. In addition, it reduces heat gain during the day and keeps the interiors cooler, and more comfortable [31].

Table 3: Thermal conductivity results

		Sample 1						Sample 2					
Mix Name	Time (mins)	0	2	4	6	8	10	0	2	4	6	8	10
A	Temp (°C)	38	41	49	57	65	73	38	42	50	58	66	74
B		43	51	58	65	77	90	43	52	59	66	77	91
C		39	43	48	55	63	70	39	41	47	56	65	72
F		39	42	48	52	60	67	39	41	47	53	61	68

Conclusions

The research findings indicate that laterite bricks made with sawdust alone exhibit lower compressive strength compared to those incorporating both sawdust and eggshell powder. The addition of 2% ESP enhanced the strength of bricks containing 5% SWD, while the highest compressive strength was obtained for the sample with laterite, cement and ESP without the SWD. Water absorption increased with an increase in SWD content, which peaks at 10% SWD due to the porous and hydrophilic nature of sawdust. In contrast, the inclusion of ESP lowers the water absorption capabilities, thereby enhancing its durability. Although bricks with 10% SWD showed lower thermal conductivity and better insulation, incorporating ESP into bricks resulted in the best insulation among the bricks evaluated. Overall, the study concludes that the inclusion of SWD alone effectively reduces the strength and insulation properties of laterite bricks, the addition of ESP counteracts those effects and provides a balanced optimal strength and insulation enhancements. ESP also enhances the durability of bricks by reducing the water absorption ability. It is recommended that future studies explore the use of lower percentages of sawdust with higher proportions of ESP to evaluate variations in strength, water absorption and thermal conductivity properties. Other sustainable non-ash pozzolans may also be explored in combination with sawdust for further exploration of brick performance.

References

[1] A.A. Ganiyu, M.O. Abdulkareem, W.O. Ajagbe, Insights on Sarooj: Oman's Traditional Pozzolan, in: International Conference on Advances in Cement and Concrete Research, 2023, Ladoke Akintola University of Technology,, Ogbomoso, Nigeria, 2024.

[2] S.I. Adedokun, A.A. Ganiyu, M.A. Adedokun, Effect of marble dust and steel slag on consistency limits and compaction characteristics of lateritic soil, in: IOP Conference Series: Materials Science and Engineering, IOP Publishing, 2019, pp. 012026. https://doi.org/10.1088/1757-899X/527/1/012026

[3] G.S. Kumar, P.K. Saini, R. Deoliya, A.K. Mishra, S.K. Negi, Characterization of laterite soil and its use in construction applications: A review, Resources, Conservation & Recycling Advances, 16 (2022). https://doi.org/10.1016/j.rcradv.2022.200120

[4] I. Bachheti, R. Rawat, B. Pratap, GGBS-Incorporated Lightweight Geopolymer Concrete: A Comprehensive Review on Its Sustainability and Performance Characteristics, Journal of Structural Design and Construction Practice, 31 (2026). https://doi.org/10.1061/JSDCCC.SCENG-1910

[5] A. Ganiyu, A. Al Abdulsalaam, H. Al Alawi, A. Badr, Impacts of alkaline activated slag blended with Omani Sarooj on expansive soil, in: The 75th Canadian Geotechnical Conference, GeoCalgary 2022., The Canadian Geotechnical Society., Calgary, AB, Canada., 2022.

[6] W.O. Ajagbe, R.O. Okpanachi, A.A. Ganiyu, M.O. Abdulkareem, Strength properties of self-compacting concrete incorporating iron ore tailings, in: Advanced Topics in Mechanics of Materials, Structures and Construction, 2023, pp. 415-421. https://doi.org/10.21741/9781644902592-43

[7] S. Maqsood, L.S.S. Eddie, Effect of using calcined eggshells as a cementitious material on early performance, Construction and Building Materials, 318 (2022). https://doi.org/10.1016/j.conbuildmat.2021.126170

[8] X.F. Li, S.I. Doh, P.J. Ramadhansyah, M.R. Hainin, W.I. Mohd Haziman, Effect of Eggshell Powder on the Mechanical and Durability Properties of Cement Mortar, in: Proceedings of the Sustainable Concrete Materials and Structures in Construction 2020, 2021, pp. 153-163. https://doi.org/10.1007/978-981-16-2187-1_14

[9] K. Chiraz, B. Fatima Zohra, N. Ghania, H. Houria, A. Hamid, M. Karima, Utilization of Eggshell Powder as a Green Mortar Material, Materials Science, 32 (2025) 241-247. https://doi.org/10.5755/j02.ms.37855

[10] M. Adamu, H. Alanazi, Y.E. Ibrahim, M. Abdellatief, Mechanical, microstructural characteristics and sustainability analysis of concrete incorporating date palm ash and eggshell powder as ternary blends cementitious materials, Construction and Building Materials, 411 (2024). https://doi.org/10.1016/j.conbuildmat.2023.134753

[11] B.O. Orogbade, M.A. Kareem, G.M. Amusan, O.S. Olaniyan, A.O. David, O.T. Opafola, E.O. Ibiwoye, Investigation of partial replacement of cement with egg shell powder in palm oil fiber-concrete, Materials Today: Proceedings, 86 (2023) 116-120. https://doi.org/10.1016/j.matpr.2023.04.164

[12] D. Cree, P. Pliya, Effect of elevated temperature on eggshell, eggshell powder and eggshell powder mortars for masonry applications, Journal of Building Engineering, 26 (2019). https://doi.org/10.1016/j.jobe.2019.100852

[13] A.C. Ogbonna, Characterization of Coconut Shell Ash and Eggshell Powder as Supplementary Cementitious Materials in Roller Compacted Concrete Industrial Access Pavements and Parking Facilities, Romanian Journal of Transport Infrastructure, 10 (2021) 67-87. https://doi.org/10.2478/rjti-2021-0005

[14] H.M. Hamada, B.A. Tayeh, A. Al-Attar, F.M. Yahaya, K. Muthusamy, A.M. Humada, The present state of the use of eggshell powder in concrete: A review, Journal of Building Engineering, 32 (2020). https://doi.org/10.1016/j.jobe.2020.101583

[15] K. Devi, N. Lida, T. Teyi, P.B. Singh, K. Sharma, N. Saini, Utilization of Sawdust as Sustainable Construction Material, in: Proceedings of SECON'22, 2023, pp. 137-145. https://doi.org/10.1007/978-3-031-12011-4_12

[16] B.C. Olaiya, M.M. Lawan, K.A. Olonade, Utilization of sawdust composites in construction-a review, SN Applied Sciences, 5 (2023). https://doi.org/10.1007/s42452-023-05361-4

[17] P. Bishetti, S. Varadharajan, B.K. Shukla, G. Bharti, Study on Suitability of Sawdust as an Alternate for Fine Aggregate in Concrete, in: Proceedings of Indian Geotechnical and Geoenvironmental Engineering Conference (IGGEC) 2021, Vol. 2, 2023, pp. 209-217. https://doi.org/10.1007/978-981-19-4731-5_19

[18] C.M. Yun, M.R. Rahman, D. Huda, K.K. Kuok, A.C.P. Sze, J.K. Seng, M.K.B. Bakri, Sawdust as Sand Filler Replacement in Concrete, in: Waste Materials in Advanced Sustainable Concrete, 2022, pp. 133-148. https://doi.org/10.1007/978-3-030-98812-8_8

[19] ASTM International, Standard Test Methods for Liquid Limit, Plastic Limit, and Plasticity Index of Soils, in: ASTM-D4318 – 17e1, ASTM International, West Conshohocken, PA, 2017.

[20] ASTM International, Standard Test Methods for Laboratory Determination of Water (Moisture) Content of Soil and Rock by Mass, in: ASTM-D2216 – 19, ASTM International, West Conshohocken, PA, 2019.

[21] ASTM International, Standard Specification for Building Brick (Solid Masonry Units Made from Clay or Shale), in: ASTM-C62 – 23, ASTM International, West Conshohocken, PA, 2023.

[22] ASTM International, Standard Test Methods for Sampling and Testing Brick and Structural Clay Tile, in: ASTM- C67/C67M – 24, ASTM International, West Conshohocken, PA, 2024.

[23] ASTM International, Standard Test Method for Determination of Thermal Conductivity of Soil and Rock by Thermal Needle Probe Procedure, in: ASTM-D5334-22, ASTM International, West Conshohocken, PA, 2022.

[24] ASTM International, Standard Practice for Description and Identification of Soils (Visual-Manual Procedures), in: ASTM-D2488 – 17e1, ASTM International, 2023.

[25] J.E.F.M. Ibrahim, O.B. Kotova, S. Sun, E. Kurovics, M. Tihtih, L.A. Gömze, Preparation of innovative eco-efficient composite bricks based on zeolite-poor rock and Hen's eggshell, Journal of Building Engineering, 45 (2022). https://doi.org/10.1016/j.jobe.2021.103491

[26] NBRRI, National Building Code of Nigeria, in, 2006.

[27] A. Rahman, M.A.H. Khondoker, Effect of Treatment Methods on Material Properties and Performance of Sawdust-Concrete and Sawdust-Polymer Composites, Polymers, 16 (2024). https://doi.org/10.3390/polym16233289

[28] S. Dias, A. Tadeu, J. Almeida, P. Humbert, J. António, J. de Brito, P. Pinhão, Physical, Mechanical, and Durability Properties of Concrete Containing Wood Chips and Sawdust: An Experimental Approach, Buildings, 12 (2022). https://doi.org/10.2139/ssrn.4081676

[29] A. Singh, A.K. Srivastava, A. Kumar, R. Bajaj, G.K. Singh, Design and development of plentiful fly ash-based glass powder-reinforced plastic composite bricks for low water absorption and high compressive and flexural strength, International Journal on Interactive Design and Manufacturing (IJIDeM), 19 (2023) 939-948. https://doi.org/10.1007/s12008-023-01580-6

[30] S. Kahangi Shahreza, J. Niklewski, M. Molnár, Experimental investigation of water absorption and penetration in clay brick masonry under simulated uniform water spray exposure, Journal of Building Engineering, 43 (2021). https://doi.org/10.1016/j.jobe.2021.102583

[31] W. Soliman, Y.M.Z. Ahmed, A. Ghitas, A.H. El-Shater, M.A. Shahat, Green building development utilising modified fired clay bricks and eggshell waste, Scientific Reports, 15 (2025). https://doi.org/10.1038/s41598-025-87435-4

Materials Research Proceedings 63 (2026) 64-71 https://doi.org/10.21741/9781644904053-8

Production of Flamboyant Pod Nanoparticles as Emerging Reinforcements for Structural Composite Materials

Sefiu Adekunle BELLO[1,a], Maruf Yinka KOLAWOLE[2,b], Musa Opeyemi ABDULLAHI[1,c], Blessing OGBONAYA[1,d], Ayodele Gabriel IDOWU[1,e], John Olu OMOTOSHO[1,f], Abdulabassit BALOGUN[1,g], Omoniyi Banji OGUNLOWO[1], Mubarak Akanbi OMOTAYO[1,i], Halimat Anuoluwapo FAMUYIWA[1,j], Salim Olakunle ABDULRAUF[1,k], Emmanuel Oladele JOHNBULL[1,l], Damilola Ifeoluwa BOLARINWA[1,m], Samuel Akinwale OGUNMODEDE[1,n], Mohammed Kayode ADEBAYO[1,o], Suleiman Danjuman DAUDU[1,p], Aisha Mayowa AKINTOLA[1,q], Sunday Wilson BALOGUN[1,r]

[1]Department of Materials Science and Engineering, Kwara State University, Malete, PMB 1530, Ilorin and 240212, Nigeria

[2]Department of Mechanical Engineering, Kwara State University, Malete, PMB 1530, Ilorin and 240212, Nigeria

[a]sefiu.bello@kwasu.edu.ng, [b]maruf.kolawole@kwasu.edu.ng, [c]adekunle-B@yahoo.com, [d]blessingogbonnaya3011@gmail.com, [e]idowuayodele009@gmail.com, [f]johnomotosho20@gmail.com, [g]abdulbassit613@gmail.com, [i]mubarakolaide800@gmail.com, [j]anuoluwapohalimat11@gmail.com, [k]abdulraufsalim16@gmail.com, [l]johnbullemmanual333@gmail.com, [m]bholarinwadamilola@gmail.com, [n]soguns2015@gmail.com, [o]mohammed.adebayo14@kwasu.edu.ng, [p]suleiman.daudu@kwasu.edu.ng, [q]aisha.akintola@kwasu.edu.ng, [r]sunday.balogun@kwasu.edu.ng

Keywords: Flamboyant, Agricultural Waste, Concrete Filler, Ball Mill, Ecofriendly Materials, Optimisation

Abstract. Dynamic changes in environmental conditions had necessitated continuous research towards the development of innovative structural composites with enhanced properties. To produce functional concrete-based composites, there is a need for structural modification using reinforcing fillers that could improve the strength and toughness of the composites. Agricultural exercise creates many nonedible components which constitute environmental nuisances owing to their little or no economic value. However, processing such components into useful reinforcing fillers in concrete and other matrices for composite development encourages a wealth creation from waste, environmental hygiene and ecofriendly materials for engineering applications. This study focuses only on the production of nanoparticles which can be used as fillers in composite development. Nanoparticles were produced from the flamboyant pods using a disc and optimised ball milling technique. Pods were pulverised using a disc miller and the powders obtained were ball milled at varied charge ratios and velocities to produce reinforcing nanoparticles. Particles obtained in each of the milling exercises were analysed and 2 functional interaction response surface model was developed to optimise the milling parameters and particle sizes. Result obtained indicated a decrease in size of the flamboyant pod particles as milling durations and charge ratios increase. P value of the model <0.0001 which is much less than 0.05 shows that the model is significant in explaining dependence of the particle sizes on the milling parameters. Hence, the developed model affirms that a minimum size of 3.5209 nm of the flamboyant pod particles can be obtained without agglomeration of particles when milling at 10 charge ratios for 17 hours 29 minutes.

Introduction

Comminution is the breaking down of solid materials into smaller particles using mechanical force. It is an important stage in many industrial processes, which includes mining, ceramic processing, and food processing [1-4]. Besides the size reduction, it helps in liberating valuable minerals from ore. The goal is to reduce particle size and increase surface area for subsequent processes like chemical reactions as found in ceramics processing, food processing, metal extraction, and composite development [5]. Comminution involves two stages which are crushing and milling. Crushing is the first stage of the comminution process. It involves breakage of large fragments into smaller pieces. Crushed powders are subjected to an additional size reduction by the mechanical milling process [6]. Millings are classified as wet, dry or cryogenic [7]. Dry milling is the breakage of particles in the air environment. It is applied to break brittle materials into pieces without suspending the feed materials or powder in any medium. Wet Milling is done in liquid medium. It is more effective than dry milling as it keeps particles together and prevents their suspension in air unlike the dry milling. It is used in breaking soft material when the dry milling is ineffective to break the powder. Cryogenic milling is carried out in inert gas (e.g nitrogen, helium) environment. The extreme coldness of the inert gas environment makes the materials/powders very brittle which aid their breakages or refinement. It is applied in breaking very soft materials which are flattening instead of breakage during the dry milling [8]. Different works are reported regarding the comminution process. Two axis rotation ball mill was designed, and its geometric model was created using CREO parametric 2.0 software. The performance of the fabricated ball mill was examined by milling 53 µm sized alumina and 37 µm sized silica at 10 charge ratios. It was concluded that the particles sizes of the alumina/silica mixture decreased as the milling duration increased [9]. Al1xxx nanoparticles were produced by milling 53 µm sized Al(1xxx) containing 0.1% carbonised coconut shell particle as the process control agent. Samples obtained at 16, 46 and 70 hours were analysed. It was reported that wide variations in particle sizes were due to different ball impacts on the individual particles during the ball milling exercises. Effects of charge ratios (mass of balls to mass of the powder) milling duration on sizes of coconut shell particles were examined using gaussian function. It was declared that the finest particle size was obtained at charge ratio value of 10 while the gaussian regression function affirmed the minimum particle size at charge ratio value of 13.49 [10]. Furthermore, effect of milling durations on particle sizes of uncarbonised coconut shell were studied [11, 12]. Sizes of the coconut shell particles obtained at 16 and 70 hours of milling were determined using images obtained from scanning electron microscope (SEM) as reported in [12]. It was shown that the particle sizes of the uncarbonised coconut shell decreased with increased milling durations. Also, sizes of the uncarbonised coconut shell nanoparticles obtained at 16, 46 and 70 hours and 10 charge ratios were determined by images obtained from SEM analysis and x-ray diffractometer aided with Scherrer's equation. Both methods affirm the decrease in the particle sizes of the uncarbonised coconut shell nanoparticles with the increase in the milling duration and there are variations among the particle sizes determined from the SEM and XRD [11]. Breakage of palm kernel, coconut and periwinkle shells were studied. Powders of palm kernel, coconut and periwinkle shells of equal sizes were milled under the same conditions. It was displayed that coconut shell particles have a smallest average size after milling for 70 hours and charge ratio value of 10 [13]. In addition, production of coconut shell-based carbon nanoparticles was reported. Carbon flakes obtained after carbonisation were milled at different durations using tumbler ball mill. It was revealed that carbon flakes attained nanoparticle size smaller than those of the uncarbonised coconut shells at 16 hours of milling [14]. Comminution is also applied in food industries in processing food granules. Maize starch was milled using a planetary ball mill operating at 500 rpm for 1, 2 and 3 hours. It was reported according to TEM and SEM examination results that there were changes in the maize granule surfaces and cavities in the central region of the distorted granules were observed in the company with disappearance of alternate growth rings [15]. Moreover, effect of ball milling on the physicochemical and functional properties of insoluble dietary fibre obtained from three types of grain bran (prosomillet, wheat and rice), which opened a broad prospect for the by-product bran with low utilization was investigated according to the report in the article [16]. It was concluded that the ball milling operation provided an effective treatment of insoluble dietary fibres by reduction of the particle size, increase of the surface area, and more exposure of the functional groups and aiding the release of polyphenols, which improves the physicochemical and functional properties of the insoluble dietary fibres [16]. Structural, physicochemical and digestive properties of wheat starch type A and type B were examined using ball milling at 10, 20, 30 40 and 60 minutes. The milled samples were examined using SEM. It was reported that more fragmented granules existed in the starch granules as the ball milling time increased and rapidly digestible components of the milled starch samples are much more than those in the control starch samples. B type wheat granule broke easily than A type wheat granule as reported in the article [16].

Flamboyant trees are flowering plants producing pods two times, every year. The pods harbouring numerous numbers of wooden-inedible seeds. Being wooden in nature, both pods and seeds are hard materials. Their hardness makes them valuable materials for modifying structures of matrices like concrete, clay, polymer or metal for composite fabrication. Uses of flamboyant seeds and pods particles have been reported in literature [17, 18] as reinforcement in polymers to engineer new composites. Those studies indicate greater strength and ductility of the composites than their pristine matrices. The studies affirm importance of the flamboyant tree products like seeds and pods, but they provide no information on optimisation of milling parameters to produce the flamboyant seed/pod particles. This study emerged to address this gap. Attempts were made in this study to produce flamboyant pod particles by varying period of milling and charge ratios. This aims to disclose particle sizes obtained at different milling parameters. Establishing an actual charge ratio and milling durations that give minimum average size of the flamboyant pod particles through experiment and modelling is one of the focuses of this study. It is expected that the mathematical model developed in this study would guide the future researchers to predict the size of the flamboyant particles and other agricultural waste having similar hardness values when milled at a charge ratio and milling duration.

Materials and Method

The material used in this study is the flamboyant pods (Fig. 1). The dried flamboyant pods were picked after drying and falling from the trees planted for ornamental purposes within the Kwara State University, Malete. Each pod was opened into two parts to recover the harbouring seeds. The pod pieces were air blown to remove the dust and crushed to produce the flamboyant pod flakes which were subject to disc grinding to transform the pod flakes to the pod powders. Powders were sorted using a set of sieves arranging in descending order of mesh sizes with the collecting or receiving pan at the bottom of the sieve set. After pouring the powder to the top sieve and covered, they were set to vibration by sine shaker for 30 minutes. The finest powders collected in the pan were subjected to millings at different conditions. Milling speed that is the drum rotation remains constant at 100 rpm for all milling exercises. In the first milling exercise, 200g of the flamboyant pod were weighed and poured into the tumbler ball mill locally fabricated at Department of Materials Science and Engineering, Kwara State University, Malete. 1,600g of hardened steel milling balls were added which equal to 8 charge ratios (1600/200). The ball mill was tightly closed to prevent air access and milling began after switching the machine on. Milling took place for 55 hours with a stop at 25 hours to take samples for analysis. In the second milling exercise, 200g of the flamboyant powders was poured into the ball mill with an addition of 2kg of milling balls which is equal to 10 charge ratios. Milling durations are the same as in the first milling exercises.

Samples taken were observed physically and analysed using XRD, SEM, and UV-Vis spectrometer.

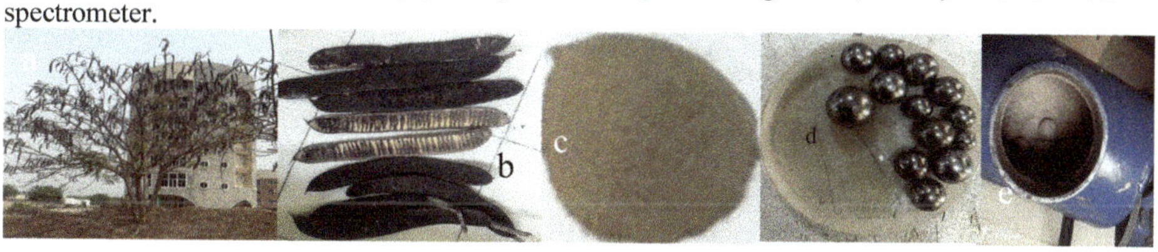

Fig. 1, Pictorial materials and milling process for producing optimised flamboyant pod particles (a) flamboyant tree (b) flamboyant pods (c) finest flamboyant powders (d) hardened stell milling balls (e) milled particles ready for discharge

X-ray Diffraction Analysis

An X-ray diffraction analysis was carried out on each milled particle to analyse the structure of crystalline materials in the particles. In the analysis X-ray beam was directed at the particles and the scattered intensities as a function of the positions (2theta) were measured. Once the beam is separated, the scatter (diffraction pattern) indicates the sample's crystalline structure. Before the analysis, particles were pressed into the sample holder at an angle of 45°. Scanning was carried out at 25°C in a continuous mode from angle 6° to 75°(2θ) at a step size of 0.0026° with the aid of Philip X-ray diffractometer set at 45 kV and 40 mA using Cu anode with Goniometer radius, distant focus divergence slit and irradiated length of 240, 91 and 4 mm respectively. Peaks were assigned with phases using X'pert High Score software. Sizes of crystalline phases detected by the X-ray were determined from wavelength (λ) of the Cu radiation used for analysing the particles, shape correction factor (K), full width half maximum (β) and cosine of angle of radiation (2θ) using Scherrer's formula in Eq 1.

$$Particle\ size = \frac{\lambda K}{\beta \cos(2\theta)} \tag{1}$$

Transmission Electron Microscopic Analysis

A multipurpose field emission transmission electron microscope (TEM), Model: JEM 2100F wit attached megapixel (2048 × 2048) bottom mount CCD camera was employed to analysis the flamboyant pod particles. Particles for the TEM analysis were prepared by drying the particles on a thin carbon film coated Cu grid. Magnifying images produced by electron beam transmitted through the particles at an accelerating voltage of 100 keV were captured by the attached CCD camera to give TEM images of the analysed particles. This description of the TEM analysis is according to the procedures published in the article [19].

Mathematical Analysis

To obtain optimal parameters for the ball milling, sizes of flamboyant pod particles determined from XRD using Scherrer's equation were analysed using historical approach 2 functional interaction response surface model (HA2FIRSM). The HA2FIRSM was used because its superior features of the modelling parameters among other functions like quadratic, cubic and linear functions. Both quadratic and cubic functions are aliased implying that it is impossible for each of both functions to generate model equation for the analysed data. Both linear and 2FI functions are significant but 2FI was chosen to build the model because of its higher Fishers' value. Moreover, HA2FIRSM was developed using milling durations and charge ratios as predictor/independent variables and the particle sizes of the flamboyant pod particles as responses/dependent variable. Minimum and maximum sizes of the particles were set as lower and upper limits, respectively. The model was developed at 95% confidence with $\alpha = 5\%$ (0.05). Model diagnostic parameters were determined to ascertain the appropriateness of the HA2FIRSM. This analysis was built according to procedures in literature [20].

Results and Discussion

X-Ray Diffraction Diffractograms of *Flamboyant* Pod Particles

Phases of flamboyant pod particles vary as it is evident in Fig. 2a that shows different phases matched with peaks at different positions (2θ). The phase differences could be attributed to chemical interactions among phases of the particles during ball milling process. Similar observation was made literatures [2, 4, 11]. Differences in phases could be responsible for colour variations observed with the particles as milling durations and charge ratios increased. Superimposed curves in Fig. 2a compares the diffractograms of the milled particles obtained at 8 and 10 charge ratios by milling for 25 and 55 hours. It is observed that intensities of the milled particles are much greater than that of the control flamboyant pod power sample. This can be explained by breakages of individual powders during milling process to produce many smaller fragments. Discrepancies in the intensities of the milling particles is also clear. This testifies to the fact that the milled particles at different charge ratios and milling durations have varied sizes. Powders used as feeds for the ball milling has an average size of 37 μm. Sizes of milled particles obtained at different milling parameters, determined using Scherrer's formula in Eq 1 are presented in Fig. 2b. This shows a plot of average sizes against milling durations at different charge ratios. It indicates that the average size of the flamboyant pod particles decreases drastically at 8 and 10 charge ratios by milling for 25 hours. The smaller average size of the milling at 8 charge ratios than that at 10 charge ratios at same milling duration of 25 hours can be attributed to the fact that minimum possible size has been attached before 25 hours in the case of milling at 10 charge ratios.

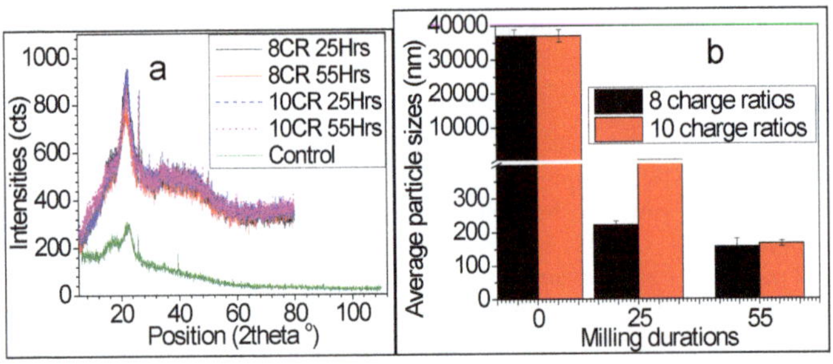

Fig. 2 , (a) superimposed diffractograms of flamboyant pod particles obtained at different milling at 8 and 10 charge ratios by milling for 25 and 55 hours (b) average sizes of the milled flamboyant particles determined using XRD aided with Scherrer's formula

Attainment of the minimum possible size is such that further milling at 10 charge ratios would result in both particle agglomeration of the finest particles due to Van der Waal's attractions and further milling of the fine particles which results in the greater average size of the particle obtained at 10 charge ratios. This explanation agrees with previous report in literature [10]. However, as milling progresses further beyond 25 hours, further particle reduction and agglomeration of ultra-fine particles accounting for slightly smaller size of the particles obtained at 10 charge ratios by milling for 55 hours than that at 10 charge ratios by milling for 25 hours. This observation is affirmed by the superimposed curves in Fig. 2a, showing much smaller intensities for the control sample. This implies that, there is no appreciable reductions in size for additional 30 hours of milling from 25 to 55 hours.

Mathematical Model

Probability (Prob>F) of the developed HA2FIRSM is <0.0001 which is smaller than $\alpha = 0.05$ (Table 1). This implies that the model (see Eq 2) is significant, and it is appropriate to explain relationship between the response (particle size) and the predictor variables (milling durations and charge ratio). Both predictor variables are significant because of their probability value <0.0001. Moreover, negative beta values of both variables predict smaller particle sizes as periods of milling and charge ratios increase whereas their product has positive beta values implying increases in particle sizes as the product of the predictors increases. This explains the greater average sizes of the particles obtained at 10 charge ratios than that obtained at 8 charge ratios as the milling progressed. Therefore, the model establishes particle agglomerations which is attributed to Van der Waal's forces. It implies that the interaction of the predictor variables causes drastic particle size reduction during ball milling to produce extremely fine particles which later resulted in physical fusion of the finest particles due to their high surface area. Greater beta value of the charge ratio compared with that of milling duration is a clue supporting its higher unique contribution to refinement of the flamboyant pod particles than that of the milling durations. Furthermore, regression coefficient of determination (R^2) of the developed HA2FIRSM is 0.91228. It testifies to goodness of the model due to closeness of its value to 100%. This affirms capacities of the model to account for 91,228 data out of every 100,000. The remnants account for the residual. High Fishers' value establishes that there is a very low chance for an error to occur in using the model for predicting flamboyant particle sizes at any values each of the milling durations and charge ratios using formula in Eq 2. Adequate precision is 97.61571. This testifies to the model's ability to navigate space. Result of model's diagnosis using leverage, residuals, outliers and Cook's distance identifies a data which is out of value range among the examined 62 inputs into the software. According to Thumb's rule, one out of 62 has nothing to do with the model. Therefore, there is no serious case that demands special attention to the model. Hence, the model is good to predict the sizes of flamboyant pod particles. Moreover, optimisation of the model using minimum particle size of the flamboyant pod particle as the target goal proposes that when milling at 10 charge ratios, maximum duration of milling to obtain minimum possible size of 3.5209 nm of the flamboyant pod particle is 17 hours 29 minutes to prevent the fusion of the finest particles which can cause particle agglomeration which can result in increasing average sizes which opposes the particle refinement which is the main purpose of the ball milling operation.

Table 1, Analysis of variance

Source	Sum of Squares	DF	Mean Square	F Value	Prob > F	R^2	Adjusted R^2	Adequate precision
Model	1.23E+09	3	4.09E+08	201.0645	< 0.0001	0.91228	0.907743	97.61571
A	2.98E+08	1	2.98E+08	146.2325	< 0.0001			
B	4.72E+08	1	4.72E+08	232.1477	< 0.0001			
AB	4.25E+08	1	4.25E+08	208.8267	< 0.0001			
Residual	1.18E+08	58	2034788					

$$p = 34311.13 - 3685.36\,A - 710.901\,B + 77.0315\,AB \quad (2)$$

p = Particle size; A = Charge ratio; B = milling durations

Transmission Electron Microscopic Images

Particles of varied geometries were obtained from each of the milling exercises as it is observed in Fig. 3a-d. Variation in sizes and shapes are linked to particles responses to impacts receiving from milling balls and drum's inner surface during the milling operations. The impact effect could be an attrition leading to particle wear, compression which causes breakage or shear. All result in particle fragmentations producing angular ultra-fine particles. It is very hard to see an isolated individual particle in each image in Fig. 3 at 20 nm resolution. This affirms physical fusion of the flamboyant pod particles during milling. Therefore, all particles exist in agglomeration. This implies that both particle refinement due to ball/drum's surface impacts and fusion due to Van der Waal's forces took place during ball milling. At the onset, when all particles are coarse, refinement supersedes agglomeration, recording high degree of particle breakages but as time goes on like above 25 hours of milling in this study, effect of particle agglomeration pronounces and opposes the purpose of the milling operation. Nanoparticle sizes obtained from XRD is supported by TEM images in Fig. 3. This also agrees with the findings from the developed HA2FIRSM in this study.

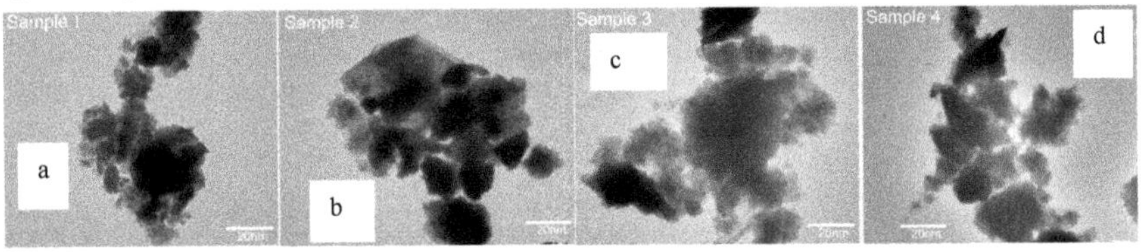

Fig. 3, Transmission electron microscopic images of flamboyant pod particles obtained at (a) 8 charge ratios, 25 hours of milling, (b) 8 charge ratios, 55 hours of milling (c) 10 charge ratios, 25 hours of milling and (d) 10 charge ratios, 55 hours of milling

Conclusions

From experimental investigation and results of this study, the following were concluded:

1. TEM indicates variation in the morphology of the particles at different milling durations and charge ratios.
2. Drastic reductions in particles sizes occurred when milling up to 25 hours.
3. Greater average sizes of the flamboyant particles obtained at 10 charge ratios could be attributed to particle agglomerations.
4. Greater beta values (coefficients of the predictor) of the charge ratio confirm its higher unique contributions to the particle refinement than milling durations.
5. Probability values of the model <0.0001 which is much less than α (0.05) affirms its significance to description of the particle size dependence on the milling durations and charge ratios
6. Moreover, optimisation of the model using minimum particle size of the flamboyant pod particle as the target goal affirms that when milling at 10 charge ratios, maximum duration of milling to obtain minimum possible size of 3.5209 nm of the flamboyant pod particle is 17 hours 29 minutes to prevent the fusion of the finest particles that may lead to particle agglomeration which can result in increasing average sizes of the particles.
7. Produced flamboyant nanoparticles are potential fillers for developing clay based or concrete based composites for building infrastructural applications.

Acknowledgement

Authors acknowledge Nigerian Geological Survey Agency (NGSA), Kaduna; Materials Fabrication laboratory, Kwara State University, Malete and G TEK (ISWEAR) Hub, Oluyole, Ibadan for supports in making this work a reality.

References

[1] E. Tarani, D. Stathokostopoulos, D. Karfaridis, L. Malletzidou, I.K. Sfampa, F. Stergioudi, G. Maliaris, N. Michailidis, K. Chrissafis, G. Vourlias, Journal of Thermal Analysis and Calorimetry, 148 (2023) 13065-13081. 10.1007/s10973-023-12428-2

[2] Z. Ling, T. Wang, M. Makarem, M. Santiago Cintrón, H.N. Cheng, X. Kang, M. Bacher, A. Potthast, T. Rosenau, H. King, C.D. Delhom, S. Nam, J. Vincent Edwards, S.H. Kim, F. Xu, A.D. French, Cellulose, 26 (2019) 305-328. 10.1007/s10570-018-02230-x

[3] S. Bhandari, R. Schierholz, R.-A. Eichel, A.L. Luna, A.K. Mechler, Advanced Energy and Sustainability Research, 5 (2024) 2400183. https://doi.org/10.1002/aesr.202400183

[4] M. de Oliveira Barros, A.L. Mattos, J.S. de Almeida, M. de Freitas Rosa, E.S. de Brito Effect of Ball-Milling on Starch Crystalline Structure, Gelatinization Temperature, and Rheological Properties: Towards Enhanced Utilization in Thermosensitive Systems. Foods. 12, 2023. 10.3390/foods12152924

[5] G.H. Jürgen, M.G. Cynthia, Introduction to Ceramics Processing,T U Clausthal,

[6] M. Herrmann,Chapter 3.4 - Ceramic Bearings and Seals, in: S. Somiya Academic Press, Oxford, 2013, 301-328. https://doi.org/10.1016/B978-0-12-385469-8.00017-4

[7] W. Wilfried, Ceramic Materials,SciyoJaneza Trdine 9, 51000 Rijeka, Croatia, 2010. http://www.issp.ac.ru/ebooks/books/open/Ceramic_Materials.pdf

[8] S.A. Bello,Particulate Hybrid Epoxy Nanocomposite: Structures, Tensile Properties, Regression Analysis, and Applications, in: S.A. Bello CRC, Taylor and Francis, USA, 2022, 231-277.

[9] P. Singh, N. Ram Chauhan, S. Rajesha, Materials Today: Proceedings, 42 (2021) 1202-1206. https://doi.org/10.1016/j.matpr.2020.12.690

[10] S.A. Bello, J.O. Agunsoye, J.A. Adebisi, S.B. Hassan, Engineering and Applied Science Research (EASR), 42 (2018) 262-272. 10.14456/easr.2018.36

[11] S.A. Bello, J.O. Agunsoye, S.B. Hassan, Materials Letters, 159 (2015) 514-519. http://dx.doi.org/10.1016/j.matlet.2015.07.063

[12] S.A. Bello, S.B. Hassan, J.O. Agunsoye, M.G.Z. Kana, R.I. A, Tribology in Industry, 37 (2015) 257-263.

[13] J.O. Agunsoye, J.E. Anyanwu, S.A. Bello, S.B. Hassan, Nigerian Journal of Technological Development, 15 (2018) 102-107. 10.4314/njtd.v15i3.5

[14] S.A. Bello, J.O. Agunsoye, J.A. Adebisi, F.O. Kolawole, B.H. Suleiman, Kathmandu University Journal of Science, Engineering and Technology, 12 (2016) 63-79. https://doi.org/10.3126/kuset.v12i1.21566

[15] T.Y. Liu, Y. Ma, S.F. Yu, J. Shi, S. Xue, Innovative Food Science & Emerging Technologies, 12 (2011) 586-593. https://doi.org/10.1016/j.ifset.2011.06.009

[16] L. Niu, Q. Guo, J. Xiao, Y. Li, X. Deng, T. Sun, X. Liu, C. Xiao, Food Research International, 163 (2023) 112263. https://doi.org/10.1016/j.foodres.2022.112263

[17] S.A. Bello, M.K. Adebayo, R.G. Adeyemo, P.A. Popoola, Iranian Polymer Journal, 33 (2024) 965-980. 10.1007/s13726-024-01307-8

[18] J.O. Agunsoye, S.A. Bello, L.O. Adetola, Journal of King Saud University - Engineering Sciences, 31 (2019) 70-77. https://doi.org/10.1016/j.jksues.2017.01.005

[19] J.O. Agunsoye, A.A. Bamigbaiye, S.A. Bello, S.B. Hassan, Arabian Journal for Science and Engineering, 45 (2020) 5087-5097. 10.1007/s13369-020-04345-5

[20] S.A. Bello, M.Y. Kolawole, R.G. Adeyemo, F.O. Kolawole, U.A. Aliu, A. Aboyeji, A. Rafiu, A.G.F. Alabi, O. Akinlabi, Journal of Alloys and Compounds, 1015 (2025) 178822. 10.1016/j.jallcom.2025.178822

Materials Research Proceedings 63 (2026) 72-84 https://doi.org/10.21741/9781644904053-9

A Comparative Analysis of the Influence of Fly Ash on the Mechanical Behavior of Self-Compacting Concrete

Dawit BEZA[2*], Muluken GETACHEW[1], Opeoluwa Rotimi DADA[2], Aschalew ZERIHUN[1] and Bolanle Deborah IKOTUN[2]

[1]Department of Construction Technology and Management, College of Engineering, and Technology, Mizan Tepi University. Tepi, Ethiopia

[2]Department of Civil and Environmental Engineering and Building Science, College of Science, Engineering and Technology, University of South Africa

23955090@mylife.unisa.ac.za

Keywords: Fly Ash, Self-Compacting Concrete, Mechanical Properties, Ordinary Portland, Cement

Abstract. Concrete is a widely used construction material composed of cement, fine and coarse aggregates, and water, mixed in specific proportions. Its strength and durability are critical for the performance of civil engineering structures, particularly in building construction. However, these properties are significantly influenced by factors such as mixed design, workmanship, and compaction methods. Therefore, research is essential to optimise and mitigate the effects of these variables, thereby enhancing the performance of concrete. This study aimed to evaluate the fresh and hardened properties of self-compacting concrete (SCC) incorporating fly ash as a partial replacement for cement. Four mix groups were prepared: a control mix with 0% fly ash, and SCC mixes with 5%, 10%, and 15% fly ash replacement. Materials were sourced locally from the Tepi and Jimma areas, apart from Class F fly ash, and chemical admixtures were acquired from Addis Ababa. The mix ratio used was 1:1.735:1.728 (cementitious material: fine aggregate: coarse aggregate), with varying cement and fly ash contents. Fresh properties were assessed using slump flow, flow table, segregation resistance, T500 mm time, and J-ring tests. Compressive strength was measured at 7, 14, and 28 days. The results demonstrated that all SCC mixes met ASTM standards for fresh properties. The compressive strength of all mixes with fly ash outperformed that of C-25 grade concrete, with the 10% replacement achieving the highest strength of 50.2 MPa. Moreover, all replacement levels up to 15% demonstrated superior performance compared to the control mix. Although the optimum replacement percentage was not definitively identified, SCC with fly ash showed improved properties across all parameters. In conclusion, the incorporation of fly ash in SCC enhances both performance and economic viability, rendering it a sustainable alternative to conventional SCC that solely uses ordinary Portland cement (OPC).

1. Introduction

Concrete is one of the most important and widely used construction materials globally (Ye et al., 2020). It is primarily composed of fine aggregate (sand), coarse aggregates (crushed stone), cement, and water, which are mixed in specified proportions to form a hard, stone-like mass after hydration. The versatility, availability, durability, relatively low cost of concrete and ability to conform to any structural shape have made it the generally preferred material for various civil works such as buildings, bridges, dams, roads, and tunnels (Letsosa et al., 2024) (Wlodarczyk & Jedrzejewski, 2016).

Over the decades, concrete has become synonymous with strength, durability, and permanence. As highlighted by (Zhang et al., 2018). Concrete has established itself as the backbone of modern infrastructure due to various factors such as the availability of raw materials, ease of fabrication on-site, and its ability to be moulded into various shapes. However, despite its strengths, the performance of conventional concrete can be significantly affected by several construction-related factors (Ye et al., 2020). These factors include improper water-to-cement ratio, poor mixing, inadequate vibration, and a lack of skilled labour during placement and finishing, which can lead to defects such as honeycombing, voids, and segregation. These defects ultimately compromise concrete strength, durability, and long-term serviceability (Alyamaç & Ince, 2009; Bhat, 2020; Campos et al., 2020). One major method to ensure the integrity of conventional concrete is to eliminate air voids and achieve maximum compaction through mechanical methods vibration. However, this process is not only labour-intensive but also dependent on the expertise of construction workers and the availability of appropriate equipment. In cases where skilled workers are unavailable, the risk of improper compaction increases. Therefore, in response to this challenge, a novel type of concrete known as Self-Compacting Concrete (SCC) was developed in the late 1980s in Japan (Kaish et al., 2018; Roig-Flores et al., 2016).

Self-Compacting Concrete (also referred to as Self-Consolidating Concrete or Self-Levelling Concrete) is a highly flowable, non-segregating concrete that spreads into place under its weight and fills the formwork even in the presence of dense reinforcement without the need for mechanical vibration. It maintains homogeneity during placement and offers a practical solution to challenges associated with compaction, especially in structures with complex geometries or congested steel arrangements. The development of SCC marked a breakthrough in concrete technology as SCC not only addresses the limitations of conventional concrete in terms of compaction but also enhances the quality and speed of construction. Self-Compacting Concrete has increasingly gained attention in both research and practice due to its benefits in terms of labour reduction, improved surface finishes, reduced construction noise (by eliminating vibrators), and better structural integrity in reinforced sections (Gesoğlu et al., 2012; Jagadesh et al., 2023; Roig-Flores et al., 2016).

Despite these advantages, SCC has not been widely adopted in some regions, largely due to limited awareness, lack of technical knowledge, and the relatively higher initial cost of materials such as chemical admixtures and supplementary cementitious materials. However, with increasing demand for faster, safer, and higher-quality construction, SCC is gradually gaining acceptance worldwide (Gesoğlu et al., 2012; Jagadesh et al., 2023; Jalal et al., 2015).

The production of SCC requires a precise mix design approach that balances flowability, segregation resistance, and strength. Unlike conventional concrete, SCC often incorporates a higher content of fine materials or fillers such as fly ash, limestone powder, silica fume, or slag. These materials enhance the fresh properties of SCC, such as flowability and viscosity, and assist in maintaining uniformity throughout placement. The selection of appropriate materials and mix proportions is essential to prevent problems like bleeding, segregation, or reduced strength (Hassan et al., 2015; Uysal & Yilmaz, 2011).

Fly ash, a commonly used mineral admixture in SCC, is a by-product of coal combustion in power plants. There are two main classes of fly ash, namely: Class C, which originates from lignite burning of coal, and Class F originates from anthracite materials and possesses pozzolana properties, which enhance the long-term strength and durability of concrete. Nevertheless, there is another type of fly ash known as Class N, which contains various materials such as volcanic ashes or pumicites, shales and clays. However, before it can be pozzolanic or cementitious, it must be calcinated. Generally, the spherical particle shape and fine texture of fly ash help to improve the workability and flow of SCC, making it an ideal material for enhancing fresh properties. Furthermore, the use of fly ash contributes to sustainability by reducing the consumption of Portland cement and utilizing industrial waste (Abd Elaty & Ghazy, 2018; Habibur Rahman Sobuz et al., 2023). In recent years, the application of SCC in the construction sector has increased, especially in complex and densely reinforced structures, precast elements, high-rise buildings, and repair works worldwide (Alhorani et al., 2022; Bhagwat et al., 2023).

This research, therefore, evaluates and compares the properties of self-compacting concrete with supplementary materials such as fly ash and without fly ash.

2. Fly Ash and Self-Compacting Concrete

2.1 Fly Ash

Fly ash, known as pulverized fuel ash, is the residue ash from the combustion of powdered coal, transported by the flue gases and precipitated electrostatically or mechanically from thermal or coal-fired power stations (Behl et al., 2021; Rivera et al., 2015). It is the most used artificial pozzolana material (Shetty, 2005). According to (ASTMC618, 2010), this material does not include ashes produced from the direct injection of lime into the boiler for sulfur removal or the burning of industrial or municipal garbage, which is also known as "incinerator ash". Fly ash mineral composition could vary from one coal-fired plant to another; however, the major chemical constituents in fly ash are silica, alumina and oxides of calcium and iron (Loeb, 2008). A typical fly ash composition is presented in Table 1. Fly ash produced from coal-fired power plants is naturally a variable material due to a range of influencing factors. One of the primary reasons for this variability is the mineralogical composition of the coal itself, which can differ significantly depending on its source and geological formation. Additionally, the degree to which the coal is pulverized before combustion affects the surface area exposed to burning, thereby influencing the characteristics of the resulting fly ash (Reta & Mahto, 2019).

The type of furnace used during the combustion process also plays a crucial role, as different furnace designs operate under varying thermal conditions and combustion dynamics. Furthermore, the oxidation conditions within the furnace, particularly the air-to-fuel ratio, significantly affect the chemical and physical properties of the fly ash. An excess or deficiency of oxygen can lead to incomplete combustion or the formation of unwanted phases in the ash. Finally, the method of collection, handling and storage before its intended usage can contribute to its variability. For example, the improper handling or exposure to moisture and contaminants during storage can alter its physical and chemical characteristics, reducing its consistency and reliability for use in industrial applications such as cement and concrete production.

Table 1: Typical chemical composition of fly ash ((Mulugeta, 2017).

Contents	CaO	SiO_2	Al2O3	Fe_2O_3	MgO	Na_2O	K_2O	TiO_2	LOI	MnO	P_2O_5	H_2O
%	0.01	42.28	26.72	3.04	0.88	2.00	0.48	0.54	16.53	0.08	0.18	5.60

2.1.1 Classification of Fly Ash

Fly ash is broadly classified into two main categories according to (ASTMC618, 2010) namely: Class C and Class F, depending on the type of coal, combustion temperature, collection method and percentage of Calcium oxide. Class C fly ash originates from sub-bituminous coal and lignite and contains higher calcium content (CaO > 20%), while Class F fly ash is typically derived from anthracite and bituminous coal and contains low calcium content (CaO < 10%). Class F fly ash is particularly suitable for use in concrete requiring long-term durability and sulfate resistance. When used as a partial replacement for cement in concrete, fly ash contributes to the strength development, improves workability, and enhances durability. Its effectiveness in concrete largely depends on its chemical and physical properties, particularly the presence of reactive silica and alumina, which react with calcium hydroxide during cement hydration to form additional calcium silicate hydrate (C–S–H), and the primary binder in concrete.

2.2 Self-Compacting Concrete

Self-compacting concrete (SCC), also known as super workable concrete or self-compactable concrete, is a high-performance concrete which can flow through and fill the gaps of reinforcement and the corners of the moulds without the need for external vibration (Prasad, 2016). It can be placed without vibration equally well in lightly or heavily reinforced sections. It has excellent deformability and segregation resistance, which result in the quality of the hardened concrete being independent of the workmanship during placing (Plank & Meyer, 2015). Originally developed to offset a growing shortage of skilled labour (EFNARC, 2002). In its plastic state, it flows silently under its weight and maintains homogeneity while completely filling the formwork of any shape, even around congested reinforcement. One of its advantages is that it does not require compaction as it self-compacts by its movement properties (Kapoor, 2012). Due to its high fluidity and resistance to segregation, it can be pumped over longer distances (Prasad, 2016)" and when transported in ready mixed concrete trucks, it minimizes transit mixtures and the turnaround time (Ghosh, 2017). Some key parameters to consider before classifying some concrete to be self-hardening include:

I. Filling Ability: SCC should be able to fill the formwork completely by going into every corner on its own.
II. Passing Ability: SCC should have the capability to pass the congested reinforcement without the need for external compaction, such as a vibrator.
III. Segregation resistance: SCC, like other concrete, is made up of several constituent materials having different specific gravities, so it is possible that at the time of placing materials having heavy mass tend to settle, causing segregation. The mix should be made such that it is able to resist segregation and maintain its homogeneity.

The properties of SCC based on performance are graphically illustrated in Figure 1.

The use of Self-compacting concrete (SCC) in the construction sector offers many benefits, such as improved productivity, reduced labour costs, enhanced aesthetics, and a safer work environment (The Engineering Community, 2011). However, it also presents challenges, including the need for precise materials monitoring, more trial batches, high sensitivity to mix variations, careful handling during transport and on-site placement, and increased formwork pressure that may cause leaching (concept concrete, 2019; The Engineering Community, 2011). Compared with conventional concrete of similar mechanical properties, the material cost of SCC is higher due to the relatively high demand of Cementation materials and chemical admixtures, including high-range water-reducing admixtures (HRWRA) and viscosity-enhancing admixtures (VEA) (Rajasekaran, 2017)

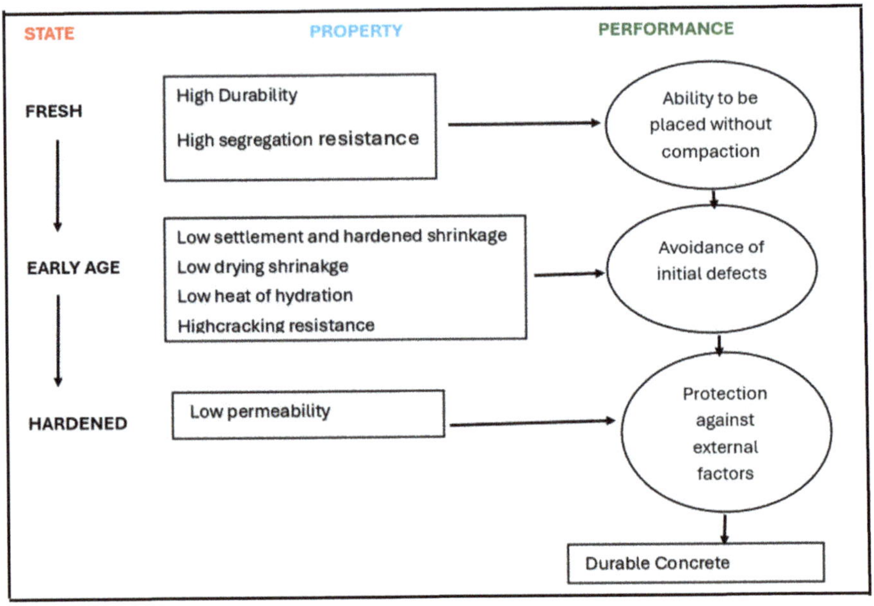

Figure 1: Graphical definition of Self-Compacting Concrete (Okamura, 2003)

Overall, the benefits of SCC, such as noise reduction, improved durability, thinner concrete sections and faster construction, outweigh those of conventional concrete, although the cost of production has been its major limitation due to larger quantities of Portland cement or supplementary cementitious materials used in SCC. Furthermore, more cementitious materials may be required to increase the fines content, to achieve stability, while admixtures such as HRWR and possibly a Viscosity Modified Admixture (VMA) will also increase the cost of SCC.

3. Materials and Methodology

3.1 Materials

The materials used for this research include ordinary Portland cement (OPC) grade 42.5N CEM, fine aggregate, water, admixture (polycarboxylic ether (PCE)), viscosity-modifying agent and fly ash.

3.2 Methodology

The methodology for this experimental research on determining the properties of self-compacting concrete with and without the inclusion of fly ash is outlined as follows: firstly, all materials used for producing concrete were carefully selected, measured, and mixed in a concrete mixer to ensure homogeneity; then, they were poured into cubic moulds measuring 15cm x 15cm x 15cm. The mix ratio used was 1:1.735:1.728 (cementitious material: fine aggregate: coarse aggregate), with varying cement and fly ash contents. When the concrete samples were demoulded, they were placed in a water-curing tank and left to cure for 7, 14, and 28 days. A total of 27 self-compacting samples were produced with 5%, 10%, and 15% fly ash inclusion, while the quantities of superplasticizer and super viscosity modifying agent used were kept constant at 1.2% and 0.4%, respectively. The Polycarboxylic ether (PCE) based superplasticizer conforms to ASTM C 494-92 in aqueous form to enhance workability and water retention, while the VMA ensures that the stability of the concrete is maintained to avoid segregation of the ingredients. Additionally, nine (9) control samples of normal concrete were made without fly ash, PCE, and VMA. These samples were cured for 7, 14, and 28 days and tested for mechanical properties, and then compared to the values obtained for self-compacting concrete. The measurement of each concrete constituent is presented in Table 2. Various tests were conducted to determine the fresh property parameters of each mix design based on fly ash addition, including the slump test, T500mm flow test, segregation resistance test, V-funnel test, L-box test, and J-ring tests. For the mechanical property test, only the compressive strength test was conducted on the concrete samples to determine the strength of the concrete under compression, especially with the inclusion of FA at different percentages. This test was done in triplicate after each curing period, and the average value recorded is presented in Figure 7.

4. Results and Discussion

Measurement of concrete constituents per mix design and the results obtained from each conducted test is presented and discussed in this section.

Concrete Constituents

The concrete composition by weight for each design mix is shown in Table 2. The cement demand decreases as fly ash content increases; however, it is maintained at a reasonable level so that the concrete's strength is not affected, while the admixture and VMA content stay constant.

Table 2: Concrete composition measurement

Concrete Composition	SCC with 5%	SCC with 10%	SCC with 15%	SCC with 0%	Normal Concrete
Cement	434.15 Kg/m^3	411.3 Kg/m^3	388.45 Kg/m^3	457Kg/m^3	370 Kg/m^3
Sand	792.92 Kg/m^3	792.92 Kg/m^3	792.92 Kg/m^3	792.92 Kg/m^3	763 Kg/m^3
Coarse Aggregate	789.96 Kg/m^3	789.96 Kg/m^3	789.96 Kg/m^3	789.96 Kg/m^3	1043.47 Kg/m^3
Fly Ash	22.85 Kg/m^3	45.7 Kg/m^3	68.55 Kg/m^3	0	-
HRWR	5.484 Kg/m^3	5.484 Kg/m^3	5.484 Kg/m^3	5.484 Kg/m^3	-
VMA	1.828 Kg/m^3	1.828 Kg/m^3	1.828 Kg/m^3	1.828 Kg/m^3	-
Water	174.07 Kg/m^3	174.07 Kg/m^3	174.07 Kg/m^3	174.07 Kg/m^3	178.6 Kg/m^3

Slump and Slump Flow Test

The slump test for normal concrete, which is used to determine the workability of concrete, was conducted by removing the slump cone from the moulded sample, and the gap between the tip of the inverted slump cone and the top of the collapsed concrete was recorded as the slump value. This is according to the ASTM C143M standard, and for normal concrete, the value was 11.3mm.

Although the ASTM C1611M standard for Slump Flow Test procedure is similar, it measures the spread diameter of SCC to determine the flowability and ability of SCC to pass through an obstruction or formwork. For this research, the Slump Flow Test result obtained for each design mix is presented in Figure 2.

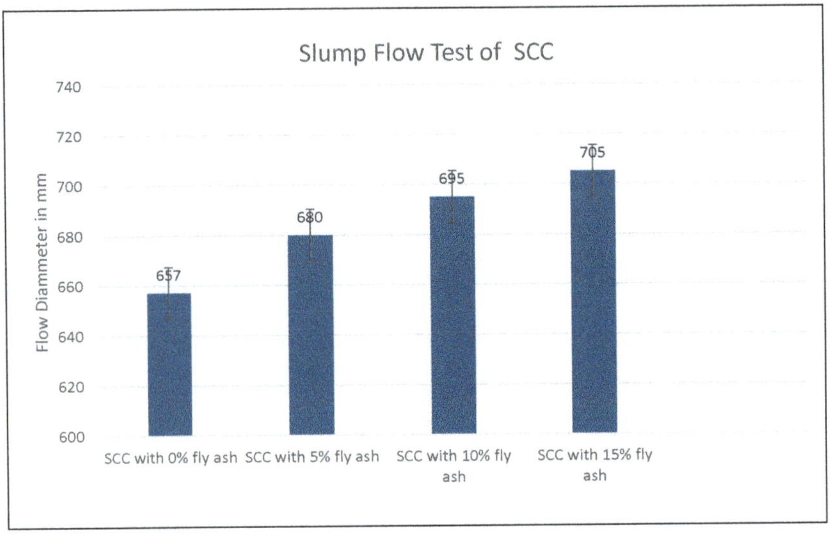

Figure 2: SCC Slump Flow Test Result

Based on ASTM C1611M standard for Slump Flow Test, a good SCC slump flow diameter ranges from 650mm to 800mm and from Figure 2, it can be seen that the more addition of fly ash to the concrete mix made the concrete the more flowable it becomes with SCC containing 15% FA exhibiting the highest flow table diameter of 705mm depicting that it has a better flowability and filling - ability properties as compared to other mix design as a result it can fill formworks faster with its weight. When compared with the control sample (SCC with 0% FA), it is evident that FA reduces the water absorption capacity of concrete, thereby contributing to the flowability properties of fresh concrete.

T_{500mm} Flow Test

The T_{500mm} flow test is a measure of the time required for SCC to flow and reach a diameter of 500mm (20 inches) during a slump flow test. This test indicates the viscosity of SCC, and it is essential to consider this factor when working with SCC, especially when the time to project completion is key. The result for the SCC design mix containing fly ash as compared to the control sample is presented in Figure 3.

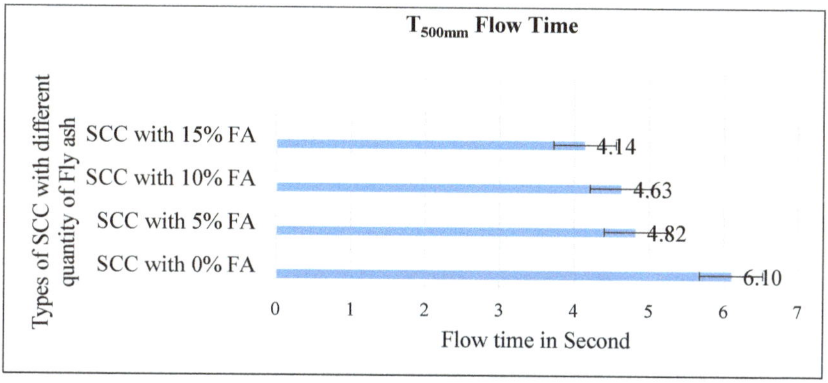

Figure 3: SCC T_{500mm} Flow Time Result

According to ASTM C1611, a good flow time for SCC is between 3-5 seconds, and from Figure 3, it can be inferred that adding FA to the concrete mix is essential for achieving this time because it increases SCC viscosity. Without FA, the time to reach a 500mm diameter is 6.10 seconds, but adding more FA to the mix increases flow speed and decreases the time needed for the concrete to flow and reach a 500mm diameter. From this research, a 15% addition of FA results in a time of 4.14 seconds to achieve a 500mm flow diameter. This test offers information on the time needed to fill a cubic metre of formwork under real-life site conditions.

Segregation Resistance of Self-Compacting Concrete

This test was conducted by pouring SCC slurry with FA on a horizontal concrete flow table to examine the free flow and stability of the concrete ingredients. As presented in Figure 4, the ingredients of the SCC mix containing fly ash are seen at all corners of the flow table. On the other hand, to achieve this in ordinary concrete, workability can only be improved by increasing the water content of the mix, which results in a reduction in the strength of hardened concrete and affects the stability of fresh concrete. However, for SCC, water is thickened by a viscosity-modifying admixture and stability is enhanced while segregation is controlled. Hence, SCC is better in segregation resistance as compared to ordinary concrete.

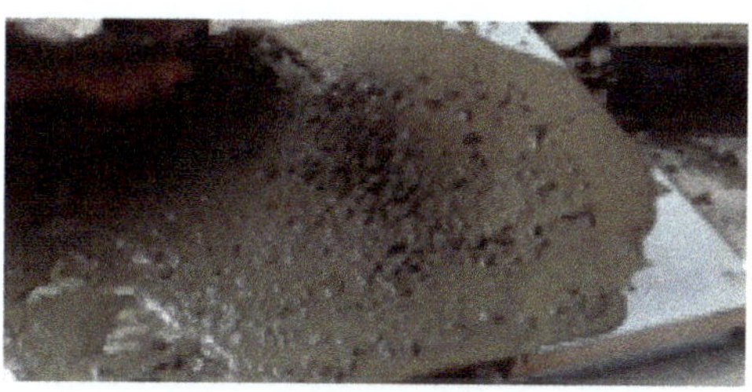

Figure 4: Showing Segregation Resistance of SCC

The J-Ring Test

The J-ring test is a quality control test for determining the passing ability of fresh SCC. The concrete passing ability of SCC slurry is obtained by measuring and calculating the difference in slump flow of the concrete with and without the J-ring apparatus, as illustrated in Figure 5. However, this property is vital in construction technology because it provides valuable information such as passing ability, concrete stability and identification of obstruction. In return, the information helps to optimize mix design and minimize structural risk.

Figure 5: Showing J-Ring Apparatus for the passing ability test of SCC

For this research, the J-Ring test was conducted according to ASTM C1621 and the values obtained for each mix design, that is, SCC with 0% FA, 5% FA, 10% FA and 15% FA are presented in Figure 6.

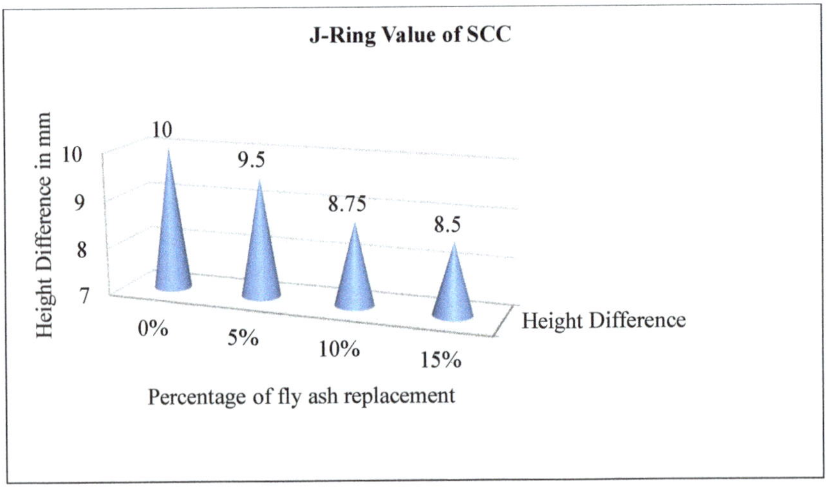

Figure 6: Result of J-Ring Test of SCC

The result of the J-ring test highlights the importance of adding fly ash to SCC, as it allows the concrete to pass more easily through obstructions or formwork compared to SCC with 0% FA. A lower J-ring value indicates a better passing ability of the SCC. Therefore, it can be concluded that SCC with 15% FA, exhibits a superior passing ability compared to SCC with 5% and 10% FA.

Compressive Strength Test

The compressive strength test was conducted on the ordinary concrete (OC) samples after curing for 7, 14 and 28 days. Similarly, the test was also conducted on SCC samples after the same curing period for concrete with 0%, 5%, 10% and 15% fly ash content to determine the strength of the concrete under compression, and to examine the effect of fly ash on the strength of SCC. The result obtained is presented in Figure 7.

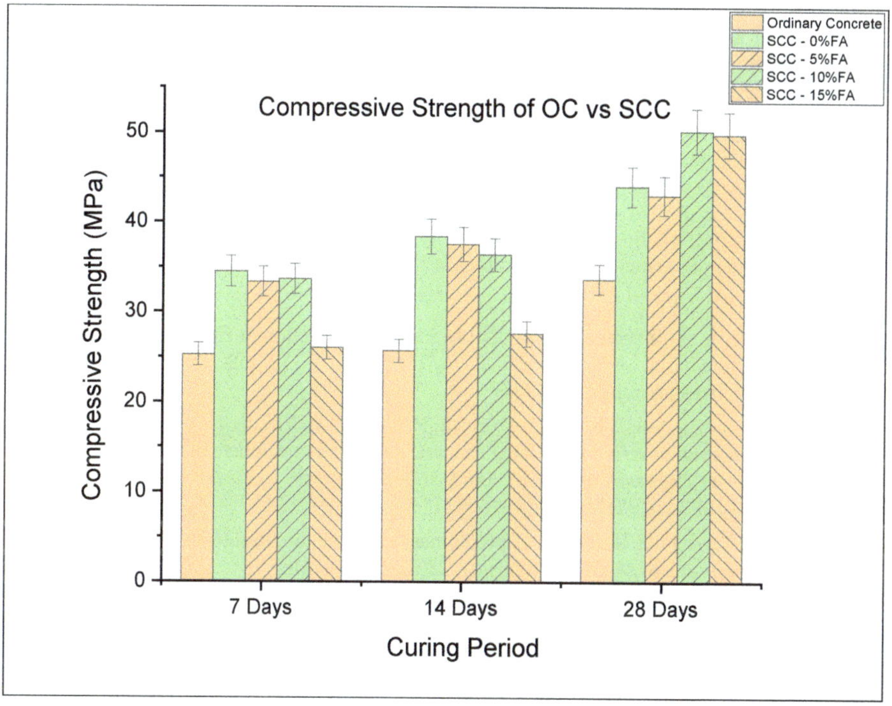

Figure 7: Compressive Strength Test of OC vs SCC

Figure 7 presents the result of the compressive strength test conducted on both ordinary concrete and self-compacted concrete. Firstly, it is observed that strength increases as the curing age progresses. Furthermore, it can be deduced that during the entire curing period, the SCC samples gained more strength than ordinary concrete, and similarly, SCC with 0% FA gained early strength between 7 and 14 days compared to SCC with the inclusion of fly ash. However, at 28 days of curing, SCC with FA content gained better strength than the one without FA, except for SCC with 5%FA. In addition, the overall strength gained by SCC was very impressive and meets the minimum strength required for many civil works; however, SCC with 10%FA presents the best compressive strength of 50.2 MPa at 28 days, followed by the design mix with 15%FA that has the strength of 49.8 MPa. Lastly, considering the results obtained from the fresh properties of SCC and the mechanical property (compressive strength), it can be inferred that the addition of fly ash is beneficial to SCC especially when inclusion percentage is optimized, and it reduces the overall cost of material purchase, as fly ash is a waste material and mostly collected from other industries at little to no cost.

Conclusion

The findings of this research have proven that the inclusion of fly ash in the design mix of self-compacting concrete has several advantages such as enhancement of flowability, filling ability, segregation resistance, and passing ability around congested reinforcement. This research also confirms that fly ash addition to SCC further enhances the concrete mechanical strength based on the compressive test result, as the highest strength of 50.2 MPa was obtained from SCC containing 10% FA, which greatly exceeds the values 33.68 MPa and 44 MPa obtained for normal concrete and concrete slurry without the addition of fly ash, respectively. Furthermore, the cost of producing SCC with fly ash is relatively lower as compared to ordinary concrete and SCC without fly ash because in most cases fly ash is a by-product of other sectors and readily available for disposal without a cost and when added to concrete mix, the amount of cement usage is reduced thereby reducing cost while promoting green construction. Finally, the adaptation of SCC in construction works helps to eliminate the need for concrete vibration or compaction, especially in tight areas, thereby reducing the time of project completion and associated cost. The scope of this research focused on the impact of fly ash on the flowability and strength of SCC; however, to determine the cost savings per cubic meter, it is recommended that a cost comparison analysis be conducted.

References

[1] Abd Elaty, M. A. A., & Ghazy, M. F. (2018). Fluidity evaluation of fiber reinforced-self compacting concrete based on buoyancy law. *HBRC Journal*, *14*(3), 368–378. https://doi.org/10.1016/j.hbrcj.2017.04.003

[2] Alhorani, R. A. M., Asaad, S. A., & Abendeh, R. M. (2022). Effect of white cement bypass dust on the degradation of roller compacted concrete subjected to sulphate attack. *International Journal of Pavement Engineering*, *23*(10), 3662–3676. https://doi.org/10.1080/10298436.2021.1913592

[3] Alyamaç, K. E., & Ince, R. (2009). A preliminary concrete mix design for SCC with marble powders. *Construction and Building Materials*, *23*(3), 1201–1210. https://doi.org/10.1016/j.conbuildmat.2008.08.012

[4] Behl, V., Singh, V., Dahiya, V., & Kumar, A. (2021). Characterization of physico-chemical and functional properties of fly ash concrete mix. *Materials Today: Proceedings*, *50*, 941–945. https://doi.org/10.1016/j.matpr.2021.06.353

[5] Bhagwat, Y., Nayak, G., Pandit, P., & Lakshmi, A. (2023). Effect of polypropylene fibres on strength and durability performance of M-sand self compacting concrete. *Cogent Engineering*, *10*(1). https://doi.org/10.1080/23311916.2023.2233783

[6] Bhat, J. A. (2020). Mechanical behaviour of self compacting concrete: Effect of wood ash and coal ash as partial cement replacement. *Materials Today: Proceedings*, *42*, 1470–1476. https://doi.org/10.1016/j.matpr.2021.01.311

[7] Campos, H. F., Klein, N. S., & Marques Filho, J. (2020). Proposed mix design method for sustainable high-strength concrete using particle packing optimization. *Journal of Cleaner Production*, *265*, 121907. https://doi.org/10.1016/j.jclepro.2020.121907

[8] concept concrete. (2019, May 25). *Current advancement and the pros and cons of self compacting concrete*. https://www.conceptconcretewa.com.au/pros-and-cons-of-self-compacting-concrete/

[9] Gesoğlu, M., Güneyisi, E., Kocabağ, M. E., Bayram, V., & Mermerdaş, K. (2012). Fresh and hardened characteristics of self compacting concretes made with combined use of marble powder, limestone filler, and fly ash. *Construction and Building Materials*, *37*, 160–170. https://doi.org/10.1016/j.conbuildmat.2012.07.092

[10] Habibur Rahman Sobuz, M., Saha, A., Akid, A. S. M., Vincent, T., Tam, V. W. Y., Yalçınkaya, Ç., Mujahid, R., & Sutan, N. M. (2023). Performance of self-compacting concrete incorporating waste glass as coarse aggregate. *Journal of Sustainable Cement-Based Materials*, *12*(5), 527–541. https://doi.org/10.1080/21650373.2022.2086936

[11] Hassan, A. A. A., Ismail, M. K., & Mayo, J. (2015). Mechanical properties of self-consolidating concrete containing lightweight recycled aggregate in different mixture compositions. *Journal of Building Engineering*, *4*, 113–126. https://doi.org/10.1016/j.jobe.2015.09.005

[12] Jagadesh, P., de Prado-Gil, J., Silva-Monteiro, N., & Martínez-García, R. (2023). Assessing the compressive strength of self-compacting concrete with recycled aggregates from mix ratio using machine learning approach. *Journal of Materials Research and Technology*, *24*, 1483–1498. https://doi.org/10.1016/j.jmrt.2023.03.037

[13] Jalal, M., Pouladkhan, A., Harandi, O. F., & Jafari, D. (2015). Comparative study on effects of Class F fly ash, nano silica and silica fume on properties of high performance self compacting concrete. *Construction and Building Materials*, *94*, 90–104. https://doi.org/10.1016/j.conbuildmat.2015.07.001

[14] Kaish, A. B. M. A., Breesem, K. M., & Abood, M. M. (2018). Influence of pre-treated alum sludge on properties of high-strength self-compacting concrete. *Journal of Cleaner Production*, *202*, 1085–1096. https://doi.org/10.1016/j.jclepro.2018.08.156

[15] Letsosa, C. M., Dada, O. R., & Ikotun, B. D. (2024). The effects of different curing conditions on the polypropylene fibre reinforcement concrete. *Journal of Physics: Conference Series*, *2888*(1). https://doi.org/10.1088/1742-6596/2888/1/012022

[16] Plank, J., & Meyer, L. (2015). New insights into physicochemical interactions occurring between polycarboxylate superplasticizers and a stabilizer in self-compacting concrete. *Journal of Sustainable Cement-Based Materials*, *4*(3), 164–175. https://doi.org/10.1080/21650373.2015.1024296

[17] Reta, Y., & Mahto, S. (2019). Experimental Investigation on Coffee Husk Ash as a Partial Replacement of Cement for C-25 concrete. *Cikitusi Journal for Multidisciplinary Research*, *6*(0975–6876), 152–158.

[18] Rivera, F., Martínez, P., Castro, J., & López, M. (2015). Massive volume fly-ash concrete: A more sustainable material with fly ash replacing cement and aggregates. *Cement and Concrete Composites*, *63*, 104–112. https://doi.org/10.1016/j.cemconcomp.2015.08.001

[19] Roig-Flores, M., Pirritano, F., Serna, P., & Ferrara, L. (2016). Effect of crystalline admixtures on the self-healing capability of early-age concrete studied by means of permeability and crack closing tests. *Construction and Building Materials*, *114*, 447–457. https://doi.org/10.1016/j.conbuildmat.2016.03.196

[20] The Engineering Community. (2011, June 11). *Self-compacting concrete (SCC) - Advantages, Disadvantages and Applications*. Engineering Articles. https://www.theengineeringcommunity.org/self-compacting-concrete-scc-advantages-disadvantages-and-applications/

[21] Uysal, M., & Yilmaz, K. (2011). Effect of mineral admixtures on properties of self-compacting concrete. *Cement and Concrete Composites*, *33*(7), 771–776. https://doi.org/10.1016/j.cemconcomp.2011.04.005

[22] Wlodarczyk, M., & Jedrzejewski, I. (2016). Concrete Slabs Strengthened with Basalt Fibres - Experimental Tests Results. *Procedia Engineering, 153*, 866–873. https://doi.org/10.1016/j.proeng.2016.08.200

[23] Ye, Y., Liu, J., Zhang, Z., Wang, Z., & Peng, Q. (2020). Experimental Study of High-Strength Steel Fiber Lightweight Aggregate Concrete on Mechanical Properties and Toughness Index. *Advances in Materials Science and Engineering, 2020*. https://doi.org/10.1155/2020/5915034

[24] Zhang, P., Han, S., Ng, S., & Wang, X. H. (2018). Fiber-Reinforced Concrete with Application in Civil Engineering. *Advances in Civil Engineering, 2018*. https://doi.org/10.1155/2018/1698905

Advances in Cement and Concrete Research - ICACCR 2025 Materials Research Forum LLC
Materials Research Proceedings 63 (2026) 85-94 https://doi.org/10.21741/9781644904053-10

Evaluation of the Current AASHTO LRFD Concrete Creep Prediction Model Using South African Materials

George C FANOURAKIS[1,a *]

[1]Department of Civil Engineering Technology, University of Johannesburg, South Africa

[a]georgef@uj.ac.za

Keywords: Creep, Concrete, Codes, Coefficients, Models

Abstract. Creep of concrete is an important design consideration. National design codes typically provide empirically based models for the estimation of creep deformation. Such models estimate a creep coefficient (ϕ), which is the ratio of the creep strain (ε_c) to the elastic strain (ε_e), which is used to predict the creep strain (ε_c) at any age. This paper assesses the accuracy of the creep coefficients (ϕ) predicted by the latest American Association of State Highway and Transportation Officials (AASHTO), Load and Resistance Factor Design (LRFD) Bridge Design Specification. The accuracy of the AASHTO Model was evaluated by comparing the predicted with the actual (measured) creep coefficients (ϕ), on a range of concretes under laboratory-controlled conditions, for six mixes. The six mixes comprised three aggregate types (quartzite, granite and andesite) and two strength grades (water-cement ratios of 0.55 and 0.4). In the case of all six mixes, at all ages considered, the AASHTO LRFD Model underpredicted the creep coefficient (ϕ). Furthermore, the predictions were most and least accurate in the case of the quartzite and andesite concretes, respectively. In the case of all six mixes, a highly significant linear relationship (maximum P = 1.4E-11 %) was found between the predicted ϕ and actual ϕ. When considering all six mixes, an overall coefficient of variation (ω_{all}) of 68.5 % was obtained. The results of this investigation were compared to those of 19 other models and the AASHTO LRFD Model was the 16th most accurate model of the 20 models assessed to date.

Introduction

Creep magnitude is an important design consideration for the durability, long-term serviceability and the load carrying capacity of structures. Creep strain can result in loss of serviceability due to unacceptable deflections or loss of prestress. Creep strain can be determined by relatively costly and time-consuming laboratory testing or estimated by means of empirically based models of various complexities. In general, the more deformation sensitive the structure, the more laboratory testing is justified. In cases where an approximation of creep is required, design code-type models suffice.

Almost all code-type creep models express creep strain in terms of the creep coefficient, (ϕ), as shown in Eq. 1.

$$\varepsilon_c(t, \tau) = \varphi(t)\ \varepsilon_{e,\tau} \quad (1)$$

Where: $\varepsilon_c(t, \tau)$ is the creep strain at any concrete age t for a concrete loaded at age τ, where $t > \tau$ and $\varepsilon_{e,\tau}$ is the elastic strain of the concrete at age τ.

The creep coefficient (ϕ) at time t can also be expressed in terms of specific creep (Cc) at time t and elastic modulus (E), as shown in Eq. 2.

$$\phi(t) = Cc(t) \times E \quad (2)$$

Where Specific creep (Cc) is creep strain (ε_c) per unit stress (σ) as shown in Eq. 3.

$$Cc = \varepsilon_c(t)/\sigma \tag{3}$$

The creep coefficient (ϕ) is empirically determined by considering one or more intrinsic and/or extrinsic variables such as age at loading or concrete stiffness. The elastic modulus used to estimate the elastic strain is estimated using an empirical equation prescribed in the relevant method.

Previous work by Fanourakis [1,2,3,4] and Fanourakis and Ballim [5] collectively assessed the accuracy of 16 code-type creep prediction models when applied to South African concretes. Fanourakis [2] investigated the correlation between the predicted specific creep (Cc) and the estimated elastic (E) and established that the most accurate creep prediction model (for the Cc), the CEB-FIP (1970) [6], was the least accurate in estimating elastic modulus (E). Furthermore, the models that yielded the most accurate estimation of elastic modulus (E) (SANS 10100, 2000 [7] and AS 3600, 2009 [8]) did not yield the most accurate estimation of specific creep (Cc).

Subsequently, Fanourakis [3] established that a highly significant (P = 0.001 %) correlation (r = 0.901) exists between the creep coefficient (ϕ) and specific creep (Cc) predicted by the 14 models considered at that time. These results justified the investigation of the accuracy of the creep coefficient (ϕ) component of code-type creep estimation models.

This paper assesses the accuracy of the creep coefficients predicted (with time) by the Latest (current) American State Highway and Transportation Officials (AASHTO) Load and Resistance Factor Design (LRFD) Model [9], when compared with the actual creep coefficients (ϕ) measured on a range of South African concretes under laboratory-controlled conditions, for a period of approximately six months. These concretes included two strength grades (w/c's of 0.56 and 0.4) and three aggregate types (quartzite, granite and andesite).

The accuracy of the AASHTO LRFD (2024) Model [9] was also compared to the accuracy of other code-type models.

The AASHTO LRFD Model

The American Association of State Highway and Transportation Officials (AASHTO), Load and Resistance Factor Design (LRFD) [9] is a Bridge Design Specification that covers bridge design, evaluation and retrofitting. This code ensures safety by factoring in uncertainties in both loads and material properties, based on the statistical understanding of loads and structural performance. This specification was preceded by the 2020 [10] and 2017 [11] versions.

In the latest (current) AASHTO LRFD code [9], the creep coefficient (ϕ) is calculated using Eq.'s 4 to 8.

$$\phi = (t,t_i) = 1.9k_s k_{hc} k_f k_{td} t_i^{-0.118} \tag{4}$$

$$K_s = 1.45 - 0.13(V/S) \geq 1.0 \tag{5}$$

$$k_{hc} = 1.56\text{-}0.008H \tag{6}$$

$$k_f = \frac{5}{1+f'_{ci}} \tag{7}$$

$$k_{td} = \frac{t}{12(\frac{100-4f'_{ci}}{f'_{ci}+20})+t} \tag{8}$$

Where,
V/S = Volume to surface ratio (inches).
H = Relative humidity (%).
f'_{ci} = Design compressive strength (ksi)

t = age of concrete since loading (days).
t_i = age of concrete at time of load application (days).

The model does not consider the aggregate type.

Experimental details

Materials. CEM I 42,5 cement, from the Dudfield factory of Alpha Cement (now AfriSam), was used for all the tests carried out in this investigation. Quartzite (Q) from the Ferro quarry in Pretoria, granite (G) from the Jukskei quarry in Midrand and andesite (A) from the Eikenhof quarry in Johannesburg (all in South Africa) were used as both the coarse and fine aggregates for the concrete. The stone was 19 mm nominal size and the fine aggregate was crusher sand.

Preparation of prisms. For each of the concrete mixes, six prisms were prepared, measuring 100 x 100 x 200 mm and cast with the 200 mm dimension vertical. After de-moulding, these prisms were continuously water cured up to an age of 28 days. After curing, three of the six prisms of each mix were used for creep tests and the remaining three were used for shrinkage measurements.

Elastic Modulus Measurements. The creep test prisms were stacked into creep loading frames and subjected to elastic strain measurements, within ten minutes of application of the loads, which were used to determine the secant moduli of the concretes.

Creep and Shrinkage Measurements. The creep tests commenced immediately after the elastic modulus measurements were taken. These tests entailed subjecting the prisms in each frame to an applied load of approximately 25 % of the 28-day compressive strength, for the 168-day period, in a room with a controlled environment (temperature of 22 ± 3° C and RH of 65 ± 5 %). Details of the creep frames are given in Fanourakis [1].

The reason for the selection of the 168-day period (approximately six months) was the premise that approximately 60% of the ultimate creep strain will occur within the first six months after loading [12].

The shrinkage (companion) prisms were placed on a rack in the same room as the creep samples. To ensure a drying surface area of the shrinkage samples equivalent to the creep samples, the two 100 mm square ends of the shrinkage prisms were dipped in warm wax to prevent drying from those surfaces.

Creep and shrinkage measurements were recorded daily for the first week, thereafter, weekly for the remainder of that month and then monthly until the culmination of the approximately six-month total loading period. The strain of each group of prisms, that is the three creep prisms or the three companion shrinkage prisms of a particular mix, was taken as the average of the strains of the prisms in that group.

The results of shrinkage measurements were subtracted from the total time-dependant strain of the loaded specimens to determine the total creep strain. The creep coefficient (ϕ) was determined from Eq. 2.

Mix details. Details of the mixes and laboratory test results are given in Table 1.

Table 1. Details of the Mixes and Laboratory Test Results

Aggregate Type	Quartzite		Granite		Andesite	
Mix	**Q1**	**Q2**	**G1**	**G2**	**A1**	**A2**
Water (l/m³)	195	195	195	195	195	195
CEM I 42,5N (kg/m³)	348	488	348	488	348	488
19 mm Stone (kg/m³)	1015	1015	965	965	1135	1135
Crusher Sand (kg/m³)	810	695	880	765	860	732
w/c Ratio	0.56	0.4	0.56	0.4	0.56	0.4
Slump (mm)	90	50	115	70	95	55
Cube Compressive Strength (MPa)	37	65	38	65	48	74
Cylinder Compressive Strength (MPa)[a]	30	53.5	30.7	53.5	38	59
Concrete Density (kg/m³)	2371	2410	2385	2432	2596	2585
Average Elastic Modulus of included Aggregate (GPa)	73		70		89	
[a] Inferred from cube strength using the conversions from EN 1992-1-1-2004 [13]						

Results and discussion

Predicted and Actual Creep Coefficient (ϕ) Values with Time. Figures 1 to 3 show the relationships between the predicted ϕ and actual ϕ values for the six mixes (Q1, Q2, G1, G2, A1 and A2) with time.

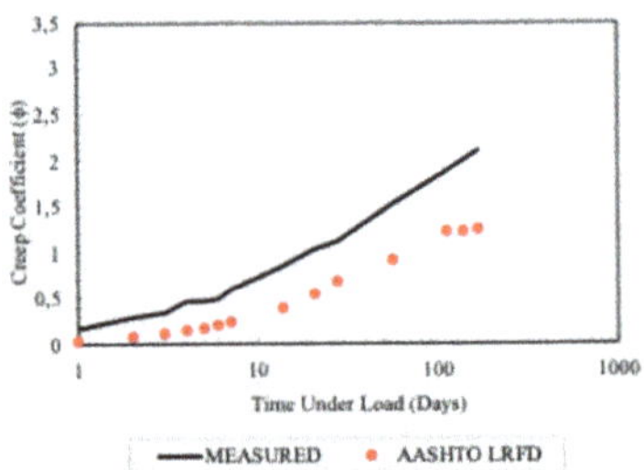

Mix Q1

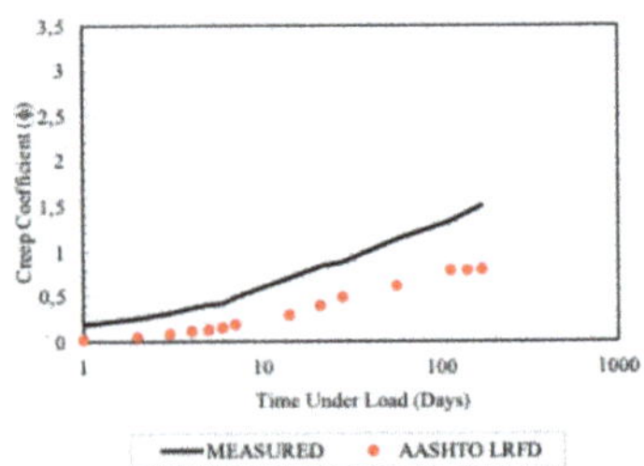

Mix Q2

Fig.1 *Figure 1. Actual and predicted creep coefficients (ϕ) with time for quartzite concretes*

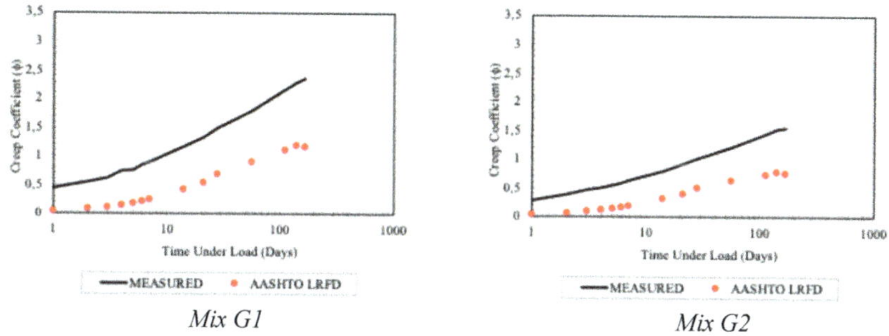

Fig.2 *Figure 2. Actual and predicted creep coefficients (ϕ) with time for granite concretes*

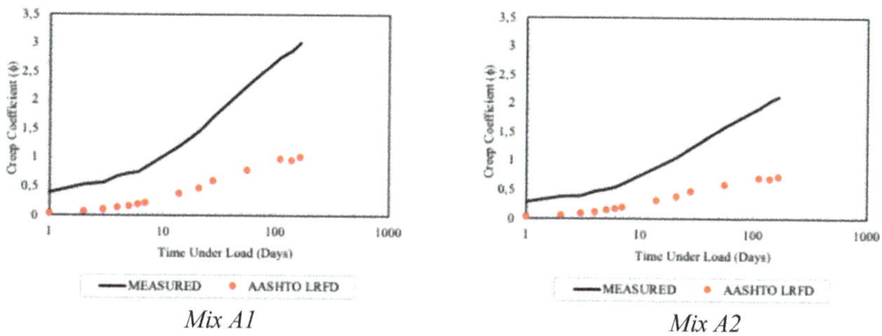

Fig.3 *Figure 3. Actual and predicted creep coefficients (ϕ) with time for andesite concretes*

As is evident from Figures 1 to 3, in the case of all the mixes, the predicted values were lower than the actual values at all ages, both of which increased with time. This underprediction agreed with the relatively recent findings of Williams et al [14], who applied the same model [9] to conventional concrete, with an average cylinder strength of 27.4 MPa at 28 days, considering a 365 day test period.

In addition, Mohebbi et al [15] who also applied the same model [9] to predict creep coefficients (ϕ) with time in the case of ultra-high-performance concrete (UHPC) specimens with strengths varying from 95 MPa to 172 MPa and loading applications times varying from 3 to 97 days. They found that this model underpredicted the creep coefficient and attributed these results to the fact that the AASHTO model was developed for conventional concrete not UHPC. However, that may not be the reason for their underpredictions, considering the similar trend of underestimations in this paper (which applied to conventional concrete).

The difference between the actual and predicted values generally increased with time, this trend being most prominent in the case of both andesite mixes (A1 and A2). Furthermore, the predicted values increased with time until an age of 112 days. Thereafter, the predicted values were essentially similar.

Materials Research Proceedings 63 (2026) 85-94 https://doi.org/10.21741/9781644904053-10

As expected, the actual and predicted creep coefficients (ϕ) were lower in the case of the higher strength mix (Q2, G2 and A2) for each included aggregate type.

Predicted versus Actual Creep Coefficient (ϕ) Values. Figures 4 to 6 show the relationships between the predicted ϕ and actual ϕ for the six mixes (Q1, Q2, G1, G2, A1 and A2). The "r = 1" line (predicted equals actual) is included in each figure to display the relative accuracy of the predicted values.

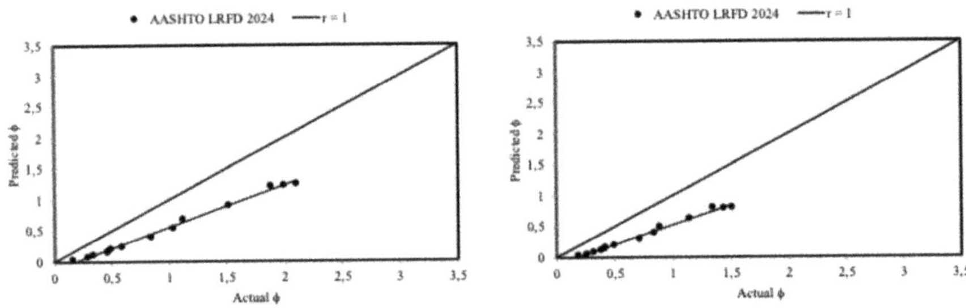

Mix Q1 *Mix Q2*

Fig.4 *Figure 4. Actual and predicted creep coefficients (ϕ) for quartzite concretes*

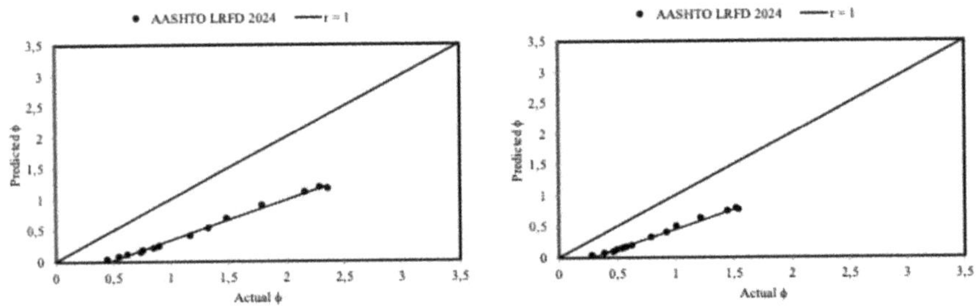

Mix G1 *Mix G2*

Fig.5 *Figure 5. Actual and predicted creep coefficients (ϕ) for granite concretes*

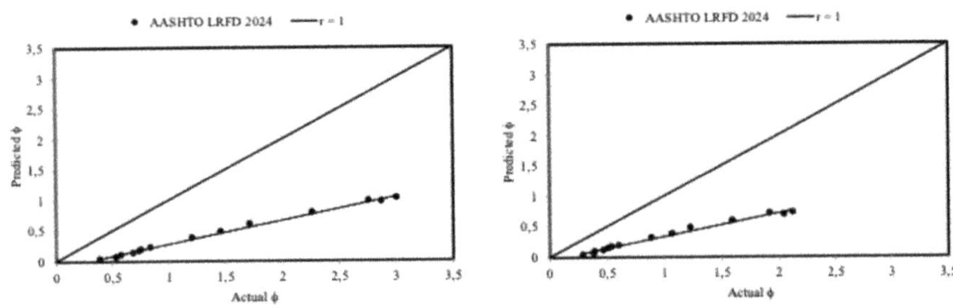

Mix A1 *Mix A2*

Fig.6 *Figure 6. Actual and predicted creep coefficients (ϕ) for andesite concretes*

From Figures 4 to 6, it is evident that the AASHTO LRFD Model [9] underpredicted the ϕ in the case of all the mixes. Furthermore, the predictions for the quartzite mixes were more accurate than the granite mixes, which were, in turn, more accurate than the andesite mixes.

The statistics pertaining to the trend lines in Figures 4 to 6, for the six mixes, are shown in Table 2.

Table 2. Statistics Pertaining to the Trend Lines

Aggregate in Mix	Correlation Coefficient (r) Range	Maximum Level of Significance (P %)
Quartzite	0.995 to 0.996	P = 1.9E-12 %
Granite	0.996 to 0.997	P = 6E-15 %
Andesite	0.993 to 0.997	P = 1.4E-11 %

From Table 2, it is evident that all the correlations established were highly significant. Furthermore, the level of significance was lowest in the case of the andesite mixes (A1 and A2).

Accuracy of the Models Assessed. To provide a statistical basis for comparing the results of creep prediction methods, Bazant and Panula [16] define a coefficient of variation of errors (ω_j) for single data sets as well for a number of data sets compared against the same prediction model (ω_{all}). The more accurate the prediction, the lower the value of ω_j. The calculated values of ω_j and ω_{all} for the different models assessed are shown in Table 3.

Table 3. Coefficients of Variation for ϕ

Coefficients of Variation (ω_j %)						ω_{all} %
Mix Q1	Mix Q2	Mix G1	Mix G2	Mix A1	Mix A2	
52.3	57.5	69.4	63.3	82.7	80.3	68.5

Comparison with the Accuracy of (ϕ) of other Models. The accuracy of the (ϕ) values predicted by the AASHTO LRDF Model [9] in this investigation was compared to the relative accuracy of the other 19 models investigated by Fanourakis [3,4,17], based on the ω_{all} values.

The ω_{all} values and relative accuracy, of all 20 models considered to date, are shown in Table 4.

Table 4. Coefficients of Variation (ω_{all}) for ϕ and Model Relative Accuracy

Prediction Method	ω_{all} [%]	+Relative Accuracy
*AASHTO LRFD (2024) [9]	68.5	16
BS 8110 (1985) [18]	29.4	2
ACI 209 (1992) [19]	45.9	13
AS 3600 (2001) [20]	44.3	10
AS 3600 (2009) [8]	82.7	17
CEB-FIP (1970) [6]	52.5	15
CEB-FIP (1978) [21]	112.5	19
CEB-FIP (1990) [22]	27.7	1
EN 1992-1-1-2004 [13]	31.7	5

FIB MC 2010 (2012) [23]	44.9	11
GL (2000) [24]	39.3	6
GL (2004) [25]	42.5	8
GZ (1993) [26]	52.3	14
HKBD (2013) [27]	45.2	12
JSCE (2007) [28]	43.1	9
JSCE HS (2007) [27]	30.6	3
RILEM Model B3 (1995) [29]	40.8	7
RILEM Model B4 (2015) [30]	103.3	18
SANS 10100 (2000) [7]	29.4	2
SANS 10100 (2000) Modified [5]	31.3	4
*Investigated in this paper. + 1 = Most accurate, 19 = least accurate;		

The comparison indicated that the AASHTO LRFD 2024 Model [9], which was investigated in this paper, was rated as the 16th most accurate relative to the other models investigated. Note that, as two of the models were rated as the second most accurate, the total number of places was 19 not 20.

Conclusions

The AASHTO LRFD Model [9] underpredicted the creep coefficient (ϕ), at all ages, when applied to South African materials.

The predictions for the quartzite mixes were more accurate than the granite mixes, which were, in turn, more accurate than the andesite mixes.

When considering the correlations between predicted ϕ and actual ϕ, all six mixes indicated highly significant linear relationships, with r values ranging from 0.993 to 0.997. The lowest (worst) probability of these relationships was P = 1.4E-11 %, in the case of the high strength andesite concrete.

The coefficient of variation (ω_{all}) for the six mixes ranged from 52.3 % to 82.7 %. The overall coefficient of variation (ω_{all}) was 68.5 %.

A comparison of the predicted (ϕ) values, with those of other investigations, revealed that the AASHTO LRFD Model was the 16th most accurate model, of the total of 20 models considered. As such the model is relatively inaccurate.

References

[1] G.C. Fanourakis, The Influence of Aggregate Stiffness on the Measured and Predicted Creep Behaviour of Concrete. MSc (Eng) Dissertation, University of the Witwatersrand, Johannesburg (1998).

[2] G.C. Fanourakis, Validation of international concrete creep prediction models by application to South African concretes, Journal of the South African Institution of Civil Engineering, 53(2), (2011). 23-30.

[3] G.C. Fanourakis, Evaluation of the creep coefficients of international concrete creep prediction models, fib (CEB-FIP) Symposium 2016, Performance-Based Approaches for Concrete Structures, Cape Town, South Africa, 21 to 23 November 2016.

[4] G.C. Fanourakis, Evaluation of the creep coefficients of the FIB 2010 and RILEM B4 concrete creep prediction models, 12th Central European Congress on Concrete Engineering,

2017, Innovative Materials and Technologies for Concrete Structures, Tokaj, Hungary, 31 August to 01 September 2017.

[5] G. C. Fanourakis, Y. Ballim, Y, An assessment of the accuracy of nine design models for predicting creep in concrete, Journal of the South African Institution of Civil Engineering, 48(4), (2006), 2-8.

[6] CEB-FIP (1970), International recommendations for the design and construction of concrete structures, principles and recommendations, FIP Sixth Congress, Prague, 27-28. Comité Euro-International du Béton - Fédération Internationale de la Précontrainte, Lausanne, Switzerland.

[7] SANS 10100 (2000), The Structural use of concrete - Part 1: Design. South African Bureau of Standards.

[8] AS 3600 (2009), Concrete structures - AS 3600-2009. Standards Association of Australia, Sydney.

[9] AASHTO (2024), AASHTO LRFD Bridge Design Specification, 10th Ed., Washington, DC.

[10] AASHTO (2020). AASHTO LRFD Bridge Design Specification, 9th Ed., Washington, DC .

[11] AASHTO (2017). AASHTO LRFD Bridge Design Specification, 8th Ed., Washington, DC .

[12] D.E. Davis, M.G. Alexander, Properties of aggregates in concrete, Part 2. Sandton, South Africa: Hippo Quarries Technical Publication, 1992, 1-27, 42-43 & 46-47.

[13] EN 1992-1-1-2004, Eurocode 2: Design of concrete structures - Part 1-1: General rules and rules for buildings.

[14] M. Williams, D. Menon, A. Meher Prasad, Creep and shrinkage in prestressed concrete beams: An experimental study, Structural Concrete, (25), (2024), 2400-2419. https://doi.org/10.1002/suco.202300625

[15] A. Mohebbi, Z. Haber, B. Graybeal B, Evaluation of AASHTO Provisions for Creep and Shrinkage of Prestressed UHPC Girders, Proceedings, 2nd International Interactive Symposium on Ultra-High-Performance Concrete, Albany, NY, 2019. https://doi.org/10.21838/uhpc.9706

[16] Z.P. Bazant, L. Panula, Practical prediction of time dependent deformations of concrete - Parts I-VI. Materials and Structures, 12, (1979),169-183. https://doi.org/10.1007/BF02494247

[17] G.C. Fanourakis, Evaluation of the Creep Coefficients of the Hong Kong (HKBD) and Japanese (JSCE) Code Concrete Creep Prediction Models, fib (CEB-FIP) Symposium 2020, Concrete Structures for Resilient Society, Shanghai, China, 22 to 24 November 2020.

[18] BS 8110 (1985), Structural use of concrete - Part 2. Code of Practice for Design and Construction. British Standards Institution, London.

[19] ACI (1992), American Concrete Institute (ACI) Committee 209, Subcommittee II. Prediction of creep, shrinkage and temperature effects in concrete structures. Report ACI 209R-92, Detroit, (1992), 1-12.

[20] AS 3600 (2001), Concrete Structures - AS 3600-2001. Standards Association of Australia, Sydney

[21] CEB-FIP (1978), International system of unified standard codes of practice for structures Volume II - CEB-FIP model code for concrete structures. 3rd ed. 56, 331-344. Comité Euro-International du Béton - Fédération Internationale de la Précontrainte, Lausanne, Switzerland.

[22] CEB-FIP (1990), CEB-FIP Model Code 1990, First Draft, Lausanne, Mar., pp. 2-3, 2-28 to 2-40 (Information Bulletin No. 195). Comité Euro-International du Béton, Lausanne, Switzerland.

[23] CEB-FIP (2012), CEB-FIP Model Code 2010 (2012) Final Draft, Bulletins 65 & 66, 125-155. Fédération Internationale du Béton (fib), Lausanne, Switzerland.

[24] N.J. Gardner, M.J. Lockman, Design provisions for drying shrinkage and creep of normal strength concrete, ACI Materials Journal, 98(2), (2001), 159-167. https://doi.org/10.14359/10199

[25] N.J. Gardner, Comparison of prediction provisions for drying shrinkage and creep of normal strength concretes, Canadian Journal for Civil Engineering, 31(5), (2004), 767-775. https://doi.org/10.1139/l04-046

[26] N.J. Gardner, J.W. Zhao, Creep and shrinkage revisited, ACI Materials Journal, 90(3), (1993), 236-246. https://doi.org/10.14359/3875

[27] HKBD (2013), Code of practice for structural use of concrete. Hong Kong Buildings Department, Hong Kong.

[28] JSCE (2007), Standard specifications for concrete structures - 2007. Design. Number 15 in JSCE Guidelines for concrete. Subcommittee on English version of standard specifications for concrete structures - 2007, Japan Society of Civil Engineers (JSCE), 2010.

[29] RILEM Model B3 (1995), Creep and shrinkage model for analysis and design of concrete structures - Model B3. Draft RILEM Recommendation, prepared by Bazant, Z. P. and Baweja, S. Materials and Structures, 28, 357-365, 415-430, 488-495, with Errata in 29 (1996), 126. https://doi.org/10.1007/BF02473152

[30] RILEM Model B4 (2015), Model B4 for creep, drying shrinkage and autogenous shrinkage of normal and high strength concretes with multi-decade applicability. Draft Recommendation: TC-242-MDC Multi-Decade creep and shrinkage of concrete: Material model and structural analysis (2015), prepared by Bazant, Z. P. Materials and Structures, 48, 753-770. https://doi.org/10.1617/s11527-014-0485-2

Performance Evaluation of RHA-Blended Cements: A Pathway toward Sustainable Cement Standards

BELLO Taofiq[1,a*], OGUNTOLA Isaiah Oluwafemi[1,b], Caesar Msughter SHAWON[1,c], Ajani Oludayo AKINTOLA[2,d], Emmanuel Babatunde KOKU-OJOMU[1,e], Akeem Ayinde RAHEEM[3,f]

[1]Nigerian Building and Road Research Institute, Km10, Idiroko Road, Ota, Nigeria

[2]Department of Civil Engineering, Ladoke Akintola University of Technology, Nigeria

[3]Department of Building, Ladoke Akintola University of Technology, Ogbomoso, Nigeria

[a]taofiqbello4@gmail.com, [b]oguntolaisaiah@yahoo.com, [c]shawoncaesar7@gmail.com, [d]oludayo.akintola@gmail.com, [e]kokuojumubabatunde@nbrri.gov.ng, [f]aaraheem@lautech.edu.ng

Keywords: RHA-Blended-Cement, Clinker-Replacement, Sustainable-Cement, Strength, Cement Grades

Abstract. The construction industry's heavy reliance on Portland Lime Cement (PLC) contributes significantly to global CO_2 emissions, highlighting the need for sustainable alternatives. This study evaluates the performance of Rice Husk Ash (RHA)-blended cement as viable substitutes for PLC. RHA was sourced from rice industry waste and processed for use as a pozzolanic material. Cement blends (M0-M50) were produced by inter-grinding RHA and clinker at 0-50% partial replacements while maintaining gypsum at 5%. Samples were tested for consistency, setting time, soundness, and compressive strength up to 56 days in accordance with EN 196-1. Results showed increasing water demand (27 to 40.6%) and extended setting times (130 to 270 minutes) with higher RHA content, while all blends maintained 0 mm soundness. At 28 days, M10 matched the strength of PLC (59.4 MPa), and by 56 days, M0–M20 reached comparable strengths (~64.4 MPa), indicating ongoing hydration. Based on EN 197-1 classifications, M0 was rated as grade 52.5, M10-M30 as 42.5, and M40-M50 as 32.5, aligning with high-, medium-, and low-strength applications, respectively. These findings demonstrate the technical feasibility of RHA as a clinker substitute with M20 blend showing optimal performance, thereby supporting the standardization of RHA-blended cements and offering a practical pathway to lower carbon emissions in the construction sector.

Introduction

The global construction industry is under increasing pressure to reduce its environmental impact, with Portland Lime Cement (PLC) production alone contributing about 8% of global CO_2 emissions, mainly due to the energy-intensive calcination of limestone [1,2]. With growing infrastructural demands, particularly in developing regions, the need for sustainable, low-carbon cement alternatives is urgent [3]. One promising solution is the use of Supplementary Cementitious Materials (SCMs) derived from agricultural and industrial by-products [4]. Rice Husk Ash (RHA), a silica-rich residue from rice processing, has shown strong pozzolanic reactivity when properly processed, improving the strength and durability of cementitious composites [5,6,7,8,9].

Studies have demonstrated that partially replacing PLC with RHA enhances mechanical performance and durability, reduces permeability, and supports sustainability goals [10]. Optimum replacement levels around 10% have been consistently reported for improved compressive strength in various mix designs [11,12,13]. However, existing research often substitutes RHA during mixing, rather than integrating it directly during cement production. This limit understanding of RHA's full effects as a clinker replacement, particularly on hydration behaviour and microstructural development [3]. Furthermore, the transition from laboratory findings to practical applications and standards remains underexplored [3].

This study addresses these gaps by using industrial RHA to partially replace clinker at the factory level. It evaluates strength development of RHA-blended mortars up to 56 days compared to normal PLC, providing empirical insights to support the integration of RHA in sustainable cement standards. The findings aim to provide robust empirical evidence supporting the viability of RHA as a sustainable clinker replacement, informing ongoing efforts to integrate RHA-blended cements within emerging sustainable cement standards, advancing both environmental goals and construction performance.

Materials and Methods

Materials: Rice Husk Ash (RHA), sourced from a rice mill dumpsite in Nigeria, was collected from 0 to about 50 mm depth of a freshly dumped RHA. Samples were sealed in airtight bags and transported to the laboratory, where they were milled and sieved to <75 μm. This fine fraction was used for oxide composition via X-Ray Fluorescence (XRF) and mineralogical analysis via X-Ray Diffraction (XRD) to assess pozzolanic potential. Other materials included PLC clinker, gypsum, and standard sand conforming to BS EN 196-1 (2005). Clinker was partially replaced with RHA at 10-50% by weight, maintaining 5% gypsum across all blends, yielding six cement types: one control (M0: 95% clinker + 5% gypsum) and five RHA-blended cements (M10-M50).

Performance Characteristics: Water-to-binder ratio was determined via consistency tests per BS EN 196-3 (2005). RHA-blended mortar adopted consistency-based w/b ratios to maintain workability [10]. Initial and final setting times were measured using the Vicat apparatus in accordance with EN 196-3 (2016), while soundness was evaluated using the Le Chatelier method to assess dimensional stability post-setting.

Mortar Production and Curing: Mortar was prepared according to BS EN 196-1 (2005). A fixed w/b ratio of 0.5 was used for M0, while RHA-blended mortars used the consistency-derived ratios. Specimens (40 × 40 × 160 mm) were cast and cured in water at 20 °C for 2, 7, 14, 28, and 56 days, representing early to later-age performance.

Flexural and Compressive Strength Testing: Mechanical performance was assessed through compressive and flexural strength tests at each curing age, following EN 196-1 (2016). Mortar specimen in (40 × 40 × 160 mm) prisms were tested under one-point mid-span loading to determine flexural strength while the compressive strength test made use of one part of the specimen resulting from flexural test. Strength values represent the average of three replicate specimens per mix/age.

Performance Classification Based on Standards: Cement performance was benchmarked against EN 197-1 (2021), using compressive strength at 2, 7, and 28 days to determine alignment with strength classes (e.g., 32.5N, 42.5R). The "N" and "R" designations denote normal and rapid early strength, respectively. Initial setting time and soundness were also evaluated to confirm compliance with the standard's thresholds (Initial Setting Time (IST) minimum limits; soundness ≤10 mm). This performance-based approach provided a basis for comparing the RHA-blended cements to the conventional PLC, with the ultimate goal of evaluating their suitability for classification within the EN 197-1 (2021) framework. The results from these tests offer insights into whether industrial-scale RHA-clinker blending can produce cementitious materials that meet the minimum strength and durability requirements for standardized cement products.

Results and Discussion

Oxide Composition and Loss on Ignition (LOI) of RHA. The chemical composition of the RHA is presented in Table 1. The RHA has a high silica (SiO_2) content of 84.14%, with accompanying amounts of alumina (Al_2O_3, 3.35%) and ferric oxide (Fe_2O_3, 1.02%) whose combination amounted to 88.51% and surpasses the minimum threshold of 70% stipulated by ASTM C618 (2019) for Class N pozzolans. This indicates that the RHA possesses strong pozzolanic potential and can contribute to strength and durability enhancement in cementitious systems [18]. The loss on ignition (LOI) was measured at 8.22%, which is also within acceptable limit of 10% according to ASTM C618 (2019). The relatively low SO_3 content (1.01%) indicates that sulfate-related durability risks are minimal.

Table 1. Oxide composition of RHA

SiO_2	Al_2O_3	Fe_2O_3	CaO	MgO	SO_3	TiO_2	K_2O	Na_2O	TiO_2	P_2O_5	LOI	$\sum$SAF
84.14	3.35	1.02	0.43	0	1.01	0.4	0.37	0.1	0.41	0.71	8.22	88.51

Overall, the oxide profile confirms that the RHA is a reactive SCM, with high silica content capable of reacting with calcium hydroxide from cement hydration to form additional calcium silicate hydrate (CSH) gel for enhanced strength property. This was corroborated by the high amorphous silica content revelaed by the XRD shwon in Figure 1.

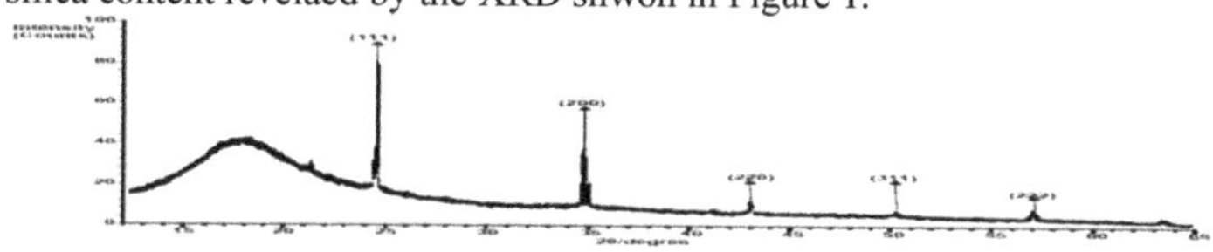

Fig 1: X-Ray Diffraction of RHA

Oxide Composition of PLC Clinker. The oxide composition of the PLC clinker used in this study is provided in Table 2. Calcium Oxide (CaO) was the dominant constituent at 65.26%, followed by silica (SiO_2, 21.04%), alumina (Al_2O_3, 4.61%), and ferric oxide (Fe_2O_3, 4.35%). These values are within the typical ranges specified by EN 197-1 (2021) for CEM I clinker and confirm the clinker's suitability for high-strength hydraulic applications. The low LOI of 0.96% reflects a well-calcined clinker with minimal residual carbon or volatiles. Minor oxides, including MgO, SO_3, Na_2O, and K_2O, were within acceptable limits (ASTM C150, 2022), suggesting good stability and compatibility when blended with pozzolans. This compositional baseline is essential for evaluating the effects of RHA substitution, as reduction in CaO and increase in SiO_2 from RHA influences the cement's hydration behavior and performance characteristics.

Table 2. Oxide composition of PLC clinker compared with EN 197-1 (2021) specifications

Compound	SiO_2	Fe_2O_3	Al_2O_3	CaO	MgO	SO_3	Na_2O	K_2O	LOI
Content (%)	21.04	4.35	4.61	65.26	2.12	0.79	0.37	0.20	0.96
EN-197-1 (%)	19-23	2-6	4–8	60-67	0.5-4	1-3	0.5-1	0.5-1	0-5

Oxide Composition of PLC and RHA-Blended Cements. The oxide compositions of the control cement (M0) and the RHA-blended cements (M10- M50) are summarized in Table 3. The results indicate a progressive increase in SiO_2 content with increasing RHA replacement from 19.93% in M0 to 53.20% in M50 reflecting the high silica content of the RHA. Conversely, the CaO content decreased steadily from 62.06% in M0 to 30.91% in M50, due to the partial replacement of clinker that is the primary source of CaO. These effects suggest a major change in cement composition, hence, cement grade variation. Minor decrease in Al_2O_3 and Fe_2O_3 were also observed with increased RHA levels indicating that PLC clinker is richer in these two compounds. While these reductions may reduce the initial cementitious reactivity, they are partly compensated by the pozzolanic contribution of the added silica. Increase in K_2O content was observed across the mixes, attributable to the intrinsic potassium content in RHA. This could affect alkali-silica reaction (ASR) risk and should be considered in durability evaluations [19].

The calculated cement quality indices show a decreasing lime saturation factor (LSF), which corresponds with the reduction in calcium-bearing phases. In contrast, both the silica ratio (SR) and alumina ratio (AR) increased with higher RHA content, highlighting a shift in chemical balance and clinker potential. However, these effects are desirable for strength enhancement [17] and resistance to sulfate attack in blended cements [18,21].

Table 3: Oxides Compositions of the Control and RHA Blended Cement

Compounds	M0	M10	M20	M30	M40	M50
	Content (%)					
SiO_2	19.93	21.43	33.36	38.19	46.52	53.20
Al_2O_3	5.16	4.83	4.69	4.47	4.30	4.20
Fe_2O_3	3.10	2.92	2.62	2.40	2.20	2.01
CaO	62.06	56.18	50.32	43.94	37.45	30.91
MgO	1.89	1.75	1.52	1.32	1.18	1.04
K_2O	0.30	0.37	0.45	0.55	0.61	0.69
LSF	0.09	0.09	0.06	0.05	0.04	0.03
SR	2.41	2.77	4.56	5.56	7.16	8.56
AR	1.66	1.65	1.79	1.86	1.95	2.09

Consistency, Setting Time, Soundness, and Loss on Ignition. The average results of standard consistency, setting time, soundness, and loss on ignition (LOI) for the control cement (M0) and the RHA-blended cement samples (M10-M50) are presented in Table 4. The standard consistency increased progressively with RHA content from 27.0% in the control cement (M0) to 40.6% in M50. This increase is attributable to the high specific surface area and porous nature of RHA, which elevates the water demand required to achieve a normal paste consistency increased with higher RHA replacement [7]. M0 recorded the shortest initial setting time of 130 minutes and final setting time of 200 minutes. These values extended to 270 and 340 minutes, respectively, in M50. The extended setting durations are linked to the partial replacement of clinker phases particularly C_3A and C_3S and the slower pozzolanic reaction of RHA. This behavior can be beneficial in hot climates or when prolonged workability is desired. All cement types showed zero expansion in the Le Chatelier soundness test, indicating excellent volumetric stability. The incorporation of RHA did not compromise dimensional stability, which is consistent with the low SO_3 and MgO contents observed in chemical composition of the RHA. The LOI values increased steadily with increasing RHA content, from 3.50% in M0 to 5.37% in M50. This trend reflects the residual unburnt carbon or organic matter inherent in RHA. However, all LOI values remained below the 6% limit prescribed by ASTM C618 (2019), affirming the suitability of the RHA as a pozzolan replacer for clinker.

Table 4: Average Consistency, Setting Time, Soundness and LOI of PLC and RHA Blended Cement

Cement Type	Consistency (%)	Setting Time (Mins)		Soundness (mm)	Loss on Ignition (LOI)
		Initial	Final		
M0	27.0	130	200	0	3.50
M10	28.0	175	225	0	3.70
M20	31.0	205	285	0	3.94
M30	33.0	230	290	0	4.11
M40	37.0	255	330	0	4.46
M50	40.6	270	340	0	5.37

Mechanical Properties of RHA-Blended Cement Mortars.

Figure 1 presents the flexural strength of PLC and RHA-blended cement mortars from 2 to 56 days. The results showed that the flexural strength of M0 mortar had the highest early strength 5.6, 6.8 and 6.8 MPa at 2, 7 and 14 days respectively and was closely followed by the M10 mortar yielding 4.9, 5.8 and 6.3 MPa. The M20 and M30 mortars had slightly lower early strength behavior. While the M20 mortar yielded 3.2, 5.5 and 6.2 MPa, the M30 recorded 2.8, 5.3 and 6.8 MPa respectively for 2, 7 and 14 days. The M40 and M50 mortars showed similar strength behavior at early ages, with M40 having slightly higher flexural strength values of 2.4, 4.6 and 5.6 MPa compared to the M50's 2.0, 4.1 and 5.2 MPa for 2, 7 and 14 days respectively. This is due to higher clinker reduction, limiting early-age strength development. Additionally, pozzolanic reaction is delayed at early ages, as pozzolans gain significant strength at later hydration ages.

The 28-day flexural strength of M0 mortars was 7.90 MPa, similar to M10 and M20 mortars due to lower RHA content. However, flexural strength decreased at M30-M50 and this could be attributed to reduced alite (C_3S) content. A general strength improvement occurred at 56 days, particularly on RHA blended cement due to pozzolanic reaction. The strength gain between 28 and 56 days was 2.15%, while the control (M0) showed gains of 3.67, 16.46, 17.14, 13.51, and 17.14%. The highest strength gain was observed at 17.14% in M30 mortars.

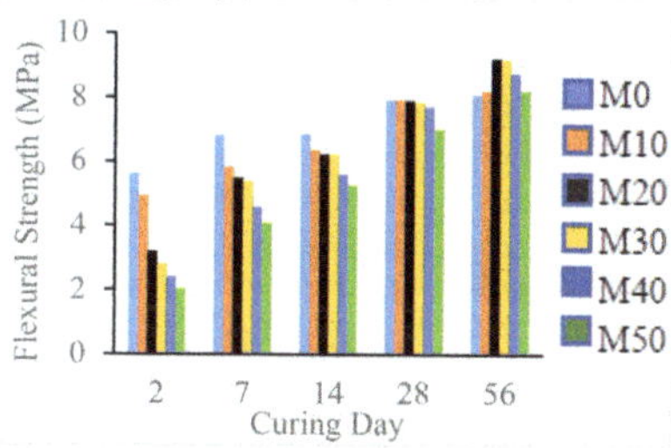

Fig 1: Flexural Strength Results

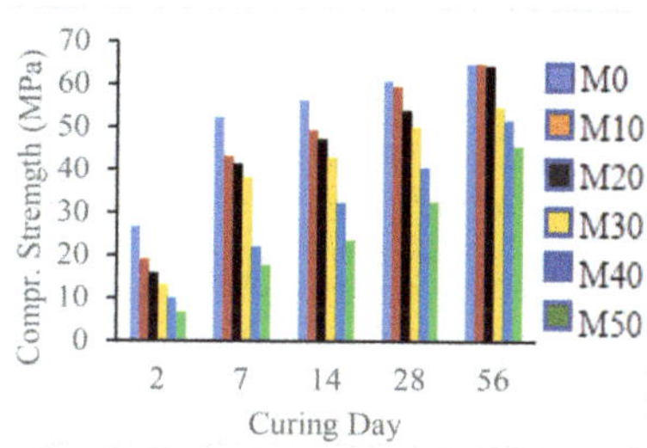

Fig 2: Compressive Strength Results

Figure 2 shows the compressive strength development of the six cement mortars during a 2- to 56-day period. All mixtures showed a general pattern of increasing strength with curing time, though the rate and magnitude of development differed depending on the level of RHA inclusion. At early ages, PLC mortar (M0) shows a significantly greater compressive strength than all RHA blended cement mortars. M0 obtains 26.6 MPa after 2 days, while M10, M20, M30, M40, and M50 have reduced strengths of 19.0 MPa (28.6% lower), 15.9 MPa (40.2% lower), 12.9 MPa (51.5% lower), 10.0 MPa (62.4% lower), and 6.6 MPa (75.2% lower), respectively. The RHA blended cement (M10-M50) shows low early strength due to clinker replacement with RHA, causing slower hydration due to low clinker phases like alite. By 7 days, M0 attains 52.2 MPa, while M10, M20, recorded 43.1 MPa, 41.5 MPa respectively, M30 reach 37.9 MPa. Higher RHA replacements result in a more pronounced strength reduction. However, noticeable strength gains on M10 and M20 mortars occur from 2 to 7 days. At 14 days, strength increases across all mixes, with M0 reaching 56.2 MPa, M10 at 49.2 MPa, M20 at 47.3 MPa, and M30 at 42.9 MPa. The gap between M0 and M10-M20 narrows, indicating hydration due to low clinker replacement. M40 and M50 experience a significant lag in strength development.

The compressive strength of PLC mortar at 28 days is 60.7 MPa, with M10 having a slightly lower strength of 59.4 MPa. The compressive strength of M20 and M30 mortars follows, with M40 and M50 significantly lower. 10% RHA replacement maintains competitive strength with PLC at 28 days, possibly due to low clinker replacement and the presence of hydration compound C_3S. Higher clinker/RHA replacements result in significant strength reduction.

At 56 days, the strength-gain trend of M0, M10, and M20 showed similar results, with M10 reaching 64.7 MPa and M20 at 64.4 MPa. The M30 mortar, however, showed a 15.6% decrease in compressive strength, indicating that pozzolanic reaction compensated for clinker replacement, enhancing later strength gain in 10-20% RHA replacement. M30 mortar also showed significant improvement, reaching 54.6 MPa, 10.5% higher than 28-day strength. The results align with existing literature where moderate RHA levels were found to improve strength through secondary C-S-H formation [10,17,22]. The long-term strength gain suggests improved durability potential, an important consideration for sustainable construction.

The study indicates that moderate replacement of clinker at 20% with RHA can improve the machanical performance of conventional PLC mortar at 56 days due to continuous pozzolanic reaction, resulting in denser and stronger mortar matrices. This aligns with previous studies like [23,24].

Classification of RHA Blended Cement. The study compares the early compressive strength, standard compressive strength, IST, and soundness values of control cement (M0) with BS EN 197-1 (2021) as described in Table 5. The control cement (M0) is classified as 52.5N due to its 2-day compressive strength above 20 MPa according to BS EN 197-1 (2021), while the M10, M20, and M30 cements are classified as 42.5N (structural application cement) and 32.5N (non-structural application cement). The use of RHA partially replaces clinker in cement production, achieving grades of 42.5 and 32.5. The study demonstrates that 10-30% PLC clinker replacement with RHA identifies with grade 42.5, suitable for structural applications. However, higher replacement levels compromise early and standard strength, limiting their application to low-strength or non-structural uses like mass concrete construction. These findings align with recent research, which recommends up to 20% RHA replacement for optimal performance [25].

Table 5: Classification of the PLC and RHA Blended Cement Based on BS EN 197-1 (2021)

Strength Class	Compressive Strength (MPa)				Initial Setting Time (Min)	Soundness (mm)	Classification of OPC and RHA Blended Cement
	Early Strength 2 Days	Early Strength 7 Days	Standard Strength 28 Days				
32.5N	-	≤16	≥32.5	≤52.5	≥75	≤10	M40, M50
32.5R	≥10	-					
42.5N	≥10	-	≥42.5	≤62.5	≥60	≤10	M10, M20, M30
42.5R	≥20	-					
52.5N	≥20	-	≥52.5	-	≥45	≤10	M0
52.5R	≥30	-					

Conclusion and Recommendation

This study assessed the performance of RHA blended cements as sustainable alternatives to PLC, focusing on performance properties and mechanical strength development. The conclusion made from the study are;

i. RHA from rice processing mill contains high content of reactive silica (84.14%), and the PLC clinker showed high conformity with relevant standards, thereby validating its quality and reliability as a baseline material for studying the incorporation of RHA as a sustainable cementitious material,
ii. The PLC aligns with global specifications (EN 197-1, 2021)/ASTM C150, 2022), exhibiting a consistency of 27%, initial setting time of 130 minutes and LOI of 3.5%.

Likewise, the RHA blended cement variants exhibited performance characteristics consistent with ASTM C595 (2021),

iii. The mechanical performance reduced from 20% RHA replacement level. RHA can effectively replace a portion of clinker in cement production without compromising strength performance at 20% at 56 days,

iv. High RHA contents (>30%) led to reduced early-age strength but exhibited gradual strength gain, making them suitable for specific applications where delayed strength is acceptable.

v. RHA blended cement produced both grade 42.5N cement (M10-M30) capable of structural applications and grade 32.5N cement (M40 and M50) for low strength applications. The study supports the integration of RHA as an SCM, and establishes a technical foundation for adopting RHA in sustainable cement production, offering a practical pathway to decarbonize the construction industry.

References

[1] J.C. Minx, W.F. Lamb, R.M. Andrew, J.G. Canadell, M. Crippa, N. Dobbeling, P.M. Forster, D. Guizzardi, J. Olivier, G.P. Peters, J. Pongratz, A. Reisinger, M. Rigby, M. Saunois, S.J. Smith, E. Solazzo, H. Tian, A comprehensive and synthetic dataset for global, regional, and national greenhouse gas emissions by sector 1970–2018 with an extension to 2019, Earth Syst. Sci. Data. 13 (2021) 5213–5252. https://doi.org/10.5194/essd-13-5213-2021

[2] IPCC, Mitigation of Climate Change. Contribution of Working Group III to the Sixth Assessment Report of the Intergovernmental Panel on Climate Change, Cambridge University Press, Cambridge, UK and New York, NY, USA, 2022. https://doi.org/10.1017/9781009157926

[3] M.M. Ahmed, A. Sadoon, M.T. Bassuoni, A. Ghazy, Utilizing agricultural residues from hot and cold climates as sustainable SCMs for low-carbon concrete, Sustainability. 16 (2024) 10715.

[4] A. Ahmed, J. Kamau, J. Pone, F. Hyndman, H. Fitriani, Chemical reaction in pozzolanic concrete, Mod. Approaches Mater. Sci. 1 (2019) 128–133.

[5] J. Kamau, A. Ahmed, K. Ngong, Sulfate resistance of Rice Husk Ash concrete, MATEC Web Conf. 199 (2018) 02006. https://doi.org/10.1051/mateconf/201819902006

[6] G. Mounika, R. Baskar, J.S.K. Rama, Rice husk ash as a potential supplementary cementitious material in concrete solutions towards sustainable construction, Innov. Infrastruct. Solut. 7 (2022) 51. https://doi.org/10.1007/s41062-21-00643-5

[7] A.A. Raheem, M.A. Anifowose, Effect of ash fineness and content on consistency and setting time of RHA-blended-cement, Mater. Today Proc. 86 (2023) 18–23.

[8] C. Fapohunda, B. Akinbile, A. Shittu, Structure and properties of mortar and concrete with rice husk ash as partial replacement of ordinary Portland cement-A review, Int. J. Sustain. Built Environ. 6 (2017) 676–689.

[9] I. Sanou, M. Sawadogo, M. Seynou, L. Zerbo, R. Ouedraogo, Study of the mechanical behaviour of mortars modified with rice husk ash, J. Miner. Mater. Charact. Eng. 7 (2019) 373-384. https://doi.org/10.4236/jmmce.2019.76025

[10] S.K. Tulashie, P. Ebo, J.K. Ansah, D. Mensah, Production of Portland pozzolana cement from rice husk ash, Materialia. 16 (2021) 101048

[11] A.Z. Seyed, A. Farshad, D. Farzan, A. Mojtaba, Rice husk ash as a partial replacement of cement in high strength concrete containing micro silica: Evaluating durability and mechanical

properties, Case Stud. Constr. Mater. 7 (2017) 73–81. https://doi.org/10.1016/j.cscm.2017.05.001

[12] D.N.L. Botchway, R.O. Afrifa, C.Y. Henaku, Effect of partial replacement of ordinary Portland cement (PLC) with Ghanaian Rice Husk Ash (RHA) on the compressive strength of concrete, Open J. Civ. Eng. 10 (2020) 353–363. https://doi.org/10.4236/ojce.2020.104027

[13] M.M. Rashwan, K.A. Abbas, M.A. Ali, Assessment of using an optimal dose of superplasticizer with the optimum replacement of rice husk ash on concrete compressive strength, Int. J. Civ. Eng. Technol. 11 (2020) 9–21.

[14] EN 196-1:2016, Methods of testing cement – Part 1: Determination of strength, European Committee for Standardization, Brussels, 2016.

[15] EN 196-3:2016, Methods of testing cement – Part 3: Determination of setting times and soundness, European Committee for Standardization, Brussels, 2016.

[16] EN 197-1:2021, Cement – Part 1: Composition, specifications and conformity criteria for common cements, European Committee for Standardization, Brussels, 2021.

[17] A.A.K. Al-Alwan, M. Al-Bazoon, F.I. Mussa, H.A. Alalwan, M.H. Shadhar, M.M. Mohammed, M.F. Mohammed, The impact of using rice husk ash as a replacement material in concrete: An experimental study, J. King Saud Univ.–Eng. Sci. 36 (2024) 249–255.

[18] A.A. Raheem, M.A. Kareem, Chemical composition and physical characteristics of rice husk ash blended cement, Int. J. Eng. Res. Afr. 32 (2017) 25–35.

[19] W. Zhiwen, B. Li, A. Othman, Z. Zhang, Endogenous Alakli Silica reaction Evaluation of Rice Husk Ash in Mortar, J. Build. Eng. 75 (2023) 106970.

[20] ASTM C618-19, Standard Specification for Coal Fly Ash and Raw or Calcined Natural Pozzolan for Use in Concrete, ASTM International, West Conshohocken, PA, 2019. https://doi.org/10.1520/C0618-19

[21] S. Barbhuiya, J. Nepal, B.B. Das, Properties, compatibility, environmental benefits, and future directions of limestone calcined clay cement (LC3) concrete: A review, J. Build. Eng. 79 (2023) 107794.

[22] W.S. De Jesus, T. Van Mullem, N. Alderete, D. Baeyens, N. De Belie, Influencing factors to the capillary water uptake of (un)cracked cementitious materials, MATEC Web Conf. 378 (2025) 02028.

[23] B. Bayrak, H.G. Alcan, M. Tanyildizi, G. Kaplan, S. Ipek, A.C. Aydin, E. Guneyiysi, Effect of silica fume and rice husk ash contents on the engineering properties and high-performance resistance of slag-based prepacked geopolymers, J. Build. Eng. 92 (2024) 109746.

[24] S.M. Ubayi, C.V. Mathur, B.S. Abubakar, H. Idris, Rice husk ash in concrete: A review of cement replacement proportions and mechanical properties, Int. J. Eng. Sci. 1 (2024) 1–15.

[25] M.D. Yeshiwas, M.D. Yehualaw, B.T. Habtegebreal, W.M. Nebiyu, W.Z. Taffese, Rice husk ash and waste marble powder as alternative materials for cement, Infrastructures. 10 (2025) 78. https://doi.org/10.3390/infrastructure10040078

Advances in Cement and Concrete Research - ICACCR 2025 Materials Research Forum LLC
Materials Research Proceedings 63 (2026) 103-111 https://doi.org/10.21741/9781644904053-12

Mechanical and Morphological Properties of Rice Husk Ash-Calcium Nitrate Blended Cement Concrete

Isaiah Oluwafemi OGUNTOLA[1,a], Taofiq BELLO[1,b*], Caesar Msughter SHAWON[1,c], Oludayo Ajani AKINTOLA[2,d], Babatunde Emmanuel KOKU-OJOMU[1,e],

Akeem Ayinde RAHEEM[6,f]

[1]Nigerian Building and Road Research Institute, Km10, Idiroko Road, Ota, Ogun State, 112212 Nigeria

[2f] Department of Building, Ladoke Akintola University of Technology, Ogbomoso, 210214 Nigeria

[a]oguntolaisaiah@nbrri.gov.ng, [b]bellotaofeeq@nbrri.gov.ng, [c]shawoncaesar7@gmail.com, [d]oludayo.akintola@gmail.com, [e]kokuojumubabatunde@nbrri.gov.ng, [f]aaraheem@lautech.edu.ng

Keywords: Rice Hush Ash, Calcium Nitrate, Compressive Strength, Flexural Strength, Splitting Tensile Strength

Abstract. Blended cement concrete involves the use of supplementary cementitious materials (SCMs) which have gained significant attention in recent years due to its potential to improve various properties of concrete. Rice Husk Ash (RHA) had been used as SCM in concrete and the properties are promising. This study investigates the mechanical and morphological properties of RHA-calcium nitrate (CN) with a view to enhance its properties. RHA was sourced from Hillcrest Rice Mill in Offa, Kwara State. CN was purchased in Osogbo, Osun State. Portland Limestone Cement (PLC), fine and coarse aggregates were purchased in Sango Ota, Ogun State. The chemical composition of RHA, CN and PLC were analyzed using XRF. RHA at 15% and CN at 1,2,3,4 and 5% were replaced by weight of PLC. Two control specimens were established: Control 1 comprised solely of PLC, while Control 2 incorporated RHA without the addition of CN. The compressive strength was assessed from 7 to 90 days of curing. Flexural and splitting tensile strength were evaluated from 7 to 28 days. Morphological evaluation was conducted using scanning electron microscopy. The SiO_2, Al_2O_3 and Fe_2O_3 present in RHA were 69.32, 3.41 and 2.83%, respectively. The workability (slump and compacting factor) ranged from 14-34 and 0.72-0.84 mm, respectively. The compressive, flexural and splitting tensile strength of RHA-CN concrete varied from 20.84 to 45.8, 1.4 to 4.9 and 1.6 to 2.9 MPa, respectively. The microstructure displayed fewer pores as CN percentage increases. Further studies are recommended to increase the quantity of CN.

Introduction

As the demand for sustainable and cost-effective construction solutions grows, attention has increasingly turned toward blended cements due to their lower environmental impact and reduced production costs resulting from decreased clinker and energy usage. Researchers are actively investigating alternative materials and admixtures that can enhance the mechanical characteristics of traditional concrete. Among these, Rice Husk Ash (RHA), a silica-rich by-product of rice processing, has proven to be a highly reactive pozzolan. When incorporated into cement, RHA reacts with calcium hydroxide during hydration, forming additional calcium silicate hydrate (C-S-H), thereby improving strength and durability [1, 2].

Additionally, calcium nitrate (CN) acts as a chemical accelerator, enhancing early strength gain and reducing setting time through its influence on hydration kinetics [3]. The combined use of RHA and CN has been shown to significantly improve compressive, tensile and flexural strength, while also refining the microstructure of the cement matrix. RHA not only minimizes cement usage, thereby lowering carbon emissions, but also reduces porosity and strengthens the interfacial transition zone between aggregate and paste [4]. Research has confirmed that RHA meets the chemical standards required for supplementary cementitious materials [5], while CN is widely recognized for its accelerating properties in concrete production standards [6]. This study investigates the joint effect of RHA and CN on concrete's mechanical performance and microstructure, aiming to establish their suitability in producing durable and environmentally responsible construction materials.

Materials and Methods

The study utilized RHA obtained from Hillcrest Rice Mill, Offa, Kwara State (Latitude: N8°14′45″, Longitude: E4°48′42″) as shown in Plate 1. Portland Limestone Cement (PLC) was sourced from a vendor in Sango Ota, Ogun State. River sand and crushed granite were collected from a quarry in the same area. CN was purchased from Safideb Chemical Nigeria Ltd., Osogbo, Osun State as shown in Plate 2. Potable water used for mixing was supplied from the Nigerian Building and Road Research Institute (NBRRI) laboratory in Ota, Ogun State, where the experimental work was conducted.

Plate 1: RHA *Plate 2: CN*

Testing of Materials

X-ray fluorescence (XRF) was employed to analyze the oxide compositions of RHA, CN, and PLC, including SiO_2, Al_2O_3, Fe_2O_3, CaO, MgO, and SO_3. The tests were conducted at Afe Babalola University and NBRRI, following the procedures outlined in [7, 8]. A 1:2:4 concrete mix with a 0.6 water-cement ratio was adopted, as recommended in [9]. A 15% RHA replacement was chosen as optimal, as suggested in [10, 11]. Mixing was done manually as described in [12]. Table 3.1 presents the batching: Control 1 (no RHA/CN), Control 2 (15% RHA), and test mixes with 15% RHA and CN varying from 1% to 5%.

Table 1. Batching of Material for Production of Concrete Specimen

Samples Identification	Specimen	Cement [kg]	RHA [kg]	CN [kg]	FA [kg]	CA [kg]
A	100% OPC (Control 1)	50	0	0	100	200
B	85% OPC+15RHA (Control 2)	42.5	7.5	0	100	200
C	84% OPC + 15% RHA + 1%CN	42	7.5	0.5	100	200
D	83% OPC + 15% RHA + 2%CN	41.5	7.5	1.0	100	200
E	82% OPC + 15% RHA + 3%CN	41	7.5	1.5	100	200
F	81% OPC + 15% RHA + 4%CN	40.5	7.5	2.0	100	200
G	80% OPC + 15% RHA + 5%CN	40	7.5	2.5	100	200

Specimen Preparation
Compressive strength tests were conducted on 100 mm concrete cubes at 7, 14, 21, 28, 56, and 90 days of curing. Three specimens per mix were tested to obtain the average strength, totalling 126 cubes. Testing was performed using a 2000 kN Control Model Universal Testing Machine at NBRRI, following [13] standards.

Flexural strength was evaluated using 100 × 100 × 500 mm beam specimens at 7, 14, 21, and 28 days. Three beams per mix were tested, amounting to 84 specimens. The indirect test method was adopted using the same machine, in accordance with [14].

Splitting tensile strength was measured on 100 × 300 mm cylindrical specimens at 7, 14, 21, and 28 days. A total of 84 cylinders were tested (three per mix) using the indirect method, with loading applied through a 2000 kN Universal Testing Machine at NBRRI, as specified in [15].

Characterization of Materials
Table 2 presents the results of the XRF analysis. The RHA sample qualifies as a Class F pozzolan, as the combined content of SiO_2, Al_2O_3, and Fe_2O_3 is 75.36%, which exceeds the 70% minimum threshold, and the loss on ignition (LOI) is 1.5%, well below the 6% limit specified by [16]. This classification aligns with findings reported by [7]. Furthermore, the XRF data for CN and OPC reveal CaO contents of 81.95% and 68.76%, respectively, indicating greater cementitious properties compared to RHA. The SO_3 content in RHA is 0.5%, which complies with the 5% maximum value recommended for cementitious materials [17].

Table 2. Chemical Composition of RHA, CN and PLC

	Percentage Composition [%]		
Chemical Constituent	RHA	CN	PLC
SiO_2	69.32	2.05	18.65
Al_2O_3	3.41	0.60	3.24
Fe_2O_3	2.63	5.01	3.24
CaO	5.62	81.95	68.76
MgO	0.73	5.61	1.82
Na_2O_3	0.32	0.43	0.37
K_2O	6.16	1.10	1.65
P_2O_5	0.61	0.05	-
SO_3	0.50	1.60	2.90
LOI	1.50	1.49	1.73
$SiO_2 + Al_2O_3 + Fe_2O_3$	75.36		

Particle Size Distribution
Figure 1 presents the particle size distribution (PSD) results. According to the Unified Soil Classification System (USCS), river sand with a coefficient of uniformity (Cu) of 3.3 and a coefficient of curvature (Cc) of 1.1 is classified as well-graded. Granite, having Cu of 2.0 and Cc of 1.09, is more uniformly graded. RHA was sieved through sieve No. 200, confirming its fine nature. These characteristics indicate that river sand, granite, and RHA are appropriate for concrete production [18,19,20].

Specific Gravity
Table 3 shows that the specific gravity of RHA (2.27 g/cm³) is lower than that of granite (2.71 g/cm³) and river sand (2.62 g/cm³), suggesting its potential to produce lightweight concrete

with better thermal insulation. RHA qualifies as a lightweight aggregate [21]. While granite and sand fall within the 2.5–3.0 g/cm³ range for normal-weight aggregates [22,23,24,25],

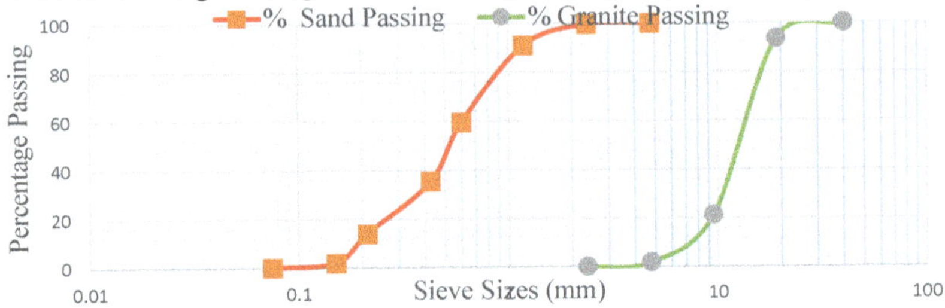

Fig. 1: Particle Size Analysis of Sand and Granite

Table 3. Specific gravity of RHA, PLC, Sand and Granite

Materials	Specific Gravity
RHA	2.27
PLC	3.10
Fine Aggregate (Sand)	2.62
Coarse Aggregate (Granite)	2.71

Workability

This section discusses the workability of the control samples and varying RHA-CN blended mortar.

Slump

Figure 2 presents the slump test results, revealing a decline in workability as RHA and CN contents increase. Control samples A and B recorded slump values of 39 mm and 36 mm, respectively. When RHA-CN was added at 1–5%, slump values dropped from 34 mm to 14 mm. This downward trend indicates that increasing CN dosage significantly reduces the mix's fluidity. This behavior aligns with the findings of [26], which confirm that increasing CN lowers concrete workability. Despite reduced slump, RHA-CN blends promote sustainability by cutting cement usage and utilizing agricultural waste. These mixes are especially suitable for low-workability applications such as road pavements, rigid foundations, and precast components. Engineers must account for reduced slump in mix design and site implementation to ensure performance consistency when using such eco-efficient alternatives.]

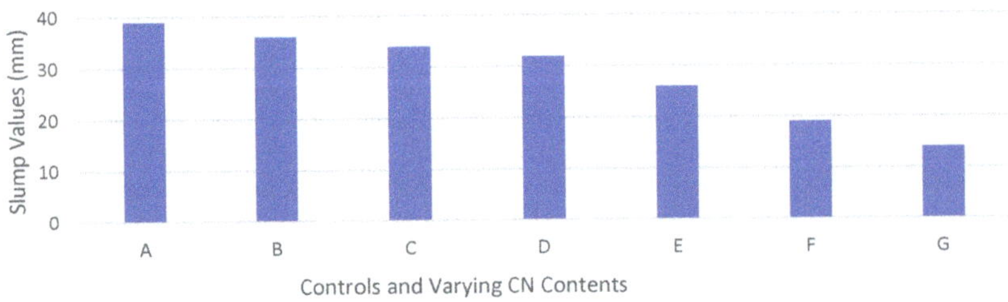

Fig. 2: Slump of RHA-CN Concrete

Compacting factor

Figure 3 presents the compacting factor results, where samples A and B recorded values of 0.92 and 0.89, respectively, classifying them within the *medium workability* range, ideal for general reinforced concrete applications like slabs and beams. As CN content increased from 1 to 5%, the compacting factor values progressively declined (0.84 to 0.72), indicating reduced workability. This trend aligns with [27], which classifies compacting factors as medium (0.85–0.92), low (0.78–0.84), and very low (<0.78) per BRE standards. Samples C and D, in the *low workability* category, may require mechanical vibration, making them suitable for lightly reinforced or simple formwork elements. Samples E to G, with *very low workability*, are best suited for mass concrete applications like pavements and foundations where fluidity is less critical. The decrease in compacting factor with higher CN dosage suggests the potential need for added water or chemical admixtures to maintain workability.

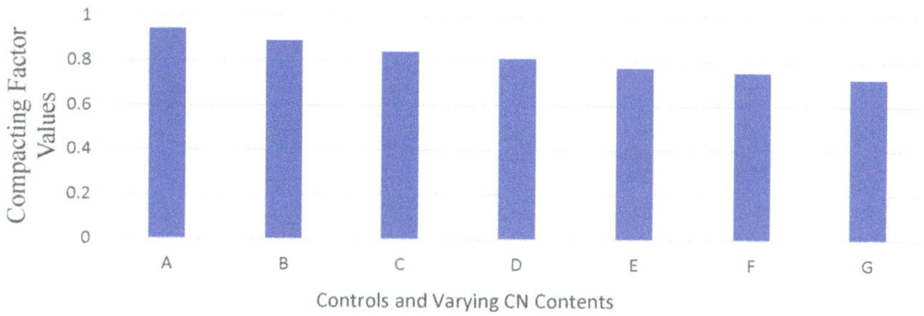

Fig. 3: Compacting factor of RHA-CN Concrete

Compressive Strength of RHA-CN Blended Cement Concrete

Figure 4 illustrates the compressive strength development of RHA-CN blended cement concrete over time. At 7 days, Sample A showed a strength of 20.02 MPa, slightly higher than Sample B at 18.01 MPa. Strength increased progressively with higher CN content, reaching up to 25.31 MPa at 5% CN. Similar growth trends continued at 14, 21, 28, 56 and 90 days, with RHA-CN mixes showing compressive strengths up to 45.84 MPa at 90 days. Notably, beyond 3% CN content, the strength of RHA-CN concrete exceeded that of Sample A, demonstrating that CN significantly enhances long-term strength. These results indicate that the optimum CN content is around 5%, and strength gains improve steadily with curing age aligning with [28], who reported that combining pozzolanic materials with chemical accelerators improves both early and long-term concrete strength by enhancing hydration and microstructure.

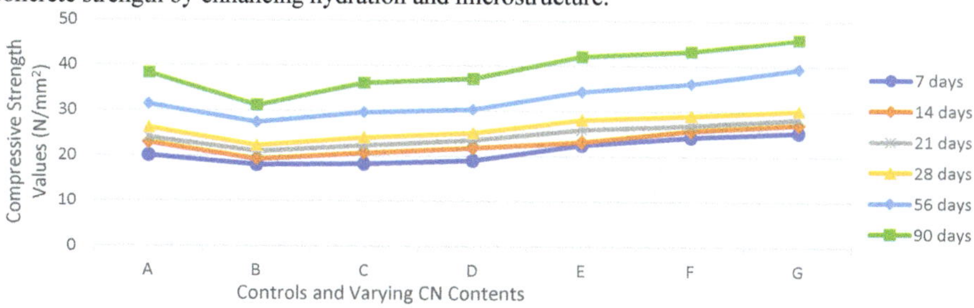

Fig. 4. Compressive Strength of RHA-CN Blended Cement Concrete.

Flexural Strength of RHA-CN Blended Cement Concrete

The flexural strength results in Figure 5 reveal that Sample A consistently outperforms Sample B at all curing ages. As CN content increases from 1% to 5%, flexural strength improves across Samples C to G. At 7 days, values range from 1.56 to 3.19 MPa, increasing further at 14, 21, and 28 days to peak values of 4.86 MPa. This progressive strength gain with higher CN dosage confirms enhanced tensile resistance over time, consistent with [27]. These findings agree with [10,11], who noted that chemical accelerators improve the flexural and tensile performance of pozzolanic concrete by enhancing hydration and matrix density. Thus, RHA–CN blended concrete is suitable for structural elements like beams and slabs, offering reduced reinforcement and cost-effective construction.

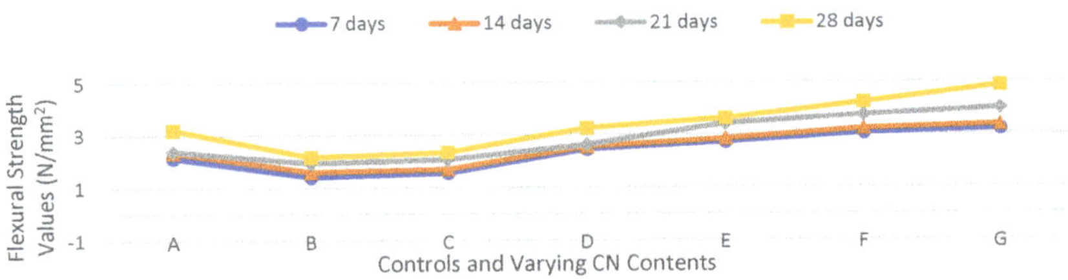

Fig. 5. Flexural Strength of RHA-CN Blended Cement Concrete

Splitting Tensile Strength of RHA-CN Blended Cement Concrete

Figure 6 illustrates the splitting tensile strength of RHA-CN blended concrete. Sample A consistently outperforms Sample B at all curing stages. From Sample C onward, strength increases with CN content, starting at 1.80 MPa and peaking at 2.38 MPa by Sample G (5% CN) at 7 days. This upward trend continues at 14, 21, and 28 days, reaching 2.95 MPa at 28 days for 5% CN. These results show that tensile strength improves with CN dosage and curing time, supporting findings by [27].

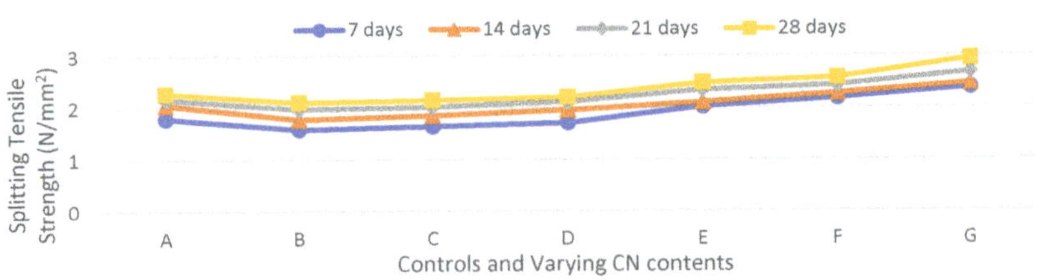

Fig. 6: Splitting Tensile Strength of RHA-CN Blended Cement Concrete

Morphology of RHA-CN Blended Cement Concrete

Sample A to G illustrate the SEM observations of RHA-CN concrete samples. Sample A exhibited a uniform porous matrix with the lowest bulk density (0.36 g/cm³) and smaller particle size (18.43 µm), indicating lower compaction. In contrast, Sample B showed a slightly denser but still porous blocky structure. As CN content increased from Samples C to G, particle sizes and bulk densities rose (up to 20.38 µm and 0.57 g/cm³, respectively), with microstructures transitioning from irregular, porous formations to denser, interconnected networks with fewer voids. This densification, attributed to CN's effect on hydration and matrix development, enhances particle bonding and reduces porosity. These changes contribute to improved strength and durability, with better resistance to moisture and chemical ingress. The trend confirms the synergistic role of RHA and CN in producing high-performance concrete, aligning with findings by [29].

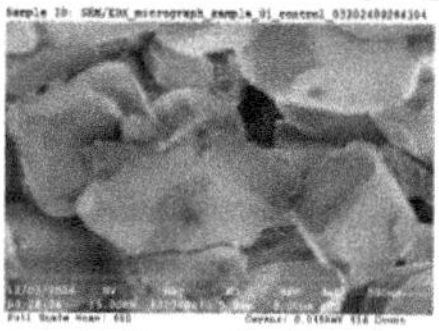
Sample A

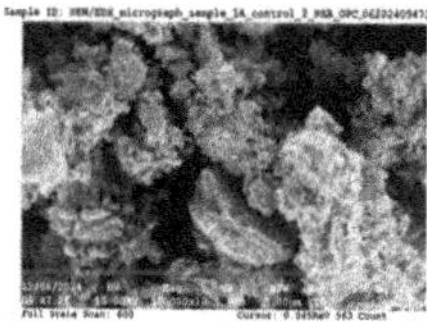
Sample B

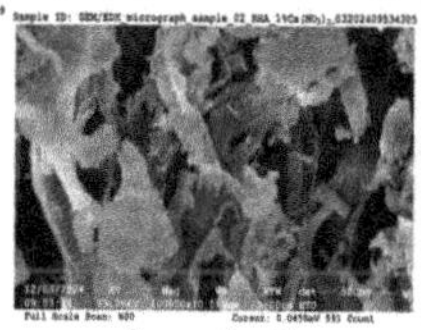
Sample C

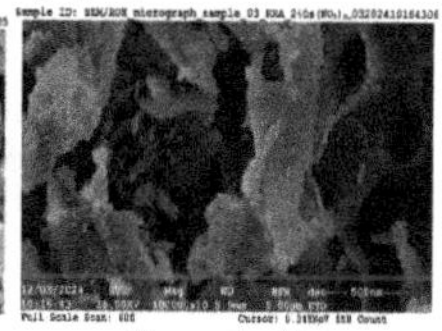
Sample D

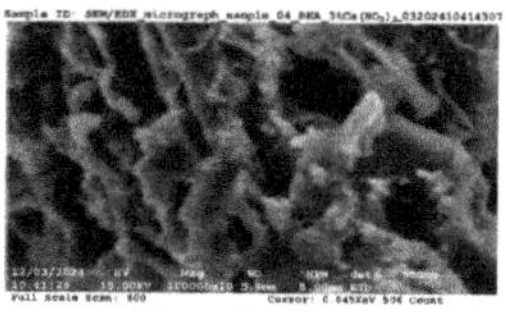
Sample E

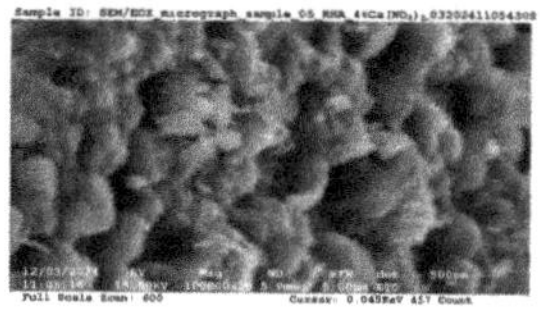
Sample F

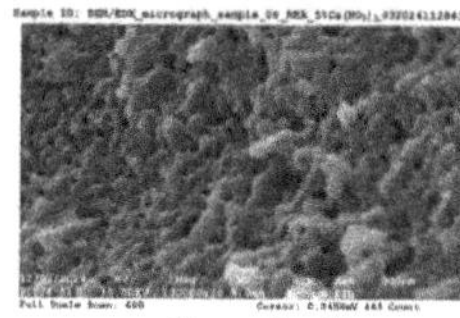
Sample G

Conclusions

i. RHA qualifies as a pozzolanic material with a combined SiO_2, Al_2O_3, and Fe_2O_3 content of 75.36%, meeting the required standard.
ii. The inclusion of calcium nitrate (CN) enhances the workability of RHA-based concrete.
iii. The addition of CN significantly improves the compressive, flexural, and tensile strengths of blended cement concrete.
iv. Increased CN content results in a denser concrete microstructure with reduced porosity.

References

[1] A.A. Raheem, A.A. Soyingbe, E. Emenike, Effect of Curing Methods on the Compressive Strength of Concrete with Rice Husk Ash as Admixture, Journal of Sustainable Development, 6(8) (2013) 107–112.

[2] P.K. Mehta, P.J.M. Monteiro, Concrete: Microstructure, Properties, and Materials, fourth ed., McGraw-Hill Education, 2014.

[3] M.D.A. Thomas, M.H. Shehata, S.G. Shashiprakash, Use of Supplementary Cementing Materials in Concrete: Basics, Benefits, and the Influence of Calcium Nitrate on Workability, ACI Materials Journal, 109(6) (2012) 767–776.

[4] D.G. Nair, K.S. Jagadish, A.L.A. Fraaij, Reactive Pozzolanas from Rice Husk Ash: An Alternative to Cement for Rural Housing, Cement and Concrete Research, 36(6) (2006) 1062–1071.

[5] ASTM C618-19, Standard Specification for Coal Fly Ash and Raw or Calcined Natural Pozzolan for Use in Concrete, ASTM International, West Conshohocken, PA, 2019.

[6] BS EN 197-1, Cement – Part 1: Composition, Specifications and Conformity Criteria for Common Cements, British Standards Institution, London, 2011.

[7] A.A. Raheem, M.A. Kareem, Optimal Raw Material Mix for the Production of Rice Husk Ash Blended Cement, International Journal of Sustainable Construction Engineering and Technology, 7(2) (2017a).

[8] A.Z. Seyed, A. Farshad, D. Farzan, M. Mojtaba, Rice Husk Ash as a Partial Replacement of Cement in High-Strength Concrete Containing Micro Silica, Case Studies in Construction Materials, 7 (2017) 73–81.

[9] K.O. Oriola, A.A. Raheem, M.A. Kareem, R. Abdulwahab, Assessment of Workability and Compressive Strength of Rice Husk Ash-Blended Palm Kernel Shell Concrete, Journal of Civil and Environmental Studies, 7(1) (2021). https://doi.org/10.36108/laujoces/1202.70.0101

[10] A.A. Raheem, M.A. Kareem, Chemical Composition and Physical Characteristics of Rice Husk Ash Blended Cement, International Journal of Engineering Research Africa, 32 (2017b) 25–35.

[11] R. Swetha, M.S. Rani, U. Raju, Experimental Works on Self-Compacting Concrete by Partial Replacement of Rice Husk Ash Subjected to Acid Attack, International Journal of Innovative Technology and Exploring Engineering, 9 (2020) 664–667.

[12] BS 1881-125, Method of Mixing and Sampling Fresh Concrete in the Laboratory, British Standards Institution, London, 2013.

[13] BS EN 12390-3, Testing Hardened Concrete – Compressive Strength of Test Specimens, British Standards Institution, London, 2019.

[14] BS EN 12390-5, Testing Hardened Concrete – Flexural Strength of Test Specimens, British Standards Institution, London, 2019.

[15] BS EN 12390-6, Testing Hardened Concrete – Splitting Tensile Strength of Test Specimens, British Standards Institution, London, 2009.

[16] ASTM C618-21, Standard Specification for Coal Fly Ash and Raw or Calcined Natural Pozzolan for Use in Concrete, ASTM International, West Conshohocken, PA, 2021.

[17] ASTM C618-12, Standard Specification for Coal Fly Ash and Raw or Calcined Natural Pozzolan for Use in Concrete, ASTM International, West Conshohocken, PA, 2012.

[18] ASTM D2487, Standard Practice for Classification of Soils for Engineering Purposes, ASTM International, West Conshohocken, PA, 2011.

[19] A.A. Raheem, B.D. Ikotun, Incorporation of Agricultural Residues as Partial Substitution for Cement in Concrete and Mortar – A Review, Journal of Building Engineering, 31 (2020) 101428. https://doi.org/10.1016/j.jobe.2020.101428

[20] J.E. Bowles, Engineering Properties of Soils and Their Measurement, fourth ed., McGraw-Hill, New York, 1992.

[21] D.C.L. Teo, M.A. Mannan, V.J. Kurian, C. Ganapathy, Lightweight Concrete Made from Oil Palm Shell (OPS): Structural Bond and Durability Properties, Building and Environment, 42 (2007) 2614–2621.

[22] BS EN ISO 10545, Ceramic Tiles – Determination of Water Absorption, Apparent Porosity, Apparent Relative Density and Bulk Density, British Standards Institution, London, 1997.

[23] S. Kazmi, S. Minhaj, S. Abbas, M.J. Munir, A. Khitab, Exploratory Study on the Effect of Waste Rice Husk and Sugarcane Bagasse Ashes in Burnt Clay Bricks, Journal of Building Engineering, 7 (2016) 372–378.

[24] A.M. Neville, Properties of Concrete, fifth ed., Pearson Education, 2011.

[25] ASTM C33/C33M, Standard Specification for Concrete Aggregates, ASTM International, West Conshohocken, PA, 2018. https://www.astm.org/c0033_c0033m-18.html

[26] E.B. Ogunbode, I.O. Hassan, Effect of Addition of Calcium Nitrate on Selected Properties of Concrete Containing Volcanic Ash, Leonardo Electronic Journal of Practices and Technologies, 19 (2011) 29–38.

[27] A.M. Neville, J.J. Brooks, Concrete Technology, Pearson Education, 2010.

[28] BS ENV 206, Concrete – Performance, Production, Placing and Compliance Criteria, British Standards Institution, London, 1992.

[29] U. Bulham, C.A. Abdulkadir, Analysis of Fly Ash Concrete with Scanning Electron Microscopy and X-Ray Diffraction, Advances in Science and Technology Research Journal, 13(4) (2019) 100–110.

Influence of Carbonated Calcium Generated from Waste Concrete Sludge on the Strength and Microstructural Properties of Portland Cement Mortar

Mustapha Bamidele JAJI[1,a *], Adewumi John BABAFEMI[2,b], Babatunde OLADIPO[3,c] and Tunde OJUMU[3,d]

[1]Department of Civil Engineering, University of KwaZulu-Natal, Durban 4041, South Africa

[2]Department of Civil Engineering, Stellenbosch University, Stellenbosch 7602, South Africa

[3]Department of Chemical Engineering, Cape Peninsula University of Technology, Cape Town 7535, South Africa

[a]jajim@ukzn.ac.za, [b]ajbabafemi@sun.ac.za, [c]oladipob@cput.ac.za, [d]ojumut@cput.ac.za

Keywords: Carbonation, Waste Concrete, Calcium Carbonate, Portland Cement, Mortar

Abstract. It is imperative to identify solutions to lower the carbon footprint of construction materials. One of the ways is to recycle and reuse waste concrete materials through indirect CO2 mineral carbonation. This study replaces Portland cement (PC) in mortar composites with up to 15% carbonated calcium (CC) generated from waste concrete at 5% intervals (i.e, 0%, 5%, 10%, 15%) by mass of PC. The slump and flow of the mortar composites (MC) developed (MC0, MC5, MC10, and MC15) were determined using the mini-slump cone and flow table tests. 50 × 50 ×50 mm cubes and 160 mm × 40 mm × 40 mm beam specimens were cured in a climate-controlled room at 23 ± 2 °C and a relative humidity of 65 ± 5% for 24 hours before being water cured for 7 and 28 days, respectively. The bulk density (BD), dry density (DD), compressive strength (CS), and flexural strength (FS) were determined. Additionally, the morphology of the hardened mortar was examined using scanning electron microscopy (SEM). The slump decreases from 8 mm in MC0 to 4 mm in MC15, and the BD is 2272–2462 kg/m^3. The DD of the hardened mortar increases from 2104–2290 kg/m^3. At 28 days, MC0, MC5, MC10, and MC15 exhibit CS of 16.8, 19.2, 22.8, and 24.9 MPa, respectively, and FS of 3.4, 3.7, 3.9, and 4.3 MPa. The microstructural analysis shows that the rhombohedral-shaped CC matrix is formed around the cement particles, depicting a whitish hydration product termed calcium silicate hydrate, with minimal visible porosity as CC increases in the SEM images. Conclusively, 15% CC inclusion in MC exhibited a 50% decrease in slump. Also, BD, DD, CS, and FS increased by 7.7%, 8.1%, 33%, and 2.1%, respectively, at optimum CC content due to the dense microstructure revealed in the SEM.

1. Introduction

The environmental sustainability of cement production is under increasing scrutiny, especially as technological innovations continue to spring up across the construction and concrete industries. Despite significant advancements in concrete technology, cement manufacturing remains one of the most energy-intensive and CO_2-emitting industrial activities, contributing substantially to global environmental challenges such as climate change [1]. Consequently, alternative materials, including fly ash [2], silica fume [3], ground granulated blast-furnace slag (GGBS) [4], cassava peel ash [5], and rice husk ash [6] have been investigated for their potential to improve strength, durability [7], and sustainability [8] when used as supplementary cementitious materials in the development of Portland cement (PC) composites.

The cement industry also generates large volumes of solid waste—such as concrete debris, kiln dust, and waste ready-mix concrete sludge (i.e., by-product from washing concrete mixers, containing fine aggregates, cement, and water)—which can degrade land and water resources if not properly managed [9]. In South Africa and many developed nations, the disposal of such waste in landfills is prohibited, driving the need for alternative management strategies [10]. Conventional methods like stockpiling and incineration have significant drawbacks: stockpiling consumes valuable land resources, while incineration releases harmful pollutants [11]. Moreover, these practices fail to promote material recovery and reuse, underscoring the need for more sustainable solutions.

Recent research has increasingly focused on utilising solid waste from cement and concrete production by leveraging their chemical reactivity [12]. Many of these waste materials, such as kiln dust and demolished concrete, contain reactive calcium compounds that can be recycled into new products [13]. Waste ready-mix concrete, for instance, consists of excess or leftover concrete that is typically discarded [14]. A promising recycling approach is mineral carbon capture and utilisation (MCCU), which exploits the reactivity of these wastes to capture and store CO_2 while producing valuable by-products such as calcium carbonate (CC) that can be reused in construction [15]. MCCU can occur via direct or indirect carbonation: in direct carbonation, CO_2 reacts directly with the solid material, while in indirect carbonation, reactive species such as calcium are first extracted into solution before carbonation. Indirect carbonation is generally preferred because it yields more crystalline products and achieves greater process efficiency, making it suitable for industrial-scale production of high-quality CC [15].

An expanding body of literature has examined the incorporation of CC into sustainable concrete and mortar systems, demonstrating its potential to enhance environmental performance [16] as well as rheological, mechanical, and microstructural properties [17]. Antoni et al. [18] reported the use of FA (10%), SF (2.5%), natural fine size CC (0-15%), and PC (87.5%) in MC. The study combines pozzolanic materials (FA and SF) with natural CC up to 15% CC at 5% intervals. The result shows that 15% CC inclusion with pozzolans exhibits an increase in workability and improved CS by up to 5.2% (early age) and 4% (later age). Conversely, other studies observed reductions in compressive strength when using carbonated recycled concrete aggregates [19]. Safaei et al. [20] found that 2% nano-CC improved compressive strength and 1.5% enhanced tensile strength, with higher additions leading to a reduction in strength. Similarly, replacing PC with 1% nano-CC yielded an optimum compressive strength of 42 MPa compared to 35 MPa for the control concrete mix [21]. Nejad et al. [22] reported a concrete compressive strength of 54 MPa at 0.5% nano-CC, followed by decreases to 51 MPa and 48 MPa at 0.7% and 1%, respectively. Collectively, these findings indicate that the physico-structural and mechanical properties of CC are highly dependent on the synthesis conditions, which in turn influence their performance in construction applications.

In the present study, CC derived from the indirect mineral carbonation of waste concrete sludge was used as a partial substitute for cement in mortar production. A distinguishing feature of this work is the use of a 1000 L capacity concrete sludge recycling pilot plant for mineral carbonation, marking a significant step beyond laboratory-scale experimentation. Operating at this scale enables a more efficient extraction of calcium ions, overcoming kinetic and equilibrium limitations typically encountered in smaller systems. This pilot-scale approach demonstrates industrial feasibility while aligning with circular economy principles through waste valorisation, reduced landfill disposal, and lower CO_2 emissions. Accordingly, Portland cement (PC) was partially replaced with up to 15% CC by mass, obtained via indirect MCCU. The study investigated the fresh properties of the mortar pastes, evaluated the hardened mechanical performance, and performed microstructural characterisation using scanning electron microscopy (SEM) to elucidate the composite behaviour.

2. Materials characterisation

The materials utilised comprised CEM II/A-L 52.5N PC and CC derived from the indirect mineral carbonation of waste ready-mix concrete (WRMC) sludge. The WRMC was obtained from a mid-tier mining and materials company based in Cape Town, South Africa. Deionized water free from soluble salts was used in all experimental procedures. High-purity carbon dioxide (99.95%) was supplied by a local vendor and administered at a controlled flow rate of 20 L/min. The chemical composition and morphology of the binders (PC and CC) were analyzed using Panalytical Axios X-ray fluorescence (XRF) equipment and a scanning electron microscope (SEM) JCM-7000 NeoScope, respectively, both available at the Chemical Engineering Department, Cape Peninsula University of Technology, South Africa. XRF (Table 1) indicates a high content of CaO in PC and CC. The SEM image at a magnification of 2000*X* (Fig.1) shows that particles of PC are angular and irregular, while CC exhibits a rhombohedral shape with a visible void between the particles.

Table 1: Chemical composition of PC and CC

	SiO_2	TiO_2	Al_2O_3	Fe_2O_3	MnO	MgO	CaO	Na_2O	K_2O	SO_3	LOI
PC	18.73	0.18	3.43	2.85	0.05	1.12	63.0	0.19	0.42	1.89	7.2
CC	1.08	0.01	0.13	0.08	-	-	55.23	-	-	0.72	42.01

Fig 1: SEM images of PC and CC

2.1 Mortar composition, mixture procedure, and fresh properties testing.

The mixture composition of the mortar is presented in Table 2. The control mixture contains 100% PC, labelled MC0. Thereafter, PC was replaced with CC up to 15% at 5% intervals, and labelled MC5, MC10, and MC15, respectively. A constant binder to aggregate mass ratio of 1:3, and a water to binder ratio of 0.4, was utilised in line with previous studies [23][24]. A 1% by mass of binder Chryso Premia superplasticizer (SP) was adopted to tailor rheology [25][26].

Table 2. Mortar mixture proportion

MIX ID	PC [g]	CC [g]	Sand [g]	Water [g]	SP [g]
MC0 (Control)	1300	-	3900	520	13
MC5	1235	65	3900	520	13
MC10	1170	130	3900	520	13
MC15	1105	195	3900	520	13

For the control mortar, the fine aggregate was first weighed and transferred into the 10 L Pan mixer, as shown in Fig. 2a. The PC was then added and mixed thoroughly for 3 minutes, followed by the introduction of water and the SP, and an additional 3 minutes of mixing. For the other mortar mixtures, both CC and PC were added to the aggregate immediately after placing the fine aggregate in the mixer, mixed thoroughly for 3 minutes, after which water and SP were incorporated, and mixing continued for another 3 minutes. Slump test was performed using a mini-slump cone on the fresh mortar mixtures following the procedures outlined in ASTM C0230/C0230M-20 [27]. The fresh mortar was placed into a mould with a height of 60 mm, a base diameter of 100 mm, and a top diameter of 70 mm, as illustrated in Fig. 2b. The cone was filled in two layers, each compacted with ten strokes of the tamping rod. After filling, the cone was carefully lifted and positioned beside the formed mortar, and the height difference was recorded as the slump value of the mortar. Additionally, the flow test was performed using the flow table test (Fig. 2c). The bulk density was determined by measuring the mass of mortar occupying a known volume, as illustrated in Fig. 2c. A 3-liter batch of fresh mortar was prepared for the test. The container was first half-filled with mortar using a scoop, then tilted about 30 mm to each side and dropped ten times onto a rigid surface to achieve compaction. The remaining portion of the container was filled, and the same procedure was repeated. The bulk density of the mortar was subsequently calculated using Eq. 1.

$$\rho_m = \frac{m_2 - m_1}{V_v} \tag{1}$$

where (ρ_m) in $\frac{Kg}{m^3}$ is the bulk density, m_1 is the mass of the empty container, m_2 is the mass of the container filled with mortar, and V_v is the volume of the container.

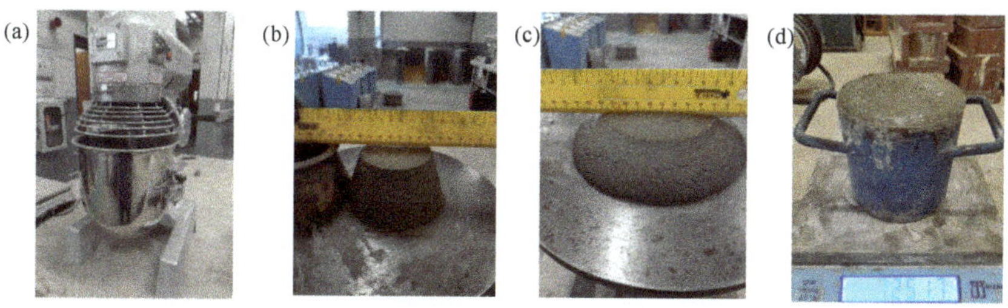

Fig. 2 (a) 10 L pan cake mixer, (b) slump, (c) slump flow, and (d) bulk density

2.2 Mechanical performance and microstructural properties testing procedures.

50 mm × 50 mm × 50 mm standard cubes and 160 mm × 40mm × 40mm beam moulds were filled with fresh mortar and compacted on a vibrating table in two layers for 3 min after each layer, as shown in Fig. 3a and placed in a climate-controlled room at 23±2 °C and a relative humidity of 65±5% for 24 h (Fig. 3b) before it was demoulded and subject to standard water curing. The Zwick Z250 Material Testing Machine was used to determine the compressive and flexural strengths of the cubes and beams after 7- and 28-day curing (Fig.3c and d). The compressive and flexural strengths of the cubes and beams were determined using Eq. 2–5.

$$f_c = \frac{F}{A_c} \tag{2}$$

where f_c is the compressive strength (MPa), F is the compressive load (N), and A_c is the compressive area of the cube (mm^3).

$$W = \frac{1}{6} b h^2 \tag{3}$$

Materials Research Forum LLC
https://doi.org/10.21741/9781644904053-13

$$M_{max} = F \times x \tag{4}$$
$$F_f = \frac{M_{max}}{W} \tag{5}$$

where F_f is flexural strength (MPa), M_{max} is the maximum moment (Nmm), W is the section modulus (mm^3), F is the maximum load (N), and x is the distance (mm).

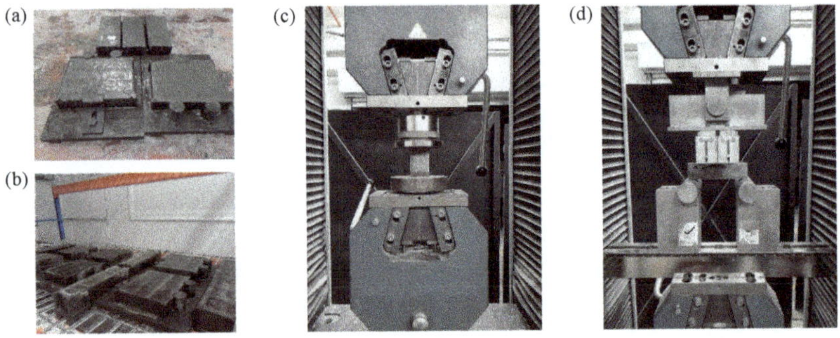

Fig. 3 (a) Specimens, (b) climate-controlled curing, (c) compression test, and (d) flexure test

3. Results and discussion

The results of slump and flow are shown in Fig. 4a. Both the slump and slump flow of MC decreased with increasing CC in a nearly linear proportion. In Fig. 4b, BD increases by 2.4, 7.4, and 7.7%, respectively, as CC content increases from 0 to 15%. A similar trend was exhibited with 2.6%, 7.3%, and 8.1% increase in DD at CC of 5%, 10% and 15% compared to the control. The increase is in agreement with what was obtained in another study [28].

The compressive and flexural strengths against mixture ID are shown in Figs. 4c&d. After 7 days of curing, compressive strengths of MC5, MC10, and M15 were enhanced by 18, 30, and 34%, respectively, compared to MC0. Also, after 28 days, 13%, 27%, and 33% increases were exhibited compared to the control. This result is supported by Gonnon and Lootens [29]. The MC kinetics reflect strength enhancement as CC increases up to 15%. The flexural strength against mixture identification is shown in Fig. 4d. After 7 days of curing, MC5, MC10, and MC15 outperformed the control samples by 4, 14, and 20%, respectively. Similarly, flexural strength was enhanced by 8, 13, and 21%, respectively, after 28 days. The improvement in flexural strength is due to the enhancement and accelerating effect of CC in the cement hydration process through its filler [30][31]. In addition, CC, as a cement filler material, was reported to have an incremental impact on flexural properties [32].

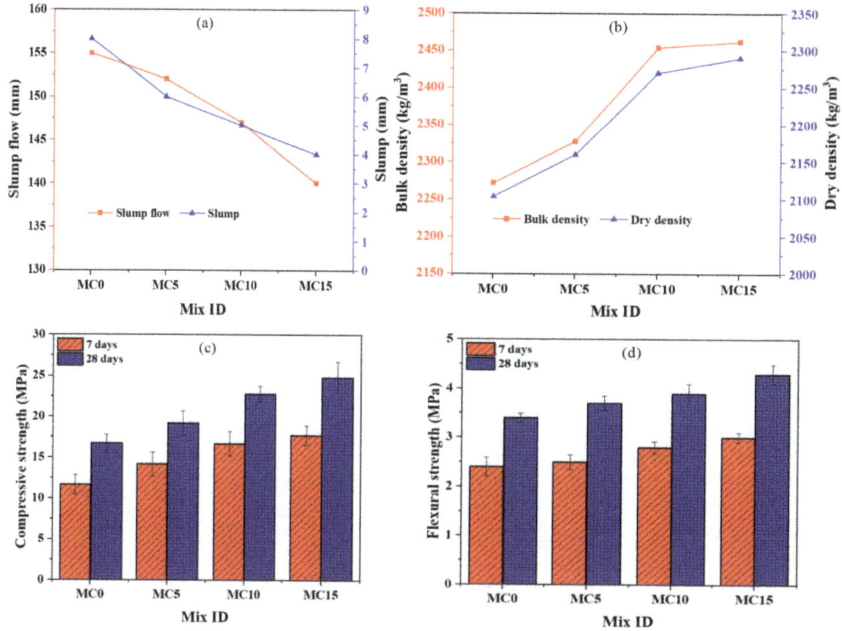

Fig. 4. (a) Slump and slump flow, (b) bulk and dry density, (c) compressive strength, and (d) flexural strength

The internal morphology of the MC using SEM images is depicted in Fig. 5a-d. The formation of hydration product, calcium silicate hydrate (C-S-H), is shown on the surface of MC in Fig. 5a. The CC matrix, exhibiting a rhombohedral shape, surrounds the cement particles and forms more whitish C-S-H (Fig. 5b), with more angular particles, which aid interlocking and improve mechanical properties. As CC increases, the MC-based structure offers high cohesion with only minor observable porosity and pore refinement (Fig. 5c). CC acts as a filler, refining the matrix and producing a relatively dense structure with improved particle packing, although some localised porosity persists. MC15 (Fig. 5d) exhibits a more refined microstructure, with the filler effect enhanced and CC particles more uniformly dispersed within the matrix. This improved particle packing substantially reduces voids, prevents susceptibility to crack propagation by filling the interstices of the MC, thereby enhancing the mechanical properties [33].

Materials Research Proceedings 63 (2026) 112-121
https://doi.org/10.21741/9781644904053-13

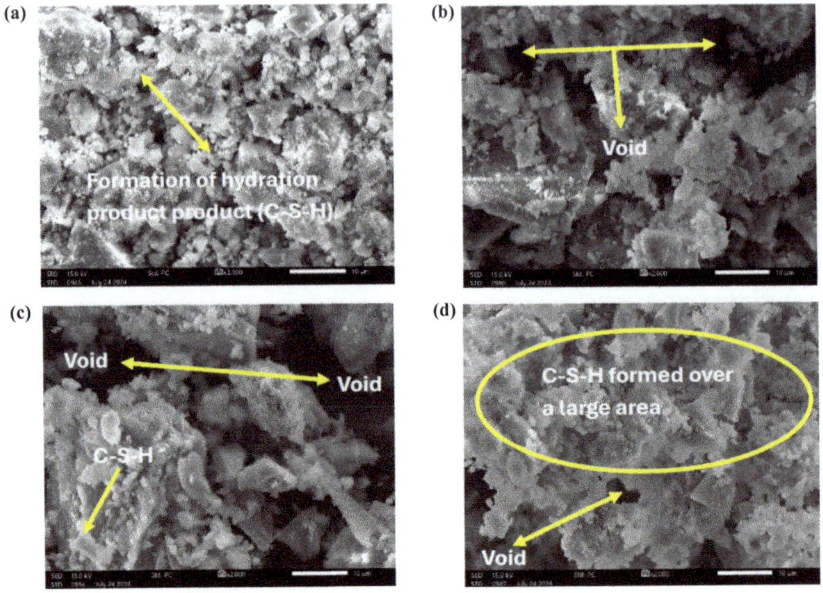

Fig. 5. SEM (a) MC0, (b) MC5 (c) MC10 and (d) MC15

4. Conclusion

The feasibility of employing CC, which is produced via indirect CO_2 mineral carbonation of waste ready-mix concrete, in mortar application was examined in this study. By encouraging circularity in building materials and lowering reliance on PC, the project benefits the environment. Up to 15% of Portland cement (PC) in mortar was substituted with CC at intervals of 5%. After examining the mortar's fresh, mechanical, and microstructural properties, the following conclusion can be drawn:

I. CC inclusion accelerates mortar setting times and reduces workability by up to 50%. Beyond 15% CC inclusion, the MC was not workable due to insufficient consistency. Hence, an optimum CC inclusion of 15% was adopted.

II. The slump of MC decreased up to 50% from MC0 to MC15. Additionally, BD and DD increased by 7.7 and 8.1%, respectively, at 15% optimum CC inclusion.

III. The CS and FS of MC15 outperformed the control (MC0) by 33 and 2.1%, respectively, after 28 days due to the densification of the composite matrix through C-S-H formation over a large area of the SEM images.

5. Recommendation

I. The long-term mechanical properties of MC incorporating CC are recommended for further studies.

II. The durability performance of MC incorporating CC when subjected to mild and aggressive environmental conditions is recommended for further studies.

References

[1] Y. Izumi, A. Iizuka, and H. J. Ho, "Calculation of greenhouse gas emissions for a carbon recycling system using mineral carbon capture and utilization technology in the cement industry," *J. Clean. Prod.*, vol. 312, no. December 2020, p. 127618, 2021. https://doi.org/10.1016/j.jclepro.2021.127618

[2] J. Yang *et al.*, "Effect of steam curing on compressive strength and microstructure of high volume ultrafine fly ash cement mortar," *Constr. Build. Mater.*, vol. 266, p. 120894, 2021. https://doi.org/10.1016/j.conbuildmat.2020.120894

[3] P. H. C. H. Caldas, A. R. G. de Azevedo, and M. T. Marvila, "Silica fume activated by NaOH and KOH in cement mortars: Rheological and mechanical study," *Constr. Build. Mater.*, vol. 400, no. July, p. 132623, 2023. https://doi.org/10.1016/j.conbuildmat.2023.132623

[4] T. Y. Nojoukambari, A. Sadrmomtazi, and R. K. Khoshkbijari, "Influence of Feldspar and Ground Granulated Blast-furnace slag as a partial cement replacement on mechanical properties and thermal durability and environmental impacts of cement mortars," *J. Clean. Prod.*, vol. 398, no. December 2022, p. 136074, 2023. https://doi.org/10.1016/j.jclepro.2023.136074

[5] G. M. Amusan, M. O. Popoola, and M. B. Jaji, "Assessment of Cassava Peel Ash Blended Cement in Concrete Production," *Int. J. Adv. Eng. Manag.*, vol. 3, no. 10, p. 224, 2021. https://doi.org/10.35629/5252-0310224227

[6] M. Thiedeitz, B. Ostermaier, and T. Kr, "Resources , Conservation & Recycling Rice husk ash as an additive in mortar – Contribution to microstructural , strength and durability performance," vol. 184, no. July 2021, 2022. https://doi.org/10.1016/j.resconrec.2022.106389

[7] M. B. Jaji, "Rheo-mechanics , Durability , and Microstructural Characterisation of Slag-modified Metakaolin-based Geopolymer Concrete for Extrusion-based 3D Printing Applications," *Unpubl. Ph.D. Diss.*, no. December, 2023, [Online]. Available: http://www.sun.ac.za

[8] M. B. Jaji, G. P. A. G. Van Zijl, and A. J. Babafemi, "Slag-modified metakaolin-based geopolymer for 3D concrete printing application : Evaluating fresh and hardened properties," *Clean. Eng. Technol.*, vol. 15, no. December 2022, p. 100665, 2023. https://doi.org/10.1016/j.clet.2023.100665

[9] M. Sandanayake, Y. Bouras, R. Haigh, and Z. Vrcelj, "Current Sustainable Trends of Using Waste Materials in Concrete — A Decade Review," 2020.

[10] W. De Villiers, M. Mwongo, A. J. Babafemi, and G. Van Zijl, "Quantifying Recycled Construction and Demolition Waste for Use in 3D-Printed Concrete," *Recycling*, vol. 9, no. 4, 2024. https://doi.org/10.3390/recycling9040055

[11] J. Waltisberg and R. Weber, "Disposal of waste-based fuels and raw materials in cement plants in Germany and Switzerland e What can be learned for global co- incineration practice and policy ?," *Emerg. Contam.*, vol. 6, pp. 93–102, 2020. https://doi.org/10.1016/j.emcon.2020.02.001

[12] H. J. Ho, A. Iizuka, and E. Shibata, "Chemical recycling and use of various types of concrete waste: A review," *J. Clean. Prod.*, vol. 284, p. 124785, 2021. https://doi.org/10.1016/j.jclepro.2020.124785

[13] A. Iizuka *et al.*, "Pilot-Scale Operation of a Concrete Sludge Recycling Plant and Simultaneous Production of Calcium Carbonate," *Chem. Eng. Commun.*, vol. 204, no. 1, pp. 79–85, 2017. https://doi.org/10.1080/00986445.2016.1235564

[14] A. Iizuka, Y. Sakai, A. Yamasaki, M. Honma, Y. Hayakawa, and Y. Yanagisawa, "Bench-scale operation of a concrete sludge recycling plant," *Ind. Eng. Chem. Res.*, vol. 51, no. 17, pp. 6099–6104, 2012. https://doi.org/10.1021/ie300620u

[15] H. J. Ho, A. Iizuka, E. Shibata, H. Tomita, K. Takano, and T. Endo, "CO2utilization via direct aqueous carbonation of synthesized concrete fines under atmospheric pressure," *ACS Omega*, vol. 5, no. 26, pp. 15877–15890, 2020. https://doi.org/10.1021/acsomega.0c00985

[16] N. Makul, Q. Hussain, A. Nawaz, P. Saingam, and G. Sua-iam, "Effect of para-wood ash and calcium carbonate on the properties of eco-friendly self-compacting mortar reinforced with electronic waste fibers," *J. Build. Eng.*, vol. 95, no. July, p. 110353, 2024. https://doi.org/10.1016/j.jobe.2024.110353

[17] P. Saingam *et al.*, "Synergizing Portland Cement, high-volume fly ash and calcined calcium carbonate in producing self-compacting concrete: A comprehensive investigation of rheological, mechanical, and microstructural properties," *Case Stud. Constr. Mater.*, vol. 21, no. October, p. e03832, 2024. https://doi.org/10.1016/j.cscm.2024.e03832

[18] Antoni, L. Chandra, and D. Hardjito, "The impact of using fly ash, silica fume and calcium carbonate on the workability and compressive strength of mortar," *Procedia Eng.*, vol. 125, pp. 773–779, 2015. https://doi.org/10.1016/j.proeng.2015.11.132

[19] J. Zhang, C. Shi, Y. Li, X. Pan, C. S. Poon, and Z. Xie, "Influence of carbonated recycled concrete aggregate on properties of cement mortar," *Constr. Build. Mater.*, vol. 98, pp. 1–7, 2015. https://doi.org/10.1016/j.conbuildmat.2015.08.087

[20] B. Safaei, E. Davodian, A. M. Fattahi, and M. Asmael, "Calcium carbonate nanoparticles effects on cement plast properties," *Microsyst. Technol.*, vol. 27, no. 8, pp. 3059–3076, 2021. https://doi.org/10.1007/s00542-020-05136-6

[21] S. W. M. Supit and F. U. A. Shaikh, "Effect of Nano-CaCO3 on compressive strength development of high volume fly ash mortars and concretes," 2014, *Journal of Advanced Concrete Technology*. doi: 10.3151/jact.12.178

[22] F. M. Nejad, M. Tolouei, H. Nazari, and A. Naderan, "Effects of calcium carbonate nanoparticles and fly ash on mechanical and permeability properties of concrete," *Adv. Civ. Eng. Mater.*, vol. 7, no. 1, pp. 651–668, 2018. https://doi.org/10.1520/ACEM20180066

[23] M. B. Jaji, K. A. Ibrahim, G. Pieter, A. Greeff, and A. J. Babafemi, "Materials Today : Proceedings Thixotropic characterisation of slag modified 3D printable metakaolin based geopolymer composite," *Mater. Today Proc.*, no. xxxx, pp. 0–6, 2024. https://doi.org/10.1016/j.matpr.2023.03.530

[24] M. B. Jaji, A. J. Babafemi, and G. P. A. G. van Zijl, "Mechanical performance of extrusion-based two-part 3D-printed geopolymer concrete: A review of advances in laboratory and real-scale construction projects," *Mater. Today Sustain.*, vol. 31, no. February, p. 101131, 2025. https://doi.org/10.1016/j.mtsust.2025.101131

[25] M. B. Jaji, G. P. A. G. Van Zijl, and A. J. Babafemi, "Fresh properties and strength evolution of slag modified fibre-reinforced metakaolin-based geopolymer composite for 3D concrete printing application," *Curr. Perspect. New Dir. Mech. Model. Des. Struct. Syst.*, pp. 1630–1634, 2022. https://doi.org/10.1201/9781003348443-267

[26] K. A. Ibrahim, M. B. Jaji, G. P. A. G. van Zijl, and A. J. Babafemi, "Influence of effective microorganisms on the rheology and fresh state properties of SCMs-based concrete for digital fabrication," *Mater. Today Proc.*, no. xxxx, 2023. https://doi.org/10.1016/j.matpr.2023.03.520

[27] ASTMC0230, "American Standard for Testing Materials," *ASTM Intl*, vol. i, no. Reapproved, pp. 3–5, 2018. https://doi.org/10.1520/C0230

[28] M. Balcikanli Bankir, M. Ozturk, U. K. Sevim, and T. Depci, "Effect of n-CaCO3 on fresh, hardened properties and acid resistance of granulated blast furnace slag added mortar," *J. Build. Eng.*, vol. 29, no. January, p. 101209, 2020. https://doi.org/10.1016/j.jobe.2020.101209

[29] P. Gonnon and D. Lootens, "Toward net zero carbon for concrete and mortar: Clinker substitution with ground calcium carbonate," *Cem. Concr. Compos.*, vol. 142, no. June, p. 105190, 2023. https://doi.org/10.1016/j.cemconcomp.2023.105190

[30] K. De Weerdt, K. O. Kjellsen, E. Sellevold, and H. Justnes, "Synergy between fly ash and limestone powder in ternary cements," *Cem. Concr. Compos.*, vol. 33, no. 1, pp. 30–38, 2011. https://doi.org/10.1016/j.cemconcomp.2010.09.006

[31] W. S. H. Q. Cheng Caoa, "Analysis of Strength Fly Ash," *Cem. Concr. Res. 30 71–75*, vol. 30, pp. 71–75, 2016.

[32] A. Marzouki, A. Lecomte, A. Beddey, C. Diliberto, and M. Ben Ouezdou, "The effects of grinding on the properties of Portland-limestone cement," *Constr. Build. Mater.*, vol. 48, pp. 1145–1155, 2013. https://doi.org/10.1016/j.conbuildmat.2013.07.053

[33] Y. Mao, S. Drissi, P. He, X. Hu, J. Zhang, and C. Shi, "Quantifying the effects of wet carbonated recycled cement paste powder on the properties of cement paste," *Cem. Concr. Res.*, vol. 175, no. October 2023, p. 107381, 2024. https://doi.org/10.1016/j.cemconres.2023.107381

Eco-Design of Sustainable Concrete Using Treated Industrial Wastewater and Bio-Based Admixtures

Mariam Iyabo ADEOBA[1,a*], Thanyani PANDELANI[1,b], Harry NGANGWA[1,c], Tracy MASEBE[2,d]

[1]UNISA Biomechanics Research Group, Department of Mechanical, Bioresources and Biomedical Engineering, College of Science, Engineering and Technology (CSET), University of South Africa, Florida 1710, South Africa

[2]Department of Life and Consumer Sciences, College of Agriculture and Environmental Sciences, University of South Africa (UNISA), Florida 1710, South Africa

[a]adeobmi@unisa.ac.za, [b]epandet@unisa.ac.za, [c]ngwanhm@unisa.ac.za, [d]masebtm.unisa.ac.za

Keywords: Bio-Based Admixtures, Eco-Design, Sustainable Concrete, Wastewater Reuse, Green Building Materials

Abstract. The construction industry faces pressing challenges related to concrete demand, freshwater scarcity, and environmental sustainability. This thematic review focuses on eco-innovative practices in sustainable concrete design, particularly the use of treated industrial wastewater (TIWW) and bio-based admixtures as alternatives to conventional materials. By analyzing some literature from 2020 to 2025, we consolidate information that investigates the function of bio-derived admixtures, including polysaccharides and lignin-based chemicals, in improving rheological properties, durability, and microstructural refinement. The review emphasizes how eco-friendly inputs can sustain or enhance mechanical strength while decreasing permeability and water absorption. These developments facilitate the creation of low-carbon, resource-efficient concrete compositions, consistent with circular economy concepts and environmental stewardship. The paper concludes by identifying research gaps and proposing future directions for scaling sustainable concrete technologies within industrial and infrastructure applications.

1. Introduction

With an annual production of over 10 billion tons and 7-8% of global CO_2 emissions, the building industry, and particularly the production of concrete, is a significant contributor to environmental degradation worldwide [1]. Significant greenhouse gas emissions are released by the energy intensive processes used to produce cement, and the production of concrete uses enormous amounts of natural resources, such as 150–200 liters of water per cubic meter, which exacerbates the freshwater shortage [2]. Furthermore, because of their toxic residues and non-biodegradable nature, synthetic chemical admixtures present environmental risks that can endanger human health and aquatic life [3]. A paradigm shift in concrete design and production is urgently needed, as the disposal of construction waste further contaminates soil and water bodies [4].

Eco-design, also known as ecological design, incorporates environmental factors into the creation of building materials with the goal of minimizing their negative effects on the environment while improving performance. This strategy incorporates innovations like low-carbon cement and recycled aggregates, encourages life cycle thinking, energy efficiency, and the use of renewable resources. The use of bio-based admixtures and treated industrial wastewater in the production of concrete is a promising path in eco-design [5]. Wastewater can be treated to reduce effluent and save potable water by being used as a resource for concrete. Because they come from natural sources, bio-based admixtures are less hazardous and biodegradable, which promotes ecological sustainability.

By combining these materials, concrete with similar or better durability can be produced, non-renewable resources can be used less frequently, and waste materials can be valued [6]. By reducing emissions and encouraging green building practices, this strategy supports sustainable development goals [7]. Eco-design provides a thorough framework for resolving environmental issues with traditional concrete, and bio-based admixtures and treated wastewater are showing promise as future sustainable concrete production methods. Therefore, this mini review analyzed peer-reviewed publications published from 2020 to 2024 that investigate the incorporation of eco-design concepts into sustainable concrete development. Significant attention was made to employing treated industrial wastewater as a partial substitute for freshwater in mixing operations, alongside the integration of biobased admixtures sourced from renewable materials to improve concrete performance and mitigate environmental effects. This analysis synthesizes current developments in these fields to emphasize developing techniques that correspond with circular economic objectives, alleviate resource depletion, and promote low-carbon construction practices.

2. Literature review

The sustainability of concrete production has emerged as a global issue owing to the significant consumption of water and natural resources, with the environmental repercussions of cement manufacturing. Recent studies have concentrated on integrating bio-based admixtures and processed industrial effluent into concrete to alleviate these effects without sacrificing performance [8]. Concrete manufacture generally takes approximately 150 liters of potable water per cubic meter. Research on the utilization of treated industrial wastewater (TIWW) as an alternative to freshwater has yielded encouraging outcomes. Research has examined several sources of wastewater, including food manufacturing, medicines, textiles, and oil refining. Essential water quality parameters, including total dissolved solids (TDS), chemical oxygen demand (COD), and heavy metal concentration, are evaluated to determine reuse viability. For instance, treated textile effluent might sustain or improve the compressive strength of concrete mixtures [9]. Although the use of wastewater in the production of concrete yields encouraging results, questions still surround the long-term viability and environmental effects of using different types of wastewaters. Untreated effluents can add chloride and sulfate ions, thereby exacerbating the corrosion of steel reinforcement. The utilization of wastewater also affects setting time, workability, and shrinkage. Properly processed TIWW can function as a sustainable substitute for potable water in concrete manufacturing.

Bio-based admixtures, sourced from renewable materials such as lignin and cellulose, are increasingly recognized as environmentally sustainable substitutes for synthetic admixtures. Lignosulfonates, for instance, improve workability and decrease water demand [10]. Alternative organic substances, such as modified starches and plant-derived oils, have demonstrated potential in enhancing concrete characteristics [11]. Notwithstanding their advantages, bio-admixtures may demonstrate performance variability attributable to variations in chemical composition and extraction techniques [12]. Standardization and characterization are vital for the uniform application of bio-based admixtures and treated wastewater in construction, yet current international standards (e.g., ASTM, ACI) lack specific guidelines for these materials. Other challenges include insufficient data on long-term durability, inadequate regulations for wastewater treatment, and variability in effluent composition. Additionally, misconceptions and regulatory hurdles impede adoption, compounded by concerns over the availability and compatibility of bio-based admixtures. Without government support and financial incentives, the construction industry may be reluctant to adopt these sustainable practices, despite promising results in controlled.

2.1 Environmental Impact of Traditional Concrete

Concrete is vital for contemporary infrastructure. Nonetheless, its manufacture significantly harms the environment, chiefly due to cement manufacturing, excessive freshwater use, and the use of chemical admixtures [13,14]. The manufacturing of cement, especially ordinary Portland cement (OPC), is a significant contributor to CO_2 emissions, accounting for roughly 7–8% of worldwide greenhouse gas emissions. The energy-intensive process exhausts fossil fuels and contributes to air pollution, while the concrete industry utilizes around 1.7% of the global freshwater supply each year, placing pressure on water supplies in arid areas. The utilization of chemical admixtures, frequently sourced from petrochemicals, prompts environmental apprehensions owing to their possible toxicity and bioaccumulation within ecosystems. Inadequate disposal of wash water from building activities can adversely affect aquatic organisms and modify local water quality. The lifecycle of these admixtures, encompassing manufacture to disposal, significantly influences their environmental impact, underscoring the necessity for more sustainable alternatives. In addition to these concerns, typical concrete's environmental impact encompasses habitat degradation due to aggregate extraction, pollution from building waste, and the exacerbation of the urban heat island effect caused by concrete surfaces. Although concrete's endurance is frequently regarded as beneficial, it is eclipsed by the environmental damage incurred over its existence.

To tackle these difficulties, the concrete industry must implement sustainable practices, including the exploration of alternative materials, the utilization of industrial waste, water conservation, and the development of low-toxicity admixtures (See Table 1). Re-evaluating concrete manufacturing through ecological principles is essential for safeguarding environmental health during infrastructure development [15].

Table 1: Synthesis of Geographic Coverage, Wastewater Strategies, and Biobased Admixtures Use for sustainable concrete

Geographic Scope	Wastewater Integration	Biobased Admixtures	Key Findings	Reviewer-Friendly Commentary	Reference
Global	Indirect; advocates marginal gains across water-energy nexus	SCMs (e.g., fly ash, slag)	Up to 50% GHG reduction via value chain optimization	Foundational review; supports ecodesign framing and marginal improvement strategies	Habert et al. [2]
Brazil	100% replacement with secondary treated wastewater	Not specified	>40 MPa compressive strength; minimal impact on fresh properties	Strong empirical support for wastewater reuse; relevant for circular water strategies	Almeida & Tonetti [16]
Nigeria	Not covered	Citrus peel, lignin, bacteria-based BBAs	Enhanced durability and strength; compatibility challenges noted	Comprehensive review of BBAs; useful for identifying viable natural admixtures	Ekop et al. [17]
Ethiopia	Not covered	Plant-, animal-, and microbe-derived admixtures	Improved rheology and mechanical performance	Valuable classification of BBAs; aligns with renewable material sourcing goals	Bedada et al. [18]

Geographic Scope	Wastewater Integration	Biobased Admixtures	Key Findings	Reviewer-Friendly Commentary	Reference
India	Not covered	Wood ash, rice husk ash, corn cob granules	Reduced CO_2 footprint; improved thermal performance	Highlights the practical feasibility of bio-waste integration; addresses durability trade-offs	Muskan & Nitu (2025)
Germany	Partial replacement with treated greywater	Lignin-based plasticizers	Maintained workability and strength	Demonstrates the feasibility of dual substitution (water + admixtures) in lab-scale trials	Proske et al. (2023)
USA	Modeled reclaimed water scenarios	Natural polymers (e.g., starch)	Eco-efficiency indices improved	Offers quantifiable metrics for ecodesign evaluation; supports LCA integration	Kourehpaz & Miller (2022)
France	Simulated wastewater use	Citrus peel extract	Enhanced freeze-thaw resistance	Innovative use of food waste derivatives supports durability enhancement	Berodier et al. (2020)
UK	Indirect reference in low-cement mixes	Not specified	Feasibility of low-clinker concrete	Material flow analysis supports resource efficiency and clinker reduction	Shanks et al. (2021)
Brazil	Life cycle water inventory including industrial reuse	Not covered	Identified water hotspots in concrete production	Strong LCA foundation; supports justification for wastewater substitution	Mack-Vergara & John (2021)

2.2 Treated Industrial Wastewater in Concrete

Using treated industrial wastewater (TIWW) in the production of concrete is a sustainable substitute that can lower freshwater consumption and encourage circular resource use. The type of wastewater, the effectiveness of treatment, and the effect on concrete performance determine whether TIWW is appropriate [19]. Different kinds of industrial wastewater are produced by different industries, such as textiles, food processing, thermal power, and petrochemicals. For example, wastewater from food processing contains organic matter and fats, while the textile industry produces alkaline effluent that is rich in salts and dyes. Thermal power plants produce effluents containing heavy metals and salts that can cause corrosion in concrete, and petrochemical wastewater is frequently toxic and needs to be thoroughly treated. Before being used in concrete, industrial wastewater must be effectively treated using physical, chemical, and biological techniques. While chemical processes like oxidation and neutralization deal with heavy metals and pH levels, physical treatments like filtration and sedimentation remove suspended particles. Although biological techniques, such as membrane bioreactors, lower the organic content of concrete, they must be properly controlled to prevent microbial contamination. Guidelines for using non-potable water in concrete that adhere to performance and safety requirements are provided by ASTM C1602-18. Concrete's workability, strength, and durability can all be impacted by the chemical characteristics of treated wastewater. While poorly treated wastewater can weaken concrete because of harmful ions, well-treated wastewater can produce compressive strengths that are on par with those made with potable water [20]. Long-term durability is an issue, especially when it comes to efflorescence and steel reinforcement corrosion, which can be signs of more serious problems with porosity and water penetration [21]. According to advanced testing, TIWW can change the microstructure of concrete, which could result in performance problems like decreased freeze-thaw resistance and increased permeability. If appropriate treatment procedures are followed and adherence to ASTM C1602 and local regulations is maintained, reusing TIWW in the production of concrete promotes sustainability in the building industry. For eco-efficient concrete practices to be widely adopted, it is imperative to address long-term risks and promote research and trials [22].

2.3 Bio-Based Admixtures in Concrete

Interest in bio-based admixtures as more ecologically friendly alternatives to synthetic chemical admixtures in concrete has surged due to the demand for sustainable building materials. These natural materials, derived from agricultural and plant sources, can reduce the environmental impact of concrete while increasing its functionality [23]. Bio-based substitutes promote the principles of the circular economy because they are low in toxicity, renewable, and biodegradable.

Common bio-based admixtures include polysaccharides, lignosulfonates, and tannins, each of which has unique characteristics that influence the characteristics of concrete. Paper industry byproducts called lignosulfonates enhance workability and strength by serving as dispersants and water reducers. Tannins, found in many plant sources, function as corrosion inhibitors and set retarders, while polysaccharides enhance viscosity and internal curing. Sustainability can be advanced by employing waste resources to create bio-based admixtures from forestry and agricultural byproducts. Their success in concrete is attributed to mechanisms such as internal curing, plasticizing, water reduction, and synergistic interactions with cement hydration. These processes improve concrete's durability, reduce shrinkage, and make it more resistant to environmental influences [24]. Bio-based admixtures pose fewer pollution risks than their synthetic counterparts because of their low toxicity and biodegradability. Economically speaking, they are usually sourced locally, which lowers transportation costs and strengthens rural economies. Their use can also lead to long-term savings by improving concrete performance and reducing maintenance needs [25]Bio-based admixtures offer sustainable development in concrete technology that has advantages for the economy, performance, and the environment. Their use in concrete production is consistent with the global trend toward ecological responsibility and resource efficiency, making them crucial components of future sustainable concrete design.

2.4 Synergistic Integration of Wastewater and Bio-Admixtures

To enhance sustainability while maintaining structural performance, researchers are exploring the synergistic use of bio-based admixtures and treated industrial wastewater (TIWW) in concrete production [26]. This approach improves the ecological footprint of concrete through waste valorization and the incorporation of renewable additives. Research indicates that combining TIWW with bio-admixtures such as lignosulfonates and starches enhances durability and mechanical properties while mitigating the adverse impacts of wastewater contaminants.

Research demonstrates that lignosulfonates enhance hydration in concrete mixtures utilizing saline-treated wastewater, consequently augmenting compressive strength [27]. Moreover, starch-modified concrete has demonstrated reduced water absorption and enhanced resistance to chloride diffusion [28]. Bio-admixtures regulate evaporation and preserve moisture, hence contributing to the mitigation of shrinkage. The integration of these elements addresses significant sustainability concerns, such as the carbon footprint of synthetic admixtures and the substantial water usage in concrete manufacturing [28,29]. This approach converts garbage into valuable construction materials, so promoting a circular economy. Numerous case studies have demonstrated practical uses. For instance, a pilot project in Gujarat, India, employed lignin-based superplasticizers and treated textile effluent to enhance the characteristics of concrete. Similarly, research in Brazil has shown that the use of wastewater from a fruit-canning facility combined with a starch-based additive enhanced durability and reduced water absorption [30]. The integration of bio-admixtures and TIWW is a feasible strategy for achieving environmental sustainability in the construction sector, potentially enhancing durability, reducing carbon emissions, and conserving water [31,32].

Further study is necessary to develop scalable implementation methodologies and standards for this innovative approach.

2.5 Challenges and Limitations

Concrete with bio-based admixtures and treated industrial wastewater (TIWW) has substantial environmental advantages, such as water conservation and less carbon emissions. Nonetheless, its actual implementation encounters obstacles like quality control, heterogeneity in wastewater composition, and variable efficacy of bio-admixtures [33]. The variability in the chemical composition of TIWW can negatively impact concrete performance, since fluctuations in pH, total dissolved solids, and other parameters may result in durability problems. Bio-based admixtures, originating from agricultural or industrial waste, demonstrate inconsistent performance attributable to the variability of their raw ingredients [34]. The inconsistency, together with restricted commercial availability and the absence of defined production procedures, impedes their scalability and acceptance in high-performance applications. The lack of established quality control procedures and performance criteria for both TIWW and bio-admixtures hampers their incorporation into concrete. Current standards, including ASTM C1602, insufficiently address the intricacies of industrial contaminants or bio-admixture efficacy, resulting in liability issues for stakeholders. An urgent need exists for the standardization of water quality indicators, bio-admixture classifications, and pollutant treatment regulations to overcome these challenges. Addressing these difficulties is crucial for realizing the full potential of eco-concrete in sustainable construction.

2.6 Policy, Regulation, and Implementation Strategies

The adoption of sustainable concrete technologies, including bio-based admixtures and TIWW requires robust policy frameworks, regulatory standards, and coordinated strategies to facilitate the transition to greener materials in the construction sector [35]. Standardization by organizations such as CEN, ISO, and ASTM is essential for ensuring the quality and reliability of these materials, while national standards organizations are encouraged to align with international sustainability goals [36]. Economic incentives and green building certification programs can promote the use of sustainable practices, particularly in regions with developing regulations like Nigeria. Collaboration among stakeholders, including government, industry, and research institutions, is critical for fostering innovation and implementing effective policies [37].

Challenges such as material variability and insufficient lifecycle assessments hinder the widespread adoption of TIWW and bio-admixtures. Future research should focus on advanced treatment technologies, engineered bio-admixtures, and comprehensive lifecycle assessments to quantify environmental benefits and optimize performance [38]. Addressing these areas will support the construction industry's alignment with global sustainability goals and enhance the resilience of the built environment. Prior studies indicate that well treated industrial wastewater can sustain or even improve compressive strength compared to potable water, as evidenced by Almeida & Tonetti, whose assessment documents strengths above 40 MPa with complete wastewater substitution [16]. Likewise, textile-effluent concrete demonstrated enhanced durability performance in the study of Shanmugasundaram et al. [9]. Lignosulfonate-based additives in bio-based admixtures have demonstrated a reduction in water demand and an enhancement of workability in diverse cementitious systems [10], whereas polysaccharide and starch-derived admixtures markedly improve durability and decrease water absorption [11,28]. Recent studies emphasizing synergy suggest that the integration of TIWW with bio-admixtures like tannic acid or lignin derivatives enhances hydration dynamics and reduces chloride-induced degradation [27–29]. In contrast to previous isolated studies, the integrated TIWW + bio-admixture framework presented in this review exhibits a more pronounced cumulative effect, with empirical evidence from multiple studies [16,27–30] consistently indicating simultaneous improvements in strength, microstructural refinement, and durability that cannot be attained through wastewater or bio-admixtures independently. This comparison illustrates that the multi-material ecological method outlined here exceeds the findings of standard single-intervention studies in literature.

2.7 Future Research Directions

The incorporation of bio-based admixtures and treated industrial wastewater (TIWW) into concrete presents numerous environmental benefits for the construction industry; nevertheless, it also entails disadvantages such as material variability and performance unpredictability [38]. The erratic composition of wastewater and the efficacy of treatment provide significant obstacles, necessitating the implementation of advanced technologies such as reverse osmosis and ultrafiltration, alongside tailored treatment protocols for specific sectors. Synthetic bio-admixtures must be engineered through molecular design to mitigate natural variability. Collaboration among materials scientists, biochemists, and civil engineers is crucial for standardizing techniques and scaling solutions. To enhance sustainability in concrete production, forthcoming research must focus on region-specific data, comparative analyses, and long-term durability in lifecycle assessments [39].

Conclusion

The transformative potential of using bio-based admixtures and treated industrial wastewater in sustainable concrete design significantly reduces environmental impacts and carbon footprints. This eco-design aligns with circular economy principles, promoting resource efficiency and waste valorization, particularly benefiting rural communities. A shift in mindset and policy is essential for widespread adoption, requiring collaboration among governments, industry, and researchers to establish supportive frameworks. Ultimately, embracing eco-designed concrete is both a technical necessity and a moral obligation to ensure sustainable infrastructure for future generations.

References

[1] ALAA OMAR ABED AL RAHMAN TANASH, K. Muthusamy, Concrete Industry, Environment Issue, and Green Concrete: A Review, CONSTRUCTION 2 (2022) 01–09. https://doi.org/10.15282/construction.v2i1.7188

[2] G. Habert, S.A. Miller, V.M. John, J.L. Provis, A. Favier, A. Horvath, K.L. Scrivener, Environmental impacts and decarbonization strategies in the cement and concrete industries, Nat Rev Earth Environ 1 (2020) 559–573. https://doi.org/10.1038/s43017-020-0093-3

[3] S. Kumari, A. Rao, M. Kaur, G. Dhania, Petroleum-Based Plastics Versus Bio-Based Plastics: A Review, Nature Environment and Pollution Technology 22 (2023) 1111–1124. https://doi.org/10.46488/NEPT.2023.v22i03.003

[4] A. Cheshmehzangi, Revisiting the built environment: 10 potential development changes and paradigm shifts due to COVID-19, Journal of Urban Management 10 (2021) 166–175. https://doi.org/10.1016/j.jum.2021.01.002

[5] N. Chayaamor-Heil, V. Perricone, P. Gruber, F. Guéna, Bioinspired, biobased and living material designs: a review of recent research in architecture and construction, Bioinspir Biomim 18 (2023) 041001. https://doi.org/10.1088/1748-3190/acd82e

[6] M.G. Davidson, S. Elgie, S. Parsons, T.J. Young, Production of HMF, FDCA and their derived products: a review of life cycle assessment (LCA) and techno-economic analysis (TEA) studies, Green Chemistry 23 (2021) 3154–3171. https://doi.org/10.1039/D1GC00721A

[7] R. Horry, C.A. Booth, A. Mahamadu, P. Manu, P. Georgakis, Environmental management systems in the architectural, engineering and construction sectors: a roadmap to aid the delivery

of the sustainable development goals, Environ Dev Sustain 24 (2022) 10585–10615. https://doi.org/10.1007/s10668-021-01874-3

[8] N.S.A. Yaro, M.H. Sutanto, L. Baloo, N.Z. Habib, A. Usman, A.K. Yousafzai, A. Ahmad, A.H. Birniwa, A.H. Jagaba, A. Noor, A Comprehensive Overview of the Utilization of Recycled Waste Materials and Technologies in Asphalt Pavements: Towards Environmental and Sustainable Low-Carbon Roads, Processes 11 (2023) 2095. https://doi.org/10.3390/pr11072095

[9] V. Shanmugasundaram, A. Manoharan, K. Selvarajan, M. A, Incorporation of furnace-treated alum and textile sludge in concrete: strength, durability and environmental impact, Journal of Taibah University for Science 19 (2025). https://doi.org/10.1080/16583655.2025.2534222

[10] A. Kalliola, T. Vehmas, T. Liitiä, T. Tamminen, Alkali-O2 oxidized lignin – A bio-based concrete plasticizer, Ind Crops Prod 74 (2015) 150–157. https://doi.org/10.1016/j.indcrop.2015.04.056

[11] Y. Boutouam, M. Hayek, K. Bouarab, A. Yahia, A Comprehensive Review of Plant-Based Biopolymers as Viscosity-Modifying Admixtures in Cement-Based Materials, Applied Sciences 14 (2024) 4307. https://doi.org/10.3390/app14104307

[12] B. Aaliya, K.V. Sunooj, M. Lackner, Biopolymer composites: a review, International Journal of Biobased Plastics 3 (2021) 40–84. https://doi.org/10.1080/24759651.2021.1881214

[13] M. Jawaid, B. Singh, L.K. Kian, S.A. Zaki, A.M. Radzi, Processing techniques on plastic waste materials for construction and building applications, Curr Opin Green Sustain Chem 40 (2023) 100761. https://doi.org/10.1016/j.cogsc.2023.100761

[14] G. Li, A. Akbar, L.-W. Zhang, F. Rosei, K.M. Liew, Exploring durability and environmental impact of cementitious composites modified by fluoroalkyl-silane based additive, Constr Build Mater 370 (2023) 130665. https://doi.org/10.1016/j.conbuildmat.2023.130665

[15] C. Passoni, E. Palumbo, R. Pinho, A. Marini, The LCT Challenge: Defining New Design Objectives to Increase the Sustainability of Building Retrofit Interventions, Sustainability 14 (2022) 8860. https://doi.org/10.3390/su14148860

[16] M.E.P. Almeida, A.L. Tonetti, Treated wastewater as a sustainable alternative to concrete manufacturing: a literature review on its performance, International Journal of Environmental Science and Technology 20 (2023) 8157–8174. https://doi.org/10.1007/s13762-022-04686-8

[17] I. Ekop, I.-I.N. Etim, E. Ambrose, U.E. Edike, Bio-Based Cement Concrete Admixtures for Green Recovery in the Construction Industry: a Critical Survey, Materials Circular Economy 6 (2024) 31. https://doi.org/10.1007/s42824-024-00113-0

[18] K. BEDADA, A. NYABUTO, I. KINOTI, J. MARANGU, Review on Advances in Bio-based Admixtures for Concrete, Journal of Sustainable Construction Materials and Technologies 8 (2023) 344–367. https://doi.org/10.47481/jscmt.1328915

[19] H. Varshney, R.A. Khan, I.K. Khan, Sustainable use of different wastewater in concrete construction: A review, Journal of Building Engineering 41 (2021) 102411. https://doi.org/10.1016/j.jobe.2021.102411

[20] R. Raydan, J. Khatib, A. Jahami, A.K. El Hamoui, F. Chamseddine, Prediction of the mechanical strength of concrete containing glass powder as partial cement replacement material, Innovative Infrastructure Solutions 7 (2022) 311. https://doi.org/10.1007/s41062-022-00896-8

[21] B. Cherry, G. Warren, Corrosion and Protection of Reinforced Concrete, CRC Press, 2021. https://doi.org/10.1201/9781003081302

[22] P.A. Rasheed, S.K. Nayar, I. Barsoum, A. Alfantazi, Degradation of Concrete Structures in Nuclear Power Plants: A Review of the Major Causes and Possible Preventive Measures, Energies (Basel) 15 (2022) 8011. https://doi.org/10.3390/en15218011

[23] D. Ibañez, M. Salka, V. Guallart, S. Boeri, L. Shamir, M.L. De Marco, S. Paoli, M.C. Pastore, M. Fragiacomo, L.R. Gobakken, S. Boulet, Innovative Design, Materials, and Construction Models for BioCities, in: 2023: pp. 183–215. https://doi.org/10.1007/978-3-031-29466-2_8

[24] R. Mohanraj, S. Senthilkumar, P. Goel, R. Bharti, A state-of-the-art review of Euphorbia Tortilis cactus as a bio-additive for sustainable construction materials, Mater Today Proc (2023). https://doi.org/10.1016/j.matpr.2023.03.762

[25] C. Tan, F. Han, S. Zhang, P. Li, N. Shang, Novel Bio-Based Materials and Applications in Antimicrobial Food Packaging: Recent Advances and Future Trends, Int J Mol Sci 22 (2021) 9663. https://doi.org/10.3390/ijms22189663

[26] I. Shufrin, E. Pasternak, A. Dyskin, Environmentally Friendly Smart Construction—Review of Recent Developments and Opportunities, Applied Sciences 13 (2023) 12891. https://doi.org/10.3390/app132312891

[27] B. Şimşek, T. Uygunoğlu, Ö.F. Dilmaç, M. Uğur, İ. Sevgili, Analyzing the effect of tannic acid as a bio-additive and water purification agent in cement pastes, Mater Lett 350 (2023) 134969. https://doi.org/10.1016/j.matlet.2023.134969

[28] D. Oni, J. Mwero, C. Kabubo, The Effect of Cassava Starch on the Durability Characteristics of Concrete, The Open Civil Engineering Journal 14 (2020) 289–301. https://doi.org/10.2174/1874149502014010289

[29] M. Sheikh Hassani, J.C. Matos, Y.X. Zhang, E. Teixeira, Concrete production with domestic and industrial wastewaters—A literature review, Structural Concrete 24 (2023) 5582–5599. https://doi.org/10.1002/suco.202200637

[30] J.M. de Souza, Cassava Wastewater: An Introduction, Characterization and Potential, in: 2021: pp. 1–10. https://doi.org/10.1007/978-3-030-58782-6_1

[31] I. Oberti, A. Paciello, Bioplastic as a Substitute for Plastic in Construction Industry, Encyclopedia 2 (2022) 1408–1420. https://doi.org/10.3390/encyclopedia2030095

[32] O. Okolie, A. Kumar, C. Edwards, L.A. Lawton, A. Oke, S. McDonald, V.K. Thakur, J. Njuguna, Bio-Based Sustainable Polymers and Materials: From Processing to Biodegradation, Journal of Composites Science 7 (2023) 213. https://doi.org/10.3390/jcs7060213

[33] I.A. Bella, N. Bella, A. Asroun, S. Saddiki, Contribution on the Use of Household Waste as Bio-Admixture, in: 2022: pp. 569–574. https://doi.org/10.4028/www.scientific.net/CTA.1.569

[34] P. Pant, S. Pandey, S. Dall'Acqua, The Influence of Environmental Conditions on Secondary Metabolites in Medicinal Plants: A Literature Review, Chem Biodivers 18 (2021). https://doi.org/10.1002/cbdv.202100345

[35] A. Tiwari, Advancement of Materials to Sustainable & Green World, Adv Mater Lett (2023). https://doi.org/10.5185/amlett.2023.031724

[36] C.A. Fletcher, K. Niemenoja, R. Hunt, J. Adams, A. Dempsey, C.E. Banks, Addressing Stakeholder Concerns Regarding the Effective Use of Bio-Based and Biodegradable Plastics, Resources 10 (2021) 95. https://doi.org/10.3390/resources10100095

[37] F. Younas, A. Mustafa, Z.U.R. Farooqi, X. Wang, S. Younas, W. Mohy-Ud-Din, M. Ashir Hameed, M. Mohsin Abrar, A.A. Maitlo, S. Noreen, M.M. Hussain, Current and Emerging Adsorbent Technologies for Wastewater Treatment: Trends, Limitations, and Environmental Implications, Water (Basel) 13 (2021) 215. https://doi.org/10.3390/w13020215

[38] F.D. Gregory, B.D. Trump, The Evolving Landscape of Biotechnology Research: Challenges and Opportunities in Acquisitions, Funding, Education, and Workforce Development for a Productive Bioeconomy, in: 2025: pp. 25–50. https://doi.org/10.1007/978-94-024-2296-2_2

[39] S. Oberthür, G. Khandekar, T. Wyns, Global governance for the decarbonization of energy-intensive industries: Great potential underexploited, Earth System Governance 8 (2021) 100072. https://doi.org/10.1016/j.esg.2020.100072

Smart Water-Energy Nexus in Concrete Manufacturing: Harnessing Renewable Energy and AI for Sustainable Development

Mariam Iyabo ADEOBA[1,a*], Thanyani PANDELANI[1,b], Harry NGANGWA[1,c], Tracy MASEBE[2,d]

[1]UNISA Biomechanics Research Group, Department of Mechanical, Bioresources and Biomedical Engineering, College of Science, Engineering and Technology (CSET), University of South Africa, Florida 1710, South Africa

[2]Department of Life and Consumer Sciences, College of Agriculture and Environmental Sciences, University of South Africa (UNISA), Florida 1710, South Africa

[a]adeobmi@unisa.ac.za, [b]epandet@unisa.ac.za, [c]ngwanhm@unisa.ac.za, [d]masebtm.unisa.ac.za

Keywords: Environmental Degradation, Water Sustainability, Renewable Energy, Concrete Production, Cement

Abstract. Concrete's high energy and water requirements contribute to environmental degradation and global resource use. This mini-review employs a scoping and thematic synthesis of peer reviewed literature from 2020 to 2024, focusing on the intersection of artificial intelligence (AI), renewable energy and sustainable water-energy management in concrete manufacturing. The review explores circular economy models for integrated energy-water systems AI-driven process optimization, water conservation and reuse techniques and the deployment of renewable energy technologies such as solar PV, biomass and biogas. Key findings highlight the potential of digital twins and smart sensors for real-time monitoring, the effectiveness of treated industrial wastewater in reducing freshwater dependency, and the economic viability of bio-based admixtures. While these innovations offer promising synergies for sustainability, adoption barriers such as high initial costs, technical complexity, and regulatory gaps remain. The study underscores the need for transdisciplinary innovation and policy alignment to foster low-carbon, water-efficient construction practices.

1. Introduction

Concrete remains foundational to modern construction due to its durability and affordability, yet its environmental footprint is increasingly problematic. Central to this issue is Portland cement, which accounts for 7–8% of global CO_2 emissions, resulting from energy-intensive clinker production and the calcination process [1,2]. Moreover, concrete manufacturing consumes vast volumes of natural aggregates and freshwater, with usage reaching up to 200 litres per cubic meter pressuring water-stressed regions [3,4].

The water-energy nexus highlights the reciprocal relationship between water and energy systems, aggravating sustainability challenges. Energy is needed to purify and transport water for concrete processes, while water supports energy generation. Climate change intensifies these pressures through resource scarcity and increased demand [5,6]. To combat this, the concrete sector is embracing circular economy principles, particularly through the adoption of treated industrial wastewater (TIWW) as mixing water. TIWW reduces freshwater dependency and supports industrial symbiosis, provided safety and performance thresholds are met [7]. In parallel, bio-based admixtures offer greener alternatives to petroleum-based chemicals, further lowering emissions [7,8].

However, variability in wastewater and natural admixtures demands intelligent control systems. Artificial intelligence (AI) and machine learning provide solutions for real-time monitoring, predictive mix design, and automated quality assurance [9,10]. Coupled with renewable energy sources, e.g., solar-powered wastewater treatment or green electricity for kilns, these innovations pave the way for sustainable production.

By harnessing these integrated strategies, the concrete industry can significantly reduce its ecological impact and align with climate mitigation goals. The fusion of AI, renewable energy, and circular practices marks a vital transition toward a resilient built environment [11]. Nonetheless, despite increasing interest in sustainable practices, a significant research gap persists in the thorough integration of artificial intelligence and renewable energy within the water-energy nexus of concrete production. Current research frequently examines these elements separately, failing to provide a cohesive framework that investigates their combined potential. This assessment seeks to integrate recent improvements and pinpoint prospects for comprehensive, AI-driven, low-carbon concrete production systems. While numerous studies have explored artificial intelligence, renewable energy, and water reuse in concrete manufacturing independently, there is a lack of integrated reviews that examine their combined application within the water-energy nexus. This review addresses this gap by synthesizing recent literature (2020–2024) to evaluate how these technologies can be jointly leveraged to enhance sustainability, efficiency, and resilience in concrete productio. It highlights the role of digital twins, smart sensors, and machine learning in optimizing water and energy use, while assessing the feasibility of renewable energy sources like solar PV, biomass, and wind in concrete production. Unlike previous reviews that focus on isolated technologies, this study emphasizes interconnected solutions and their potential to reshape concrete manufacturing in line with climate goals. By doing so, it contributes to the growing discourse on transdisciplinary innovation and policy alignment for sustainable infrastructure development.

2. Method

This mini-review employed a scoping thematic synthesis approach to analyze recent peer-reviewed literature on the integration of artificial intelligence (AI), renewable energy, and water-energy management in concrete manufacturing. The review focused on publications from 2020 to 2024, a period selected to capture the most recent technological and policy developments in sustainable construction. Relevant literature was sourced from Scopus and Google Scholar, using keyword combinations such as: ("concrete manufacturing" AND "artificial intelligence", "renewable energy" AND "cement production"," water-energy nexus" AND "construction", "treated wastewater" AND "concrete", "digital twin" AND "sustainable infrastructure". An initial pool of over 250 articles was identified. The selection was narrowed down through a multi-stage screening process:

1. Title and abstract screening for relevance to the review's scope.
2. Inclusion criteria: peer-reviewed journal articles, English language, and direct relevance to AI, renewable energy, or water reuse in concrete production.
3. Exclusion criteria: duplicate studies, non-peer-reviewed sources, and articles lacking methodological transparency or empirical grounding.
4. Final selection: 40 articles were chosen based on relevance, methodological quality, and contribution to thematic insights.

Further, thematic synthesis was conducted by clustering recurring concepts across the selected literature. Themes were developed through iterative coding and expert judgment, focusing on:

- Renewable energy integration in concrete production
- AI-driven optimization of water and energy use

Materials Research Proceedings 63 (2026) 132-142 https://doi.org/10.21741/9781644904053-15

- Environmental and economic impacts
- Policy, research, and implementation strategies

While this review provides a comprehensive overview, it is limited by its reliance on English-language publications and only two databases, which may exclude relevant regional or non-English studies. Nonetheless, the selected timeframe and sources offer a robust foundation for identifying emerging trends and interdisciplinary synergies in sustainable concrete manufacturing.

3. Thematic discussions

The screening and selection of the articles are further synthesized and discussed in the following sections.

3.1 Harnessing Renewable Energy in Concrete Manufacturing

The global construction industry faces significant pressure to reduce its carbon footprint while supporting urbanization and infrastructure growth, particularly in concrete production, which is energy-intensive and polluting, especially during cement manufacturing [12,13]. Utilizing renewable energy sources is essential for minimizing the carbon footprint of concrete manufacturing. This study examines relevant renewable energy sources, their applications in concrete production, and the associated benefits and drawbacks.

Table 1 shows key renewable energy options including solar energy, wind energy, and biomass. Solar photovoltaic (PV) systems can power equipment in concrete batching and mixing facilities, while solar thermal technologies can provide heat for cement kilns and curing processes. Wind turbines can generate electricity, particularly in remote locations, enhancing the sustainability of mobile batching plants. Also, as shown in Table 1, solar power offers the highest CO_2 reduction potential (up to 90%) among renewable sources [17], though its adoption remains limited due to high capital costs. Biomass can be converted into biogas or thermal energy, offering a carbon-neutral alternative to fossil fuels in cement kilns and on-site operations. In developing countries, mobile concrete batching plants often rely on diesel generators, presenting an opportunity to integrate renewable energy sources, reducing environmental impact and reliance on fossil fuels. However, challenges such as financial barriers, technical expertise requirements, and the intermittency of renewable energy sources must be addressed. The transition to renewable energy can significantly reduce greenhouse gas emissions, enhance energy security, and promote self-sufficiency, particularly in areas with unstable grid infrastructure. Despite the benefits, high upfront costs and the need for technical expertise remain significant barriers to adoption, especially for small and medium-sized enterprises (SMEs) in the construction sector [14,15]. Addressing these challenges through integrated system design, financial innovation, and government support is crucial for the concrete industry to meet global decarbonization targets and promote sustainable industrial development. The strategic integration of renewable energy will be vital for creating a cleaner, more resilient construction industry alongside resource-efficient practices and digital technology.

3.2 AI-Driven Optimization of the Water-Energy Nexus

Given the increasing demand for environmentally friendly building materials worldwide, the concrete industry needs to address its resource-intensive operations, especially regarding energy and water consumption. Concrete manufacturing is particularly impacted by the water-energy nexus, which is the interdependent relationship between water consumption and energy use, due to the high-water demand for mixing and curing and the significant energy consumption for cement production and plant operations, as shown in Figure 1. Optimizing this relationship is essential to achieving long-term sustainability, lowering environmental effects, and improving operational effectiveness. Using Artificial Intelligence (AI) technology is one of the most promising ways to handle this complexity [16]. AI has the potential to completely transform resource efficiency in the manufacturing of concrete through automation, data analytics, and smart sensing.

Table 1: Harnessing Renewable Energy in Concrete Manufacturing

Renewable Energy Source	Application in Concrete Manufacturing	Energy Output Potential	Implementation Cost	Environmental Impact	Current Adoption Rate
Solar Power	Powering cement kilns, lighting, conveyor systems	150-400 kWh/m^2 annually	Medium ($2-4M for large plants)	85-90% CO_2 reduction	15-20% of plants [17]
Wind Energy	Operating crushing equipment, mixers, and transportation	2-3 MW per turbine	High ($3-5M initial)	80-85% CO_2 reduction	10-15% of plants [17]
Hydroelectric	Powering grinding mills, material handling	1-100 MW depending on site	Very High ($5-15M)	90-95% CO_2 reduction[3]	5-8% of plants [18]
Biomass Energy	Cement kiln fuel, steam generation	10-50 MW thermal	Medium ($1-3M)	60-70% CO_2 reduction[4]	25-30% of plants [18]
Geothermal	Process heating, facility climate control[5]	5-30 MW thermal[5]	High ($4-7M)	85-90% CO_2 reduction[5]	2-5% of plants [17]
Waste Heat Recovery	Capturing kiln exhaust, steam generation	10-30% efficiency gain	Low-Medium ($0.5-2M)	20-40% CO_2 reduction[6]	40-50% of plants [17]

Artificial intelligence (AI) encompasses a range of computing tools capable of analyzing large datasets, identifying patterns, and generating conclusions or suggestions. In concrete production, AI plays a crucial role in monitoring, forecasting, and optimizing the interconnected processes that manage energy and water usage. Smart sensors integrated into production lines collect real-time data on temperature, humidity, pressure, flow rates, and material composition, which are connected through the Internet of Things (IoT) for seamless data transfer. One of the most innovative AI technologies is the digital twin, a virtual model of a concrete plant that uses real-time sensor data and AI algorithms to predict outcomes, test optimization options, and simulate plant operations. Figure 1 illustrates the integration of AI-enabled technologies, highlighting how digital twins and smart sensors interact to optimize water and energy flows in real time. For example, a digital twin can model the impact of altering the water-to-cement ratio or cure conditions on concrete's strength and energy consumption, aiding operators in making informed decisions. For instance, a case study by Delnavaz et al. (2022) demonstrated the use of a digital twin in a precast concrete plant, achieving a 15% reduction in energy consumption through real-time optimization of curing cycles [13]. Similarly, Feola et al. (2023) reported that lignosulfonate-based admixtures derived from paper industry waste can reduce the carbon footprint of concrete mixes by up to 20% while maintaining structural performance [25]. While bio-based admixtures offer environmental benefits, their performance variability necessitates intelligent control systems, as discussed further in Section 3.3.

Despite the promise of AI and IoT in optimizing concrete production, several limitations persist. AI models require large, high-quality datasets, which may be unavailable or inconsistent across regions. Additionally, cybersecurity risks in IoT-enabled systems pose threats to operational continuity and data integrity. Over-reliance on predictive algorithms without human oversight can also lead to suboptimal decisions in dynamic construction environments [16].

Machine learning (ML) algorithms, a subset of AI, are particularly effective in identifying complex relationships between operational outcomes and input variables. ML can analyze historical and current data to forecast water demand based on material characteristics and environmental conditions, estimate energy usage, and optimize curing cycles to conserve resources. These models continuously improve by adapting to new data, ensuring resource use is optimized according to current operational conditions. AI implementation in the water-energy nexus of concrete production focuses on four key areas such as predictive maintenance, energy load forecasting, water reuse monitoring, and mix design optimization.

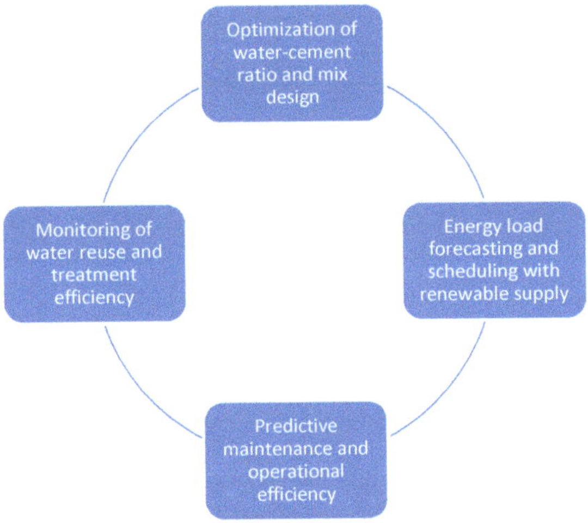

Figure 1: Overview of AI-Enabled Technologies in Concrete Production Systems

The water-to-cement (w/c) ratio, critical for concrete's strength and durability, can be dynamically adjusted by AI systems based on real-time data from embedded sensors, enhancing efficiency and performance. AI also plays a vital role in water recycling management by monitoring the pH, turbidity, and contaminants in process water, regulating chemical dosing and filtration, and identifying performance irregularities that could affect environmental safety and concrete quality. This ensures that reclaimed water meets reuse standards, reducing operational costs and environmental impact. In concrete manufacturing, which involves high energy loads, AI can predict energy demand and align it with renewable energy availability, optimizing operations to reduce reliance on the grid and lower costs [19,20]. AI can schedule energy-intensive tasks during peak solar hours and manage battery storage for renewable energy, thus promoting environmental sustainability.

Predictive maintenance powered by AI identifies potential equipment failures before they occur by analyzing data from various sensors, thereby minimizing downtime and resource waste. This approach enhances equipment lifespan and operational efficiency. Overall, AI is redefining efficiency and sustainability in concrete production by optimizing energy and water use, supporting responsible environmental practices, and improving resource management. The integration of smart sensors, IoT, and digital twins facilitate real-time data collection and simulation, enabling AI-enabled concrete plants to optimize water-cement ratios, enhance water reuse, align energy consumption with renewable sources, and reduce downtime. These advancements contribute to broader environmental goals, such as reducing carbon emissions and freshwater withdrawal, while fostering economic growth and environmental stewardship in resource-constrained developing countries [21,22]. Embracing intelligent, data-driven systems is essential for the future of sustainable concrete production.

3.3 Environmental and Economic Impacts

The construction industry is experiencing significant transformation due to increasing economic and environmental demands, particularly in concrete production. Traditional concrete manufacturing relies heavily on freshwater, fossil fuels, and non-renewable mineral aggregates, leading to substantial industrial pollution, freshwater depletion, and greenhouse gas (GHG) emissions. Innovations such as bio-based admixtures, treated industrial effluence, renewable energy integration, and intelligent process control are essential for promoting sustainable concrete manufacturing [23]. These strategies not only address environmental concerns but also enhance economic performance, demonstrating the interconnection between economic viability and environmental stewardship. The environmental impact of conventional concrete is primarily driven by freshwater usage and carbon emissions, with cement production alone contributing to 7–8% of global CO_2 emissions due to energy-intensive processes. Additionally, the concrete sector consumes significant amounts of water, between 150 and 200 liters per cubic meter for mixing and curing.

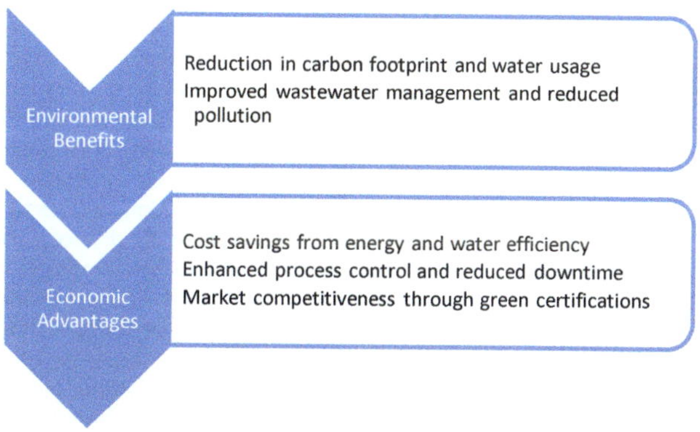

Figure 3: Environmental and Economic Impacts

The industry's reliance on freshwater resources can be significantly reduced by using treated industrial wastewater (TIWW) instead of potable water, allowing concrete plants to decrease their water footprint by up to 80% [24]. This practice is particularly impactful in areas with limited water resources where municipal and industrial water usage often compete [24]. Additionally, replacing petroleum-based chemical admixtures with bio-based alternatives derived from renewable resources like cellulose, starch, or lignin can lower embodied carbon. For example, using lignosulfonates from paper industry waste can reduce the carbon footprint of concrete mixes by up to 20% [25]. The incorporation of renewable energy sources, such as biomass, wind, and solar photovoltaic (PV), into the concrete production process can further aid decarbonization, potentially decreasing GHG emissions related to energy use by 30–70% when replacing fossil fuels. Reusing industrial wastewater not only promotes better wastewater management but also aligns with the principles of the circular economy and industrial ecology.

Investing in advanced treatment systems that produce concrete-grade water helps companies reduce pollution and create financial incentives for water reuse. Moreover, bio-based admixtures mitigate environmental risks associated with synthetic chemical admixtures, contributing to safer construction sites and cleaner manufacturing processes. The economic benefits of sustainable concrete production are compelling, as using treated wastewater can lower water procurement and disposal costs, especially in regions with strict regulations. Transitioning to renewable energy sources can also lead to significant long-term savings despite initial infrastructure costs. Additionally, bio-admixtures, often made from inexpensive agricultural waste, present a cost-effective alternative to synthetic options.

The integration of smart technology, such as AI-based process monitoring, enhances operational efficiency by optimizing additive dosage and monitoring effluent quality, thereby reducing variability and minimizing downtime. Companies adopting sustainable practices are better positioned to achieve green building certifications like LEED, BREEAM, or EDGE, which can attract environmentally conscious investors and provide financial incentives. Overall, the use of bio-based admixtures and treated wastewater in concrete production offers substantial financial and environmental benefits, supporting resource conservation and climate resilience while improving operational efficiency and market positioning. As concrete remains essential for infrastructure development, integrating sustainability into its production processes is both feasible and necessary [26,27].

3.4. Policy, Research, and Implementation Strategies

The transition to sustainable concrete production requires more than just technological innovation; it necessitates the optimization of the water-energy nexus through renewable energy and artificial intelligence (AI), alongside strong policy frameworks, skilled human resources, and strategic research investments [28,29]. Three critical pillars, policy and regulatory support, capacity building, and future research directions are essential for embedding AI-driven, resource-efficient systems into modern concrete manufacturing. Effective policies and regulations are crucial for integrating renewable energy systems and AI technologies in concrete manufacturing. Governments must create an enabling environment through financial incentives such as tax credits and subsidies, and public-private partnerships (PPPs) can help fund necessary infrastructure [30,31]. Laws that support carbon pricing can also incentivize low-emission facilities, while regulatory agencies should establish sustainability standards based on international best practices like ISO 14046 and ISO 50001. Human capital is vital for the successful integration of AI and renewable energy in concrete production. Educational institutions must develop curricula that equip engineers and technicians with the skills to manage AI-driven systems and renewable energy installations. Customized workforce development programs and community-based learning initiatives can democratize access to these technologies. To promote adoption, government organizations should provide best-practice guides and case studies that demonstrate the benefits of integrating AI with renewable energy. Continuous innovation is necessary to enhance performance and adapt to varying industrial conditions [32,33]. Research should focus on developing adaptive AI models that can improve decision-making in real time, as well as comprehensive life-cycle evaluations (LCA) to assess the environmental impacts of AI-optimized concrete systems. Localization of AI and renewable energy technologies to the specific contexts of developing countries is another important research area [[34,35]. Collaborative efforts between research institutions and local stakeholders can lead to context-specific solutions that address local development needs. Pilot projects and living labs can serve as innovation incubators for sustainable concrete solutions [36,37]. A concerted, multi-stakeholder approach is necessary to fully realize the potential of AI-driven, resource-optimized concrete manufacturing. Investing in intelligent, sustainable infrastructure is essential for a resilient and inclusive future as urbanization and climate challenges grow [38–40].

Conclusion

The construction sector is undergoing a revolutionary change towards greater resilience, efficiency, and sustainability through the integration of bio-based admixtures, treated industrial wastewater, and renewable energy sources in concrete production. This transition relies on the intelligent integration of the water-energy nexus, which enhances resource efficiency, reduces emissions, and preserves the environment. This review has examined the integration of artificial intelligence (AI), renewable energy technologies, and water reuse strategies within the context of sustainable concrete manufacturing. The findings underscore the potential of a smart water-energy nexus framework to address the environmental and operational inefficiencies. Specifically, the use of treated industrial wastewater (TIWW) can reduce freshwater consumption by up to 80% while solar photovoltaic (PV) systems offer CO_2 emission reductions of up to 90% when replacing fossil fuel-based energy sources. AI-enabled technologies including digital twins, machine learning algorithms, and smart sensors facilitate real-time monitoring, predictive mix design, and dynamic energy-water optimization. These tools, when integrated with renewable energy systems and circular resource flows, enable significant gains in process efficiency, mitigation, and resource conservation. Importantly, intelligent monitoring systems are not developed by bio-admixtures or renewable technologies but are deployed in conjunction with them to enhance system responsiveness and performance consistency. Despite these advancements, several research gaps remain. Future investigations should prioritize validation of AI models across diverse material compositions and climatic conditions, localization of renewable energy systems for off-grid or resource-constrained construction environments, and comprehensive life-cycle assessments (LCA) to quantify the long-term environmental and economic impacts of integrated systems.
To accelerate adoption, coordinated efforts are required across academia, industry, and policy domains. This includes the development of performance-based procurement frameworks, investment in pilot-scale demonstration projects, and the establishment of interdisciplinary training programs that bridge digital technologies, environmental engineering, and materials science. Embedding these innovations into mainstream concrete production is essential for achieving low-carbon, water-efficient, and resilient infrastructure aligned with global sustainability targets.

References

[1] V. Sousa, J.A. Bogas, S. Real, I. Meireles, Industrial production of recycled cement: energy consumption and carbon dioxide emission estimation, Environmental Science and Pollution Research 30 (2022) 8778–8789. https://doi.org/10.1007/s11356-022-20887-7

[2] M. Soomro, V.W.Y. Tam, A.C. Jorge Evangelista, Production of cement and its environmental impact, in: Recycled Concrete, Elsevier, 2023: pp. 11–46. https://doi.org/10.1016/B978-0-323-85210-4.00010-2

[3] S. Monkman, T. Janke, A. Hanmore, NRG COSIA Carbon XPRIZE: carbon-dioxide mineralization in recycled concrete wash water, Clean Energy 5 (2021) 553–574. https://doi.org/10.1093/ce/zkab023

[4] H. Najm, H. Wang, J. Hencken, H. Yagnik, X. Chen, X. Chen, Implementation of Porous Concrete in Sidewalks in New Jersey, New Jersey, 2022. https://rosap.ntl.bts.gov/view/dot/61819/dot_61819_DS1.pdf (accessed September 1, 2025).

[5] N. Mamassis, A. Efstratiadis, P. Dimitriadis, T. Iliopoulou, R. Ioannidis, D. Koutsoyiannis, Water and Energy, in: Handbook of Water Resources Management: Discourses, Concepts and

Examples, Springer International Publishing, Cham, 2021: pp. 619–657. https://doi.org/10.1007/978-3-030-60147-8_20

[6] M.Z. Nawaz, M. Bilal, A. Tariq, H.M.N. Iqbal, H.A. Alghamdi, H. Cheng, Bio-purification of sugar industry wastewater and production of high-value industrial products with a zero-waste concept, Crit Rev Food Sci Nutr 61 (2021) 3537–3554. https://doi.org/10.1080/10408398.2020.1802696

[7] A. Adesina, Circular economy in the concrete industry, in: Handbook of Solid Waste Management: Sustainability through Circular Economy, Springer, 2021: pp. 1–15.

[8] C.K. Purchase, D.M. Al Zulayq, B.T. O'Brien, M.J. Kowalewski, A. Berenjian, A.H. Tarighaleslami, M. Seifan, Circular Economy of Construction and Demolition Waste: A Literature Review on Lessons, Challenges, and Benefits, Materials 15 (2021) 76. https://doi.org/10.3390/ma15010076

[9] A.C. Rosa, A.W. Hammad, D. Boer, A. Haddad, Use of operational research techniques for concrete mix design: a systematic review. Heliyon 9 (4): e15362, (2023).

[10] K.C. Onyelowe, A.M. Ebid, H.A. Mahdi, F.K.C. Onyelowe, Y. Shafieyoon, M.E. Onyia, H.N. Onah, AI Mix Design of Fly Ash Admixed Concrete Based on Mechanical and Environmental Impact Considerations, Civil Engineering Journal 9 (2023) 27–45. https://doi.org/10.28991/CEJ-SP2023-09-03

[11] S. Rezvani, N. de Almeida, M. Falcão, Climate Adaptation Measures for Enhancing Urban Resilience, Buildings 13 (2023) 2163. https://doi.org/10.3390/buildings13092163

[12] S. Pamenter, R.J. Myers, Decarbonizing the cementitious materials cycle: A whole-systems review of measures to decarbonize the cement supply chain in the UK and European contexts, J Ind Ecol 25 (2021) 359–376. https://doi.org/10.1111/jiec.13105

[13] M. Delnavaz, A. Sahraei, A. Delnavaz, R. Farokhzad, S. Amiri, S. Bozorgmehrnia, Production of concrete using reclaimed water from a ready-mix concrete batching plant: Life cycle assessment (LCA), mechanical and durability properties, Journal of Building Engineering 45 (2022) 103560. https://doi.org/10.1016/j.jobe.2021.103560

[14] J. Khan, B. Johansson, Adoption, implementation and design of carbon pricing policy instruments, Energy Strategy Reviews 40 (2022) 100801. https://doi.org/10.1016/j.esr.2022.100801

[15] I. Parry, K. Zhunussova, M.S. Black, Carbon taxes or emissions trading systems?: instrument choice and design, (2022).

[16] A. Aldoseri, K.N. Al-Khalifa, A.M. Hamouda, Re-Thinking Data Strategy and Integration for Artificial Intelligence: Concepts, Opportunities, and Challenges, Applied Sciences 13 (2023) 7082. https://doi.org/10.3390/app13127082

[17] V., D.R.M., N.A., & Y.D.M. Albino, Alternative Energy Sources in Cement Manufacturing: A Systematic Review. Network for Business Sustainability., Politecnico di Bari, Italy, 2011.

[18] O.M. Fadayini, C. Madu, T.T. Oshin, A.A. Obisanya, G.O. Ajiboye, T.O. Ipaye, T.O. Rabiu, J.T. Akintola, S.J. Ajayi, N.A. Kingsley, Energy and Economic Comparison of Different Fuels in Cement Production, in: Cement Industry - Optimization, Characterization and Sustainable Application, IntechOpen, 2021. https://doi.org/10.5772/intechopen.96812

[19] D.N. Dwivedi, A. Gupta, Artificial intelligence-driven power demand estimation and short-, medium-, and long-term forecasting, in: Artificial Intelligence for Renewable Energy Systems, Elsevier, 2022: pp. 231–242. https://doi.org/10.1016/B978-0-323-90396-7.00013-4

[20] T. Mazhar, H.M. Irfan, I. Haq, I. Ullah, M. Ashraf, T. Al Shloul, Y.Y. Ghadi, Imran, D.H. Elkamchouchi, Analysis of Challenges and Solutions of IoT in Smart Grids Using AI and Machine Learning Techniques: A Review, Electronics (Basel) 12 (2023) 242. https://doi.org/10.3390/electronics12010242

[21] L.O. Cezarino, L.B. Liboni, N. Oliveira Stefanelli, B.G. Oliveira, L.C. Stocco, Diving into emerging economies bottleneck: Industry 4.0 and implications for circular economy, Management Decision 59 (2021) 1841–1862. https://doi.org/10.1108/MD-10-2018-1084

[22] T.M. Kabudi, Artificial Intelligence for Quality Education: Successes and Challenges for AI in Meeting SDG4, in: 2022: pp. 347–362. https://doi.org/10.1007/978-3-031-19429-0_21

[23] P.Y. Liew, P.S. Varbanov, A. Foley, J.J. Klemeš, Smart energy management and recovery towards Sustainable Energy System Optimisation with bio-based renewable energy, Renewable and Sustainable Energy Reviews 135 (2021) 110385. https://doi.org/10.1016/J.RSER.2020.110385

[24] A. Maury-Ramírez, D. Illera-Perozo, J.A. Mesa, Circular Economy in the Construction Sector: A Case Study of Santiago de Cali (Colombia), Sustainability 14 (2022) 1923. https://doi.org/10.3390/su14031923

[25] G. Feola, M.K. Goodman, J. Suzunaga, J. Soler, Collective memories, place-framing and the politics of imaginary futures in sustainability transitions and transformation, Geoforum 138 (2023) 103668. https://doi.org/10.1016/j.geoforum.2022.103668

[26] C. Maraveas, Incorporating Artificial Intelligence Technology in Smart Greenhouses: Current State of the Art, Applied Sciences 13 (2022) 14. https://doi.org/10.3390/app13010014

[27] N. Rane, S. Choudhary, J. Rane, Artificial Intelligence (AI) and Internet of Things (IoT) - based sensors for monitoring and controlling in architecture, engineering, and construction: applications, challenges, and opportunities, SSRN Electronic Journal (2023). https://doi.org/10.2139/ssrn.4642197

[28] O.M. Owojori, C. Okoro, The Private Sector Role as a Key Supporting Stakeholder towards Circular Economy in the Built Environment: A Scientometric and Content Analysis, Buildings 12 (2022) 695. https://doi.org/10.3390/buildings12050695

[29] K. Smith, S. Sepasgozar, Governance, Standards and Regulation: What Construction and Mining Need to Commit to Industry 4.0, Buildings 12 (2022) 1064. https://doi.org/10.3390/buildings12071064

[30] Q.A. Nisar, S. Haider, F. Ali, S. Jamshed, K. Ryu, S.S. Gill, Green human resource management practices and environmental performance in Malaysian green hotels: The role of green intellectual capital and pro-environmental behavior, J Clean Prod 311 (2021) 127504. https://doi.org/10.1016/j.jclepro.2021.127504

[31] G.U. Alaneme, K.A. Olonade, E. Esenogho, Critical review on the application of artificial intelligence techniques in the production of geopolymer-concrete, SN Appl Sci 5 (2023) 217. https://doi.org/10.1007/s42452-023-05447-z

[32] M. Ghobakhloo, M. Iranmanesh, A. Grybauskas, M. Vilkas, M. Petraitė, Industry 4.0, innovation, and sustainable development: A systematic review and a roadmap to sustainable innovation, Bus Strategy Environ 30 (2021) 4237–4257. https://doi.org/10.1002/bse.2867.

[33] S. Vinodh, J. Antony, R. Agrawal, J.A. Douglas, Integration of continuous improvement strategies with Industry 4.0: a systematic review and agenda for further research, The TQM Journal 33 (2020) 441–472. https://doi.org/10.1108/TQM-07-2020-0157.

[34] A.W. Ng, J. Nathwani, J. Fu, H. Zhou, Green financing for global energy sustainability: prospecting transformational adaptation beyond Industry 4.0, Sustainability: Science, Practice and Policy 17 (2021) 377–390. https://doi.org/10.1080/15487733.2021.1999079.

[35] P.M. Falcone, Sustainable Energy Policies in Developing Countries: A Review of Challenges and Opportunities, Energies (Basel) 16 (2023) 6682. https://doi.org/10.3390/en16186682.

[36] P. Zuñiga, L. Rubalcaba, R.C. de Fassio, Catapulting Innovation: Linking Open Innovation with Innovation Procurement, Washington, D. C., 2021. https://doi.org/10.18235/0003817.

[37] L. Katikas, S. Sotiriou, Schools as living labs for the new European bauhaus, Univers Access Inf Soc 24 (2025) 1995–2006. https://doi.org/10.1007/s10209-023-01044-4.

[38] M.A. Khan, Cities and Mega Risks: COVID-19 and Climate Change. Springer Nature, ., Springer, 2022.

[39] M. Mukim, M. Roberts, Thriving: making cities green, resilient, and inclusive in a changing climate, World Bank Publications, 2023.

[40] M. Adedeji Olawale, A.A. Ayeh, F.O. Adekola, A.S. Precious, A.O. Joshua, O.T. Oladosu, A Review on the Intersection of Artificial Intelligence on Building Resilient Infrastructure, Promoting Inclusive and Sustainable Industrialization and Fostering Innovation, International Journal of Engineering and Modern Technology 9 (2023) 44–74. https://doi.org/10.56201/ijemt.v9.no3.2023.pg44.74.

Application of Rice Husk Ash and Laterite Interlocking Bricks Sandcrete Blocks for Affordable Housing Delivery in Nigeria - A Review

Akeem Ayinde RAHEEM[1], Samuel Oyetunde AKINLOYE[2*], Favour Wuraola KOLAWOLE[2]

[1]Building Department, Ladoke Akintola University of Technology, Ogbomoso, Nigeria

[2]Civil Engineering Department, Ladoke Akintola University of Technology, Ogbomoso, Nigeria

akinloye23@gmail.com

Keywords: Affordable Housing, Compressive Strength, Laterite Interlocking Bricks, Rice Husk Ash, Sandcrete Blocks

Abstract. Nigeria's housing affordability necessitates innovative and sustainable building materials. Sandcrete block, being commonly used for load bearing and non-load bearing walls in building construction has been seriously affected by the high cost of its ingredient. The search for alternative materials to sandcrete block with the intent of reducing the cost of conventional materials has led to investigating the use of laterites blended with agricultural residues as supplementary cementitious materials thus complementing the cement. Effective use of these materials is a way of advocating for the green method of disposal of large quantity of waste, which would have caused environmental nuisance. An overview of studies carried out on laterite interlocking bricks and rice husk ash (RHA) sandcrete blocks is presented in this paper. The review considered the utilization of RHA as partial substitution for cement in sandcrete block and laterite interlocking bricks with the aim of reducing cost of conventional materials for affordable and sustainable housing unit. The chemical, physical, and mineralogical properties of these materials were described based on findings reported in previous studies, rather than through direct analysis in the present work. Also, their influence on the characteristics of sandcrete block and Laterite interlocking bricks such as; water absorptions, compressive strength, and durability (acid attack and abrasion) were discussed in details. The findings suggest that these materials can contribute to sustainable and affordable housing delivery in Nigeria, depending on project-specific requirements.

1. Introduction

The housing sector significantly drives economic growth by influencing factors like employment, savings, and investment, and it also impacts health through quality housing [1]. Affordable housing, which is essential for human needs, should be practical, affordable, and well-maintained [2] It accounts for 10–20% of a country's economy and is the largest household asset [1]. Rapid urbanization poses significant challenges to developing countries, including Nigeria, in providing adequate housing, infrastructure, and services for the growing urban population. [3]. According to Lee [4], since the COVID-19 pandemic, housing affordability has declined, with more households, especially renters and mortgage owners, spending over 30% of their income on housing. The high cost of building materials inhibits housing affordability. Sandcrete blocks, made from sand, cement, and water, account for about 22% of building expenses for load-bearing and non-load-bearing walls and depend on factors like cost and climate [5]. Rising demand for sandcrete blocks increases interest in local materials like Rice Husk Ash (RHA) and Laterite Interlocking Bricks [6]. Nigeria's Raw Materials Research and Development Council cites local materials like laterite soil and rice husk ash as affordable, sustainable options [6, 7]. Studies have looked at using rice husk ash to partially replace cement in hollow sandcrete blocks and laterite bricks for cost savings [8]. This review discusses properties and performance of sandcrete block blended with RHA and laterite bricks blended with cement, amongst are water absorption, strength, and durability.

2. Rice Husk Ash (RHA)

Rice husk, a byproduct of rice processing, is removed during harvest. According to Jalam et al[8], up to 78% of rice and 22% of rice husk are produced during milling. It is abundant in Asia and parts of Africa due to favorable conditions [9]. (RHA) is formed by burning rice husk, which increases silica reactivity at a temperature of 700°C [10,11,12]. RHA's chemical, morphological, and physical properties have been studied.

2.1 Physical, Chemical Composition and Morphology of RHA

The physical properties of rice husk ash (RHA) are influenced by various processing factors such as combustion method, temperature, duration, separation, and grinding, all of which collectively determine its effectiveness as a pozzolanic material [10]. The chemical composition of RHA after cooling is provided in Table 1. X-ray fluorescence analysis was used to determine the chemical composition. The silica content in the rice husk ash (RHA) is higher, averaging 82% among other chemical constituents, as indicated by researchers. This suggests a high potential reactivity of RHA as a pozzolanic or cementing agent [11,12]. The microstructure of RHA depends on incineration, processing, and grinding methods [37,60,62]. Scanning Electron Microscope (SEM) images, (Fig. 1), typically show that RHA particles are irregularly shaped and have a micro-porous cellular surface. Meanwhile, the XRD pattern displays both crystalline and amorphous regions with peaks between 10° and 70°. The main phases include quartz (Q), cristobalite (Cr), and tridymite (Tr). Peaks at approximately 26° and 20° indicate quartz's crystalline structure, which is a stable silica phase [37]. This study focuses on evaluating the structural integrity of RHA as a partial replacement for cement in sandcrete blocks, focusing on characterization, water absorption, compressive strength, and durability for construction purposes. This approach aims to reduce material consumption, address RHA waste disposal, conserve resources, and lessen environmental harm from improper disposal and cement emissions.

Table 1. Chemical composition of RHA

Chemical Constituent	Authors	[11]	[12]	[13]	Average (%)
SiO_2		81.04	86.51	78.87	82.14
Al_22O_3		1.80	0.61	1.61	1.34
Fe_2O_3		1.01	0.60	2.20	1.27
CaO		1.60	0.71	1.33	1.21
MgO		2.25	1.53	2.11	1.96
Na_2O		0.45	0.02	0.03	0.17
K_2O		2.35	1.89	2.03	2.09
P2O5		5.26	4.20	9.87	6.44
Total SO_2 + Al_2O_3 + Fe_2O_3		83.85	87.74	82.86	84.75

2.2 Rice Husk Ask Sandcrete Block.

The most common use of sandcrete blocks is as walling material in buildings, drainage, and other masonry projects. Over 95 percent of all walling materials consists of sandcrete blocks [15,16]. These blocks are usually rectangular hollow units with typical dimensions of 450 × 225 × 225mm and 450 × 150 × 225mm [15]. The quality of sandcrete blocks depends on several factors, including the quality of materials, the manufacturing process, the curing period, water absorption, compressive strength, and durability, especially when used for load- bearing walls [16].

Figure 1: SEM image of RHA. Source: [14]

Researchers have shown that replacing cement with RHA can significantly lower building costs and reduce greenhouse gas emissions from cement production. According to Anifowose and Raheem [17], they compared conventional sandcrete blocks with RHA sandcrete blocks, focusing on specific gravity and compressive strength. [18] studied using rice husk ash to create low- cost sandcrete blocks in Nigeria. The study concluded that compressive strength increased over time during curing and decreased as RHA percentage increased there by, recommending 20% RHA as the optimal cement replacement level. Additionally, [18] suggested replacing 17. 5% of cement with RHA for sandcrete blocks used as building units. [18] also recommended 30% RHA for cement replacement in the production of hollow sandcrete blocks in Nigeria. Meanwhile, Mayowa et al., [19] concluded that 10% rice husk ash replacement met the strength requirements for building construction and reduced costs by 5.3%. Conversely, in corresponding [20] recommends an optimal cement replacement level of 15-20% with RHA in sandcrete blocks. Furthermore, [21] evaluated how RHA impacts the water absorption properties of the blocks. Results indicate that water absorption increased with higher RHA content, correlating with increased porosity. In addition, [22] examined the water absorption coefficient of sandcrete blocks incorporating wastepaper ash (WPA) as a partial cement substitute. The density ranged from 1682 to 1872 kg/m ^{3}, while water absorption varied between 27.49 to 27.04 %, with significant values ($p < 0.05$) (Fig. 2). Whereas [23] investigated the cost benefits of using agricultural waste materials like RHA. Their results showed that the costs of RHA in sandcrete blocks are based on actual material prices and volumes, leading to approximately 24% savings in total material costs. This paper will focus on using 15% Rice Husk Ash replacement of cement based on previous research, while comparing RHASB with that of conventional Blocks in terms of strength, water absorption, density and durability.

3. Laterite

Laterite consists of residual sedimentary rocks that are reddish or brownish, soft, highly porous, and have worm-like structures. According to [24], Laterite is locally available, and its use in civil engineering, such as road construction and landfilling. Its primary industrial use is as a filling material. Since it is abundant, expanding its application could reduce building costs and promote affordable housing [25]. Additionally, laterite can serve as an eco-friendly, cost-effective fine aggregate for brick manufacturing, which can decrease block costs and support building development in Nigeria [26].

3.1 Application and use of Laterite.

According to [34] Laterite interlocking bricks, is mortar-free with dry-stack grooves, enable quick, cost-effective, eco-friendly dry stacking for low-rise load-bearing walls, suitable for two stories in residential, educational, and community buildings without cement mortar. [35].

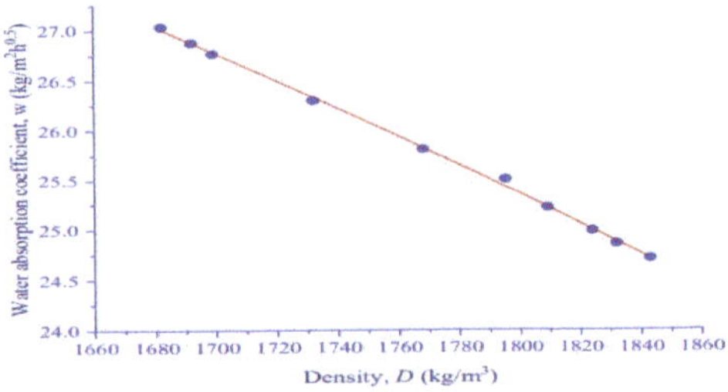

Fig. 2: Water absorption coefficient on density of the block samples [22]

3.2 Chemical properties of Laterite

Many researchers [38,39,49] state that laterites mainly contain iron, aluminum, silica, titanium, and water, with low levels of alkali and alkaline earth metals. They include water and Al_2O_3 (50-60%), the most common element. Iron and quartz levels vary inversely with alumina. TiO_2 is minor; oxides make up over 70%, sometimes around 92±5%. [24,26] suggested that laterite soil is slightly acidic, rich in iron, and amorphous, forming cementitious phases in alkaline media. According to [41], the main strategies to enhance the characteristics of laterite soil is to use an effective soil stabilizer (i.e., chemical stabilization). Previous studies have demonstrated that the strength of laterite soil has been significantly increased through the use of effective soil stabilizers in various applications.

3.3 Laterite Interlocking Bricks

The choice of walling material depends on cost, availability, durability, aesthetics, and climate [27]. Laterite bricks can be stabilized with small amounts of cement, lime, or pozzolans, and help reuse waste like rice husk ash and fly ash. They offer excellent fire and heat resistance, ideal for hot climates with interlock mortar free shown Fig 3. Lateritic bricks are durable and affordable. Several researchers [26,27,28,29] have studied Laterite Interlocking Bricks' mechanical properties, including water absorption, density, compressive strength, and durability.

3.4 Water Absorption of LIB:

Water absorption is a critical parameter for Interlock Compressed Earthen Bricks because it directly affects their durability, weather resistance, and overall performance. Generally, the less water the material absorbs, the better its quality and performance [28]. Low water absorption is preferred because it shows the material is less prone to moisture ingress, which reduces the risk of damage such as cracking, efflorescence, and freeze-thaw issues [29]. In construction materials like bricks, too much water absorption can cause structural weaknesses and aesthetic problems over time. For interlock compressed earthen bricks, only about 4% water absorption is notable, while previous studies reported a minimum of 13.20% [30]. According to [31], low absorption is due to decreased porosity from the compaction pressure during production. The compaction process reduces voids within the material, making it less permeable. Additionally, cement helps lower porosity, further increasing ICEBs' resistance to water absorption [32].

Materials Research Forum LLC
https://doi.org/10.21741/9781644904053-16

Fig 3: showing image of Laterite Interlocking Bricks.

Whereas [33] Investigate the water absorption of interlocking compressed earth brick (ICEB) units and beam bricks. ICEB units absorb water at 15.09% and 17.18%, with the half brick at 15.87% and beam brick at 13.20%. ICEBs absorb more water than normal bricks but stay below the maximum rate of 21%.

3.5 Compressive strength of Laterite Interlocking Bricks:

The mechanical properties of Laterite bricks are essential. [27] examined a comparative study of cement and lime stabilized lateritic interlocking blocks produced with laterite samples. The Nigeria Building and Road Research Institute (NBRRI, 2006) specified that the minimum 28-day strength should not be lower than 2 N/mm^2. Only cement-stabilized blocks satisfied this requirement at 10% and above stabilization levels. It was concluded that cement stabilized interlocking blocks are of better quality and cheaper than those stabilized with lime. Meanwhile, [33] used laterite as a partial replacement for fine aggregate at different percentages in interlocking blocks. The results revealed that the interlocking bricks exceeded the NBRRI standard.

3.6 Durability of LIB

According to Mohammadi et al [33,36], 28-day abrasive test results for cement and lime stabilized bricks are presented in Tables 2. The results showed that resistance to abrasion increases with both stabilizers. Unstabilized laterite interlocking blocks (the control) had a high percentage of material abraded, indicating stabilization is essential to improve durability. Cement-stabilized blocks performed better than lime-stabilized ones, reflecting higher durability. [33] Further observed that the cost of stabilizing with cement is cheaper than the cost of stabilizing with lime. This is due to the high cost of lime per bag, [33] Recommends 10% cement stabilization to meet the minimum strength. However, despite prior research on Laterite Interlocking bricks, these units haven't been used for model buildings, nor has their cost per unit meter been studied.

Table 2: Showing results of abrasiveness test for cement and lime stabilized lateritic bricks [36]

Cement Stabilization(%)	**0**	**5**	**10**	**15**	**20**	**25**
Average abraded away(%)	1.26	0.61	0.52	0.39	0.18	0.11
Lime Stabilization(%)	0	5	10	15	20	25
Average abraded away(%)	1.26	0.98	0.86	0.78	0.52	0.35

4. Conclusion

- This study concludes that RHA sandcrete blocks and LIBs can be viable for affordable housing in Nigeria, depending on context and needs. Using local materials and innovative tech can lower costs, speed up construction, and improve sustainability [25].
- RHA sandcrete blocks and LIBs serve as environmentally sustainable alternatives to conventional building materials, thereby decreasing the ecological footprint of construction endeavours. The use of indigenous waste materials such as rice husk ash and laterite further encourages sustainable construction practices and reduces waste generation [32].
- The utilisation of local materials and innovative technologies in RHA sandcrete blocks and LIBs can significantly decrease construction expenses. Additionally, LIBs, featuring their interlocking design, facilitate expedited building processes, thereby reducing both construction duration and labor costs [32].

REFERENCES

[1] J.B. Effiong and M.A. Ewah. Examination of Valuation Variance in the Use of Cost Method of Valuation in Calabar Metropolis, International Journal of Civil Engineering, Construction and Estate Management, 13,2 (2025) 1-15.

[2] E., Arinze, E.U., Reuben, Okereke., and C.E., Eucharia. Evaluation of The Role of Housing in National Development. International Journal of Progressive Research In Science And Engineering. 3, (2022) 14-21.

[3] L.C. Enekwachi-Akpa. Affordable Housing Solutions, International Journal of Civil Engineering, Construction and Estate Management, 12 (2024) 71-87

[4] S., Lee, (2023). House prices, homeownership, and household consumption: Evidence from household panel data in Korea, Economic Modelling. 126, (2023) 106 - 355.ISSN 0264-9993,https://doi.org/10.1016/j.econmod.2023.106355.

[5] E., Dorozhkina. Architectural structures for the formation of vertical landscaping of buildings. IOP Conference Series: Materials Science and Engineering. 962 (2020) 42005. 10.1088/1757-899X/962/4/042005.

[6] R.P. Amit S., Aniket R.G., Amit A,P., Onkar and N.F Uddhav. Compressive Strength And Other Properties Of Lateritic Interlocking Blocks Stabilized With Fly Ash And Cement. International Research Journal of Engineering and Technology (IRJET). 5. (2018) 1430-1434.

[7] S. Mushanga, O., Oloke, D., and D. Olukanni, D, A Review of Sustainable Housing Preferences and Affordability. IOP CONFERENCE SERIES: Earth and Environmental Science. 1342, (2024) 12030. doi:10.1088/1755-1315/1342/1/012030.

[8] U.A., Jalam, A., Jalam, S. I. Abdurrahman, O. Job. Cost Evaluation of Utilising Building Materials Derived from Agricultural Waste as Sustainable Materials for Lightweight Construction 16,(2016) 673 - 685.

[9] G.U. Fayomi, S.E. Mini, O.S.I., Fayomi, A. Ayoola, A.A. Perspective on Environmental CO2 emission and energy factor in cement industry. International Conference on Energy and Sustainable Environment. 331, (2019) 12-35. Doi: 10.1088/1755-1315/331/1/012035.

[10] A.A. Raheem, and M. Kareem. Chemical Composition and Physical Characteristics of Rice Husk Ash Blended Cement. International Journal of Engineering Research in Africa. 32, (2017) 25 - 35. 10.4028/www.scientific.net/JERA.32.25.

[11]A. Singh and B. Singh. Characterization of rice husk ash as obtained from industrial source. Journal of Sustainable Cement-Based Materials, 10(2020)193-212. https://doi.org/10.1080/21650373.2020.1789010 18

[12]S. K. Das, J. Mishra, S. M. Mustakim, A. Adesina, C. R. Kaze, and D. Das. Sustainable utilization of ultrafine rice husk ash in alkali activated concrete: Characterization and performance evaluation. Journal of Sustainable Cement Based Materials, 11,2 (2022) 100-112.

[13]D. Pranowo, A. A. Nuryono, M. R. Jumina, and F. M. C. S. Application of Silica Extracted From Rice Husk Ash For The Encapsulation Of Afb1 Antibody As A Matrix In Immunoaffinity Columns. JSM Mycotoxins, 67, 2 (2017) 77-83.

[14]K. Onyelowe, I. Obianyo, P. Onwualu, C. Chizirim. Morphology and mineralogy of rice husk ash treated soil for green and sustainable landfill liner construction.Cleaner Materials. 1. (2021) 1-12. 10.1016/j.clema.2021.100007.

[15]A. A. Raheem, and M. A. Anifowose. Effect of Ash Fineness and Content on Consistency and Setting Time of RHA-blended-cement. Materials Today: Proceedings 86, (2023) 18-23. https://doi.org/10.1016/j.matpr.2023.02.054

[16]NIS 87. "Nigerian Industrial Standard Standard for Sandcrete Blocks". Standard Organization of Nigeria (SON). Lagos, Nigeria. 2007.

[17]A.Mukaila and A.A Raheem. Review on Application of Rice Husk Ash for Production of Hollow Sandcrete Blocks: Towards Sustainable Construction. International Journal of Engineering Research in Africa. 72, (2024) 97-116. 10.4028/p- Zqj5YL.

[18]M. Abubakar and V Omotoriogun. Quality Assessment of Commercial Sandcrete Blocks in Minna Metropolis, Niger State. Nigerian Journal of Technology. 41. (2022). 222 -228. Doi:10.4314/njt.v41i2.3.

[19]S. Mayooran S. Ragavan and N. Sathiparan. Comparative Study On Open Air Burnt Low-And High-Carbon Rice Husk Ash As Partial Cement Replacement In Cement Block Production. Journal of Building Engineering, 13, (2017)137-145.

[20]H. B. Mahmud, S. Bahri, Y. W. Yee, Y. Y. Yeap. Effect of rice husk ash on the strength and durability of high strength performance sandcrete. World Acad. Sci. Eng. Technol. 10, 3 (2016) 375–380.

[21]S.P. Sangeetha. Rice husk ash sandcrete block as low cost building material. International Journal of Engineering Research and Application, 6 (2016) 46-49.

[22]O.O Samson M. Anifowose, S. Bello, Z. Giwa, and U. Wilson. Strength Properties and Microstructure of Sandcrete Blocks Incorporated with Maize Straws. Advanced Engineering Forum. 46. (2022) 71-78. 10.4028/p-kgoezq.

[23]D. Ali, U. A, Jalam and M. Balteh. Pozzolanic Characterization of Rice-Husk-Ash from Two Varieties of Rice in Gombe: A Strategy for Achieving Low-Cost Housing. International Journal of Innovative Science and Research Technology. 7, 11 (2022) 767-772.

[24]T.K. Prasad and G.R. Parthasarathy, "Laterite and Laterization – A Geomorphological Review".International Journal of Science and Research(IJSR), Issue 4. 7, (2016) 578-583.

[25]M. A. Muraleed haranetal. "Feasibility Study on Brick Using Lateritic Soil" International Journal of Engineering Science Invention Research & Development. 8. (2017) 19598 - 19601

[26] E., Ogunleye. The effects of laterite type on compressive and flexural strengths of Concrete Utilizing Laterite-Sand Fine Aggregate. Global Journal of Engineering and Technology Advances. 16. (2023) 180 - 191. Doi:10.30574/gjeta.2023.16.2.0119.

[27] O. Olopade, A., Oladimimu, K., Bolorunduro, A., Adetukasi. Comparative Strength Analysis of Laterite as Partial Replacement of Sand in Hollow Blocks and Interlocking Bricks Production. 6, (2022) 37.

[28] ASTM International, "Standard Test Methods For Particle-Size Distribution (Gradation) of Soils Using Sieve Analysis," "ASTM D6913-04R2009," 4, (2004) 1–35.

[29] J. D. Sitton, Y. Zeinali, W. H. Heidarian, and B. A. Story. "Effect of Mix Design On Compressed Earth Block Strength," Construction and Building Materials, 158,(2018)124– 131.

[30] D. Sinkhonde, R. O. Onchiri, W. O. Oyawa, and J. N. Mwero. "Durability and Water Absorption Behaviour Of Rubberised Concrete Incorporating Burnt Clay Brick Powder," Cleaner Materials, 7. (2022) 8565 Article ID 100084.

[31] T. E. Elahi, A. R. Shahriar, and M. S. Islam, "Engineering Characteristics Of Compressed Earth Blocks Stabilized With Cement And Fly Ash," Construction and Building Materials, 277, (2021) 122-367

[32] A. Yaras, M. Sutcu, O. Gencel, and E. Erdogmus, (2019)"Use of Carbonation Sludge in Clay Based Building Materials Processing For Eco-Friendly, Lightweight And Thermal Insulation," Construction and Building Materials, 224, 57–65.

[33] B. H. A. Bakar, S. Saari, and N. A. Surip, "Water Absorption Characteristic of Interlocking Compressed Earth Brick Units," in Proceedings of the International Conference of Global Network for Innovative Technology and Awam International Conference in Civil Engineering (IGNITE-AICCE'17): Sustainable Technology and Practice for Infrastructure and Community Resilience, (2017) 1892.

[34] M., Mohammadi, , T. G., Atsbha and Y. Chen. Comparison and Design of Dry- Stack Blocks with High Thermal Resistance for Exterior Walls of Sustainable Buildings in Cold Climates. Sustainability, 17,4(2025)1393. https://doi.org/10.3390/su17041393.

[35] J. Ayanlola, A. Adesope, L. Balogun, M.. Ogunjimi. Production and Evaluation of the Durability of Laterite, Interlocking Paver Block Incorporating Glass Powder. African Journal of Advances in Science and Technology Research. 19,(2025) 181-192. 10.62154/ajastr.2025.019.01020.

[36] A.A. Raheem, O., Bello, B., Tech and B.T Makinde. A Comparative Study of Cement and Lime Stabilized Lateritic Interlocking Blocks. Pac. J. Sci. Technol.. 11 (2012).

[37] J., Rego, N., Nepomuceno, A.; Figueiredo, E.; Hasparyk. Microstructure of Cement Pastes With Residual Rice Husk Ash Of Low Amorphous Silica Content. Constr. Build. Mater. 80, (2015) 56–68. [CrossRef]

[38] S.T. Owusu, C.W.W. Ng, C. Zhou, (2024). An Experimental Study of Stress-Dependent Water Retention Behaviour Of Two Lateritic Clays With Different Minerals Eng. Geol., 329. Article 107398, 10.1016/j.enggeo.2023.107398

[39] R.C. Kaze, G. L Lecomte-Nana, K.C. Elie , P. S. Camacho , A.S. Yorkshire. Microstructure and Engineering Properties of Fe2o3(Feo)-AL2o3-Sio2 based Geopolymer Composites.J.Clean. Prod., 199 ,(2018) 849-859. 10.1016/j.jclepro.2018.07.171

[40] H. Sadeghi, H. Nasiri. Hysteresis of Soil Water Retention And Shrinkage BehaviourFor Various Salt Concentrations, Geotechnique Letters, 11,1 (2021) 21-29. ISSN 2045-2543, https://doi.org/10.1680/jgele.20.00047.

[41] A.M.A., Ezreig, M.A.M Ismail, and K.I.A., Ehwailat. A State of Review: Challenges And Techniques of Laterite Soil Stabilisation Using Chemical, Economical, And Eco-Friendly Materials. Innov. Infrastruct. Solut. 7, (2022) 229. https://doi.org/10.1007/s41062-022- 00821-z

Advances in Cement and Concrete Research - ICACCR 2025 Materials Research Forum LLC
Materials Research Proceedings 63 (2026) 152-159 https://doi.org/10.21741/9781644904053-17

Effect of Restraints on Deterioration-induced Deformation in RC Structures Undergoing both Steel Corrosion and Alkali-Silica Reaction

Mike OTIENO[1,2,a*], Mebrahetu BEYENE[2,b]

[1]School of Physics, Engineering and Computer Science, University of Hertfordshire, Hatfield, United Kingdom

[2]School of Civil and Environmental Engineering, University of the Witwatersrand, Johannesburg, South Africa

[a]m.otieno@herts.ac.uk, [b]1841997@students.wits.ac.za

Keywords: Alkali-Silica Reaction, Steel Corrosion, Durability, Restraint, Reinforced Concrete Deterioration, Crack Formation

Abstract: The durability of reinforced concrete (RC) structures is compromised by various deterioration mechanisms, most notably alkali-silica reaction (ASR) and chloride-induced steel corrosion. While some studies have focused on the combined effects these deterioration mechanisms, it is necessary to understand the impact of restraint on the deterioration-induced deformations. Furthermore, the presence of restraints closely simulates most in-service RC structural members. This study presents an experimental investigation into the degradation of RC due to ASR and exposure to chloride environments, with focus on the influence of two-dimensional (2-D) restraint on deterioration-induced deformation. Six slab specimens (350 × 350 × 50 mm) were cast and subjected to exposure environments designed to accelerate the deterioration mechanisms independently and in combined situations (ASR-only, steel corrosion-only, and combined ASR and steel corrosion, respectively). For each exposure and deterioration set-up, both restrained and unrestrained conditions were used. The specimens were monitored for steel corrosion and expansion during the experimental period. Preliminary results indicate that combined ASR and corrosion resulted in synergistic and extended damage characterised by widespread map cracking of the concrete and increased steel corrosion levels particularly in unrestrained specimens. In addition to corroborating previous findings that combined deterioration processes exacerbate the risk of loss of serviceability and structural integrity, the results demonstrate that restraints can slow down the rate of deterioration and impact the geometric response of a RC member to deterioration-induced stresses.

Introduction

Durability is a crucial aspect to consider when designing and constructing reinforced concrete (RC) structures, as it directly impacts their in-service longevity and performance. Ensuring the durability of RC structures helps to minimize maintenance and repair costs, which can be significant over the lifespan of a structure. RC with high structural integrity not only improves safety and reliability but also prolongs the service life, reducing downtime and overall lifecycle expenses [1] [2]. At the design stage, key factors, concrete quality, cover depth and the exposure environment, are typically emphasized when assessing the durability of RC structures [3] [4] [5]. Several studies have indicated that the main deterioration mechanisms in RC structures are chloride-induced corrosion and ASR.

Corrosion of reinforcing steel in concrete is a worldwide problem facing both inland and coastal RC structures, with the latter being more affected [6] and demanding huge amounts of resources in the maintenance and repair of the structures. Specifically, the leading cause of durability failure in RC structures is chloride-induced corrosion [7] [8] [9] which involves ingress of chloride ions leading to the corrosion of the embedded reinforcing steel. It results in a reduction of the load-bearing capacity of the RC structure, significantly compromising its structural integrity. RC structures located near coastal areas are particularly susceptible to chloride-induced corrosion [7]. A common method used to reduce this risk is provision of sufficient concrete cover depth to delay the time to attainment of chloride threshold [10] [11].

In the case of ASR, the dissolved alkalis in the concrete pores complex with reactive silica in aggregates, if present, to form ASR gel [12]. In the presence of moisture, the gel swells, generating expansive internal stresses that can lead to extensive concrete cracking [13]. The internal pressure in the RC structure leads to cracking ultimately impacting its durability.

While concrete quality and exposure environment parameters significantly influence long-term performance, an often-overlooked aspect in durability studies is the effect of structural restraint. Under unrestrained conditions, concrete can undergo relative deformation (i.e., displacement or rotation) without necessarily developing cracks [14]. However, conditions of zero internal and/or external restraint are not practical in structural systems. Some degree of restraint is always present, limiting the ability of concrete to deform freely. Internal restraint typically arises from the presence of reinforcing steel, differential temperature gradients within the concrete mass, internal chemical reactions such as ASR or steel corrosion, and the rigidity of coarse aggregates. External restraint, on the other hand, may be imposed by adjacent structural elements, differential drying between bonded concrete surfaces, or rigid connections between structural members [14] [15]. Although restraint is often essential to ensure structural integrity, it can also induce undesirable residual tensile stresses that can result in cracking, which in turn compromise both the serviceability and durability of RC structures [16].

Experimental set-up

Coarse and fine aggregates derived from andesite and quartzite were used in the experiment. Scanning Electron Microscopy (SEM) confirmed that these aggregates contain varying amounts of reactive silica – see Table 1 and Figure 1. The aggregate with the higher reactive silica content was selected for preparing the test specimens. A summary of the properties of the aggregates is presented in Table 2.

A total of six slab specimens were prepared for this study, each uniquely defined by its restraint condition and exposure environment. The selected testing methods, such as half-cell potential measurements, are non-destructive and can be performed repeatedly on the same specimens, thereby justifying the decision to limit the total number to six. Furthermore, techniques like petrographic analysis and scanning electron microscopy (SEM) require only small core samples, minimizing the need for additional specimens. The six slabs were evenly divided into two groups: three restrained and three unrestrained specimens.

Table 1. Aggregates used to make the test specimens

Specimen	Fine aggregate (crusher sand)	Coarse aggregate (9.5 mm)
ASR and corrosion specimens	Andesite	Quartzite
Restraining RC frame	Andesite	Andesite

Table 2. Selected aggregate properties

Type of aggregate	*Nominal size (mm)*	*Relative density*	*Compacted bulk density (kg/m^3)*	*Fineness modulus*	*Si content (%)*
Andesite	9.5	2.94	2500 – 2700	2.75	32.6
Quartzite	9.5	2.70	2700 - 2800	-	47.3

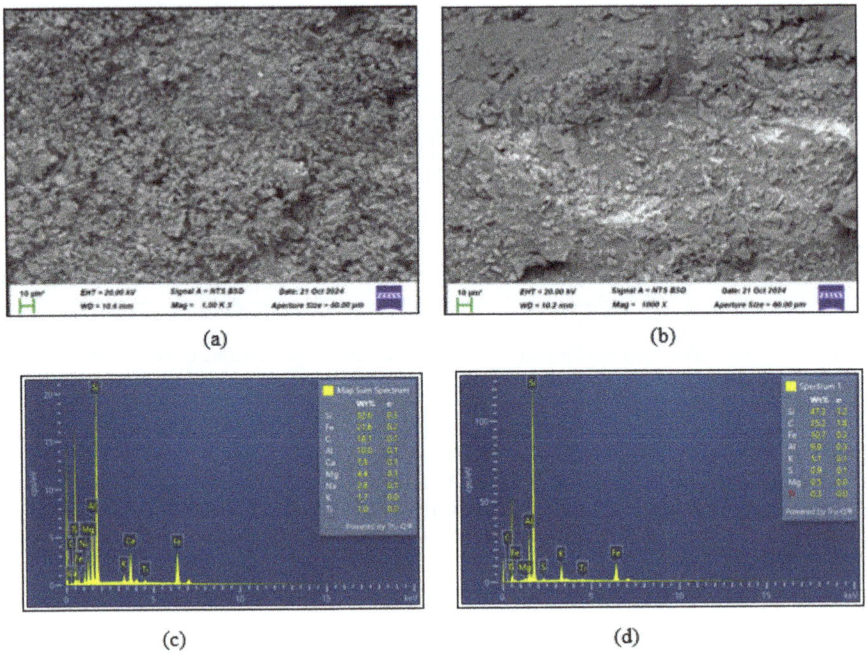

Figure 1. SEM imaging and elemental composition of aggregates, respectively:(a) & (c) for andesite, (b) & (d) for quartzite

Specimen preparation: A wooden mould was constructed for two sets of specimens – one for the restraining RC frame and the other for the experimental specimens. Details of the restraining frame and the test specimens are given in the sections below.

I. Restraining frame: a wooden formwork measuring 500 × 500 mm on the outside and 350 × 350 mm on the inside with a depth of 50 mm (see Figure 2(a)) was constructed for purposes of casting the RC restraining frames. Each RC restraining frame was reinforced with Y10 steel rebar. The use of Y10 reinforcing steel was considered adequate to resist any deformation resulting from ASR-induced expansion. To mitigate against corrosion in the frames, the reinforcing steel bars were epoxy-coated before casting. Three RC restraining frames were cast and cured for 28 days in water at 23 °C for 28 days before being used. After curing, the concrete surface was epoxy-coated to protect against moisture, chloride and alkali penetration later in the course of the experiments.

II. Test specimens: all six slab test specimens measured 350 × 350 mm and 50 mm in depth. Additionally, all test specimens were reinforced with three R8 steel bars, and each reinforcement was connected to a wire that extended out of the concrete, as shown in Figure 2(a) and (c). This setup is used for later half-cell potential measurements, as the embedded reinforcing steel is inaccessible. The formwork for the two groups of specimens (restrained and unrestrained) was prepared in two ways. The specimens that were to be restrained were cast inside the restraining frame described – see Figure 2(b), while a separate formwork was prepared for the unrestrained specimens – see Figure 2(d). The reinforcement details used for the unrestrained specimens are identical to those of the restrained specimens to

Materials Research Forum LLC
https://doi.org/10.21741/9781644904053-17

enable comparison between the two test conditions. Three unrestrained test specimens were cast inside the wooden frames and cured for 28 days in water at 23 °C.

Figure 2. Formwork setup and typical cast restrained (a & b) and unrestrained (c & d) specimens

Scanning Electron Microscopy setup: For the SEM analysis, a small section of the aggregates and concrete, measuring less than 9 mm in size, was used. Initially, a core with a diameter of 30 mm and a depth of 50 mm was drilled from each slab, then cut into smaller pieces using a precision saw to obtain suitable samples for SEM. The aggregate samples were used without cutting. Prior to imaging, the samples were mounted and placed in the SEM chamber, where vacuum evacuation was performed to remove all air and ensure appropriate imaging conditions. SEM analysis was then carried out by selecting three regions of interest on each sample. Observations were performed at various magnifications ranging from 100× to 2000×, allowing for detailed inspection of surface morphology and internal microcracks. At 1000× magnification, the crack widths were measured, and elemental composition analysis was performed using energy-dispersive X-ray spectroscopy (EDS).

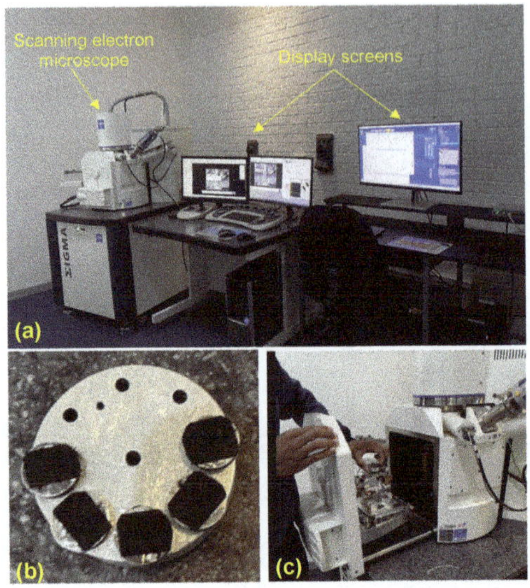

Figure 3: (a) SEM equipment (b) mounted specimens, (c) placement of specimen in SEM chamber

Results and Discussion

(i) *Half-cell potential assessment*

The corrosion probability of the embedded reinforcing steel was evaluated using two methods: indirect measurement through wires connected to the reinforcing steel and direct measurement following core extraction and exposure of the embedded steel. The latter method, involving direct half-cell potential measurements using a $Cu/CuSO_4$ electrode, was found to yield more reliable results. Specimens subjected to both ASR and chloride exposure demonstrated a synergistic effect, resulting in the most severe corrosion activity. Specifically, the ASR+SC-UR specimen recorded the most negative potential at –400 mV (see Figure 4), corresponding to a greater than 90% probability of corrosion. Similarly, SC-UR showed values ranging from –300 to –350 mV, also indicating high corrosion activity. The ASR+SC-R specimen exhibited values between –300 and –320 mV, placing it within the 50–90% corrosion probability range. In contrast, ASR-UR, ASR-R, and SC-R specimens showed potential values between –231 and –275 mV, falling into the uncertain zone where corrosion may or may not be actively occurring.

These results demonstrate that the corrosion activity of embedded reinforcing steel in concrete is significantly influenced by both the presence of restraint and the type of deterioration mechanism. Unrestrained specimens are more susceptible to corrosion, as evidenced by their consistently more negative half-cell potential values. Moreover, the combination of ASR and chloride-induced corrosion has a synergistic effect, substantially increasing the likelihood of corrosion, particularly under unrestrained conditions. The ASR+SC-UR specimens demonstrated the highest corrosion probability, exceeding 90%, highlighting the severity of combined deterioration mechanisms in the absence of restraint. Conversely, restrained specimens, while still affected, exhibited comparatively moderated corrosion activity. These findings underscore the critical importance of structural restraint and the need to consider the combined effects of deterioration mechanisms when assessing the durability and long-term performance of reinforced concrete structures.

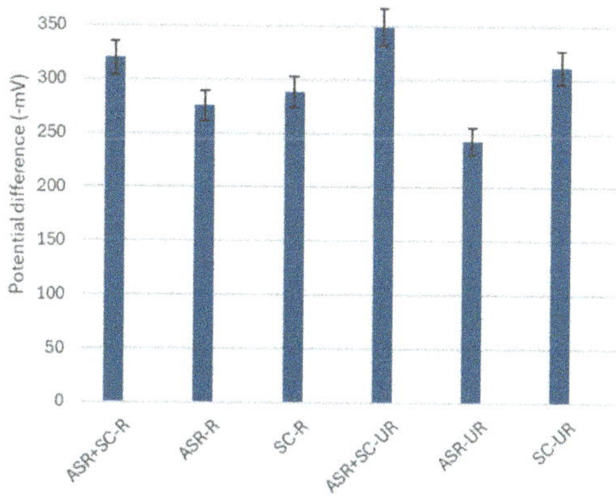

Figure 4. Average potential difference values of the specimens

(ii) *Scanning electron microscopy assessment*

SEM analysis revealed that the restrained sample showed lower expansion compared to the unrestrained counterpart. This confirmed that 2-D restraint was effective in limiting volumetric deformation, despite ongoing internal reactions. However, this mechanical constraint altered not only the extent but also the type and distribution of cracking. Petrographic analysis of ASR+SC-R showed map cracks, characteristic of ASR-induced deterioration. In contrast, ASR+SC-UR showed fewer cracks, but these were significantly wider with the largest measuring up to 22 µm (see Table 3), highlighting that the absence of restraint allowed free expansion and resulted in more severe, localized damage. The cracks patterns observed under SEM are shown in Figure 5.

This difference in cracking also influenced corrosion progression. Half-cell potential measurements recorded the most negative values in ASR+SC-UR, indicating a higher probability of active corrosion. The unrestrained condition likely facilitated greater crack openings, which in turn accelerated the ingress of chloride ions. This is consistent with the understanding that ASR-induced cracking can compromise the concrete matrix, enabling chlorides to reach the reinforcement more readily. In contrast, ASR+SC-R, though also exposed to the same solutions, showed slightly less negative half-cell potentials, suggesting that the reduced cracking from restraint helped slow chloride penetration and, consequently, corrosion.

Table 3. Surface crack size measured under ×1000 magnification

Sample ID	Crack size range (µm)
ASR + SC-R	7.9 – 8.6
ASR-R	6.2 – 11
SC-R	7.4 – 9.6
ASR + SC-UR	16 – 22.12
ASR-UR	6.9 – 8.0
SC-UR	4.4 – 5.2

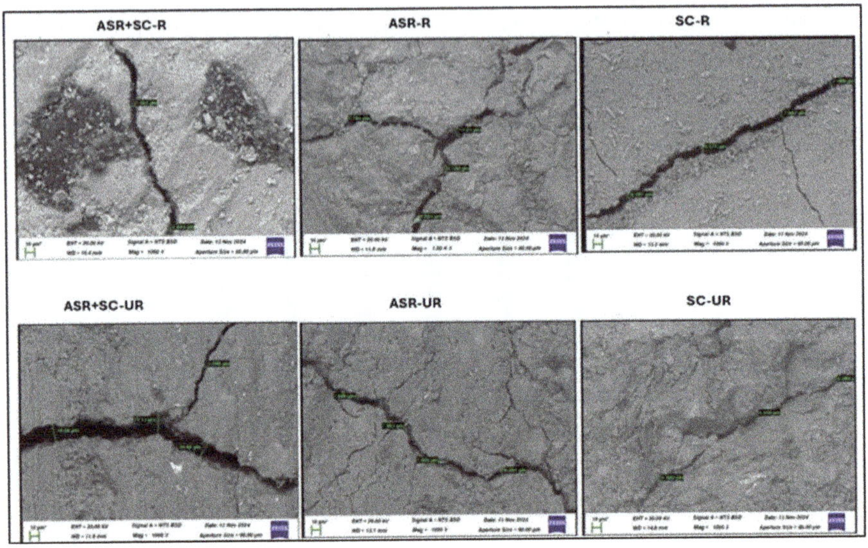

Figure 5. Crack patterns observed under a scanning electron microscope at ×1000 magnification

Conclusions

This study investigated the combined effects of ASR and chloride-induced steel corrosion on RC structures under restrained and unrestrained conditions. Six slab specimens were exposed to different deterioration mechanisms and tested using half-cell potential measurements, and scanning electron microscopy. From the results, the following conclusions can be stated:

I. Steel corrosion, assessed through half-cell potential, SEM, and petrography, was most severe in unrestrained, combined ASR+SC samples, with potential values exceeding –350 mV;

II. The presence of restraint significantly decreased deterioration due to both ASR and steel corrosion, limiting expansion, slowing crack formation, and as a result reducing the risk of corrosion since there is less cracking that allows chloride ion ingress;

III. Combined ASR and corrosion showed synergistic effects, causing more extensive damage than either mechanism alone, especially when unrestrained.

References

[1] G. L. Golewski, "Assessing of water absorption on concrete composites containing fly ash up to 30 % in regards to structures completely immersed in water," Case Studies in Construction Materials, vol. 19, 2023. https://doi.org/10.1016/j.cscm.2023.e02337

[2] S. Y. Feng Shi, T. M. Pham, R. Tuladhar and H. Hao, "Pullout and flexural performance of silane groups and hydrophilic groups grafted polypropylene fibre reinforced UHPC," Construction and Building Materials, vol. 277, 2021. https://doi.org/10.1016/j.conbuildmat.2021.122335

[3] S. Demis and V. G. Papadakis, "Durability design process of reinforced concrete structures - Service life estimation, problems and perspectives," Journal of Building Engineering, vol. 26, 2019. https://doi.org/10.1016/j.jobe.2019.100876

[4] M. B. Otieno, H.-D. Beushausen and M. G. Alexander, "Corrosion propagation in reinforced concrete structures - state of the art review and way forward," in roceedings of the 6th international conference on Concrete under severe conditions - Environment and loading , CONSEC'10, 2010. https://doi.org/10.1201/b10552-56
[5] A. Moore, H. Beushausen, M. Otieno, J. Ndawula and M. Alexander, "Oxygen availability and corrosion propagation in RC structures in the marine environment-Inferences from field and laboratory studies," Corrosion and Materials Degradation, vol. 3, pp. 363-375, 2022. https://doi.org/10.3390/cmd3030022
[6] Y. Zhang, B. M. Ayyub and J. F. Fung, "Projections of corrosion and deterioration of infrastructure in United States coasts under a changing climate," Resilient Cities and Structures, vol. 1, no. 1, pp. 98-109, 2022. https://doi.org/10.1016/j.rcns.2022.04.004
[7] C. S. Das, H. Zheng and J.-G. Dai, "A review of chloride-induced steel corrosion in coastal reinforced concrete structures: Influence of micro-climate," Ocean Engineering, vol. 325, 2025. https://doi.org/10.1016/j.oceaneng.2025.120794
[8] X. Wang, Y. Liu and S. Liu, "Effect of chloride-induced corrosion on bond performance of various steel fibers in cracked SFRC," Cement and Concrete Composites, vol. 140, 2023. https://doi.org/10.1016/j.cemconcomp.2023.105113
[9] C. Chalhoub, R. François and M. Carcasses, "Critical chloride threshold values as a function of cement type and steel surface condition," Cement and Concrete Research, vol. 134, 2020. https://doi.org/10.1016/j.cemconres.2020.106086
[10] L. Li, J. Li and C. Yang, "Theoretical approach for prediction of service life of RC pipe piles with original incomplete cracks in chloride-contaminated soils," Construction and Building Materials, vol. 228, 2019. https://doi.org/10.1016/j.conbuildmat.2019.116717
[11] Q. Liu, L. Sun, X. Zhu, L. Xu and G. Zhao, "Chloride transport in the reinforced concrete column under the marine environment: Distinguish the atmospheric, tidal-splash and submerged zones," Structures, vol. 39, 2022. https://doi.org/10.1016/j.istruc.2022.03.041
[12] I. Fernandes and M. Broekmans, "Alkali-Silica Reactions: An Overview. Part I," Metallography Microstructure and Analysis, vol. 2, no. 4, pp. 1-12, 2013. https://doi.org/10.1007/s13632-013-0085-5
[13] Y. Chen, X. Xiong, S. Shi, C. Liu, Z. Xu and Z. Ma, "Effect of glass powder on alkali-silica reaction mitigation for tunnel waste rock slag in concrete," Journal of Building Engineering, vol. 98, 2024. https://doi.org/10.1016/j.jobe.2024.111024
[14] M. A. D. Al-Gburi, "Restraint effects in early age concrete structures," Graphic production, Lulea, 2015.
[15] Martin Nilsson, M. Al-Gburi and J.-E. Jonassona, "Reducing Crack Risks Caused by Restraint in Young Concrete -A case study on Walls of Water Tanks," European Journal of Environmental and Civil Engineering, 2015.
[16] I. M. Puttipong Srimook, K. Shibuya, S. Tomita, G. Igarashi, Y. Hibino and Kazuo Yamada, "Evaluation of thermal crack width and crack spacing in massive reinforced concrete structures subject to external restraints using RBSM," Engineering Fracture Mechanics, vol. 274, 2022. https://doi.org/10.1016/j.engfracmech.2022.108800
[17] A. Leemann, M. Góra, B. L. and M. Heuberger, "Alkali silica reaction in concrete - Revealing the expansion mechanism by surface force measurements," Cement and Concrete Research, vol. 176, 2024. https://doi.org/10.1016/j.cemconres.2023.107392

Development and Characterization of Supplementary Cementitious Material from Phosphate and Sewage Sludge

Euricia MKHONTO[1,a*], Thandiwe SITHOLE[1,b] and Tebogo MASHIFANA[2,c]

[1]Department of Chemical Engineering, University of Johannesburg, P.O. Box 17011, Doornfontein 2088, South Africa

[2]Department of Chemical Engineering, College of Science, Engineering, and Technology, University of South Africa, C/o Christiaan de Wet & Pioneer Avenue, Florida Campus, Johannesburg 1710, South Africa

[a]euriciam@uj.ac.za, [b]nastassias@uj.ac.za, [c]mashitp@unisa.ac.za

Keywords: Supplementary Cementitious Material, Sewage Sludge, Sustainable Material, Waste Beneficiation

Abstract. The environmental impact associated with cement production highlights the need for alternative low-carbon binders that promote sustainability and circularity. There is limited research that investigates the combined use of sewage sludge (SS) and Phosphate sludge (P) as cement-free supplementary cementitious materials (SCM) that are normally discarded, while they contain pozzolanic elements. This study investigates the feasibility of developing composites with varied proportions of SS and P (40SS:60P, 50SS:50P, 60SS:40P) by evaluating their unconfined compressive strength (UCS) and environmental safety. The raw materials and the developed composite were characterized by their chemical composition using X-ray fluorescence (XRF), mineralogy via X-ray diffraction (XRD), microstructural analysis with a scanning electron microscope (SEM), and functional groups through Fourier-transform infrared spectroscopy (FTIR). The 60SS:40P ratio yielded the highest UCS of 0.89 MPa, with low to no metal leachability below the United States Environmental Protection Agency (USEPA). The analysis showed that the strength was influenced by the formation of C-A-F-S-H, C-S-H, and iron phosphate. The developed composite is suitable for application as backfilling and non-load-bearing blocks. The valorization of SS and P contributes to circularity and minimizes emissions from cement production while advancing SDG 12 (responsible consumption) and SDG 13 (climate action).

Introduction

Cement production accounts for approximately 8% of the world's carbon dioxide (CO_2) emissions from clinkers, a major contributor to climate change (Mirandi and Shirzad, 2025). Supplementary cementitious materials (SCM) have emerged as viable alternatives to lower CO_2 emissions while improving binder properties, such as durability, permeability, and mechanical properties (Nassiri et al., 2025). Rapid urbanization and industrialization generate a notable amount of Sewage sludge (SS) from wastewater treatment and phosphate sludge (P) metal pretreatment processes. These wastes are normally discarded into the landfills, posing a risk to groundwater pollution through heavy metal leaching, which then affects the people consuming it, leading to diseases. However, they contain a significant proportion of silica, alumina, calcium, phosphorus, and iron, which are essential constituent cementitious and pozzolanic reactions, with potential for SCM applications (Jhatial et al., 2023; Zhang et al., 2025).

Studies have shown that when the SS ash is mixed with cement or thermally treated, it exhibits pozzolanic reactivity evident through the formation of hydration products which are suitable for application in mortar, concrete, stabilized materials, and road construction (Chang et al., 2022;

Guo et al., 2022; Xia et al., 2023; Mojapelo et al., 2025). The incorporation of P waste often retards the hydration process, but it successfully produced controlled low-strength materials and ceramic bricks exhibiting good durability (Navarro-Blasco et al., 2015; Muliawan & Astutiningsih, 2018; Karoui et al., 2024; Fadili et al., 2025). Most studies focused on the individual use of SS and P with a heavy reliance on cement. However, there remains a research gap in understanding the behavior of mixing SS and P without cement as SCM. Particularly the reaction mechanism forming the binding products, the effect of mixing ratios on the UCS, and environmental performance are not sufficiently explored. Addressing these gaps presents a novel approach to multi-waste valorization, which supports the climate mitigation plan and the circular economy concept.

This study addresses this gap by investigating the feasibility of utilizing SS and P residue as blended SCM without cement. The composites were prepared at different SS:P ratios to determine the effect on UCS. The obtained composites were characterized by using X-ray fluorescence (XRF), X-ray diffractometer (XRD), Fourier-transform infrared spectra (FTIR), Scanning Electron Microscope (SEM), and Toxicity Characteristic Leaching Procedure (TCLP). The finding shows that the developed composite with a UCS of 0.89 MPa is suitable for applications such as backfilling and non-load-bearing blocks

Material preparation, mix design and methods of testing

XRF composition of the raw materials

Table 1 Chemical composition of P and SS

	SiO_2	Al_2O_3	CaO	P_2O_5	MgO	Fe_2O_3	ZnO
P	0.65	0.34	0.06	25.54	0.19	65.27	7.50
SS	36.59	18.41	20.33	7.30	1.31	6.02	0.46

The P was supplied by a local company that manufactures bolts, and the SS by a local wastewater treatment plant, both in South Africa. They are both residues filtered after leaching phosphorus. The materials were oven-dried for 24hrs at 105°C to remove the moisture. The P and SS residues were mixed with water at the proportions of 40SS: 60P, 50SS: 50P, and 60SS: 40P (wt.%). The material was mixed with water at a moisture content of 15%, which formed a paste and was cast into a 50x50x50 mm^3 mold. The hardened specimens were covered with foil and cured in the oven at 80°C for 5 days for gradual strength development while maintaining phase stability. The residue generated from P and SS leaching contained a high proportion of silica, iron, calcium, phosphorus, and alumina, as presented in Table 1. These are elements required for stabilization and development of a composite for application in building and construction (Swierczek et al., 2018).

Unconfined compressive strength testing

The UCS was determined after 5 days using the Matest C104 2.6.3 3000kN at a rate of 0.5 MPa/sec and 15% stop load was applied without shock until the specimen failed then record the maximum load. The UCS was measured on an average of three specimens for consistency and confidence in the results.

Characterization

The elemental composition of the P and SS residue was analyzed by X-ray fluorescence (XRF Rigaku ZSX Primus II). X-ray diffractometer (XRD Ultima IV 40 kV) was used to determine the mineralogical phases. The functional groups of the residue composites were determined by Fourier-transform infrared spectra (FTIR Thermo Scientific, Nicolet iS50 FT-IR). The morphology of the residue composite was studied using a Scanning Electron Microscope (High-resolution SEM; Joel JSM-7900F). The SEM was equipped with AZtec software for Energy-dispersive X-ray spectroscopy (EDS), Elemental mapping, and phase finder. They were conducted on the raw materials and on the composites after testing the UCS at 5 days.

Toxicity Characteristic Leaching Procedure (TCLP)
The test was conducted as per the United States Environmental Protection (USEPA, 1992) method. An extract solution was prepared using NaOH and CH_3OOH acid with deionized water at pH 4.93±0.005. Optimum composite and extract solution was mixed at 1:20 and the mixture was shaken for 24 h at 25°C±3. The solids were separated from the liquid by filtering with 0.45μm filter paper, and the leachate was further analyzed using ICP-OES.

Results and discussion
P and SS leaching residue composite.

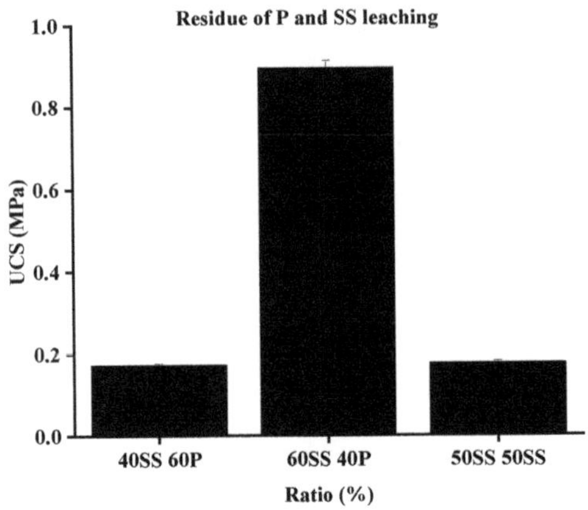

Figure 1 Composite from SS and P-waste residue

Different content of P waste and SS were mixed for stabilization at 40%, 50%, and 60% by weight and the results are presented in Figure 1. The strength of the composites increased with the increasing content/proportion of sewage sludge. The UCS increased by 4.65% when the SS content was increased from 40 to 50% by weight addition. According to ANOVA single-way analysis ($p - value = 0.75 > 0.05$) the addition of SS from 40 to 50% did not have any significant effect on strength development. A significant ($p - value = 0.0028 < 0.05$) change in UCS was observed when SS addition increased by 10% to 60% by weight, with UCS increasing from 0.18MPa to 0.89MPa. The high content of SS improved the strength as it contained a high proportion of calcium, which is responsible for strength development in composite acting as a binder (Ngayakamo et al., 2020). The developed binder is suitable for application as backfilling and non-load-bearing blocks.

Mineralogy of the developed composite P-waste and SS leaching residue
The mineralogy of the P-waste and SS sludge residue mixture from leaching is presented in Figure 2 on the diffraction pattern. The SS showed a crystalline broad diffraction peak between 25° and 30° 2(θ) which was identified as quartz (Q) and a peak of 36° 2(θ) corresponding to apatite containing calcium oxide responsible for strength development. Vivianite was detected at 22° 2(θ). The P-waste exhibits a broad diffraction peak at 20° 2(θ) which was identified as iron phosphate

with all the other peaks dominated by iron and phosphate, the same trend peaks were also observed by Goj et al. (2021) and Maarouf et al. (2021). When the residues are mixed the produced spectrum shows the peak identity of both samples. At 40SS: 60P phases with iron phosphate hydrate and silicon oxide are reported. Calcium iron phosphate broad diffraction peak between 19° and 25° 2(θ) was formed for all the mixed residues. New phases were formed at peak 31° 2(θ) namely parabutlerite, phosphoderite, and magnesium hydrate phosphate for 40, 50, and 60 % SS added in the mixture.

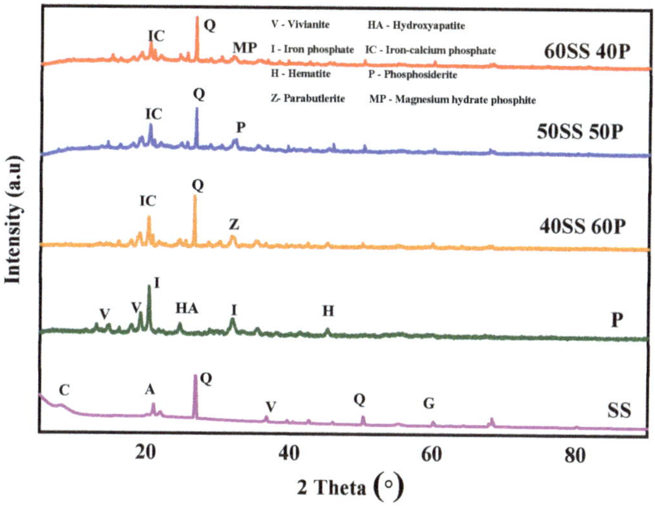

Figure 2 XRD analysis of SS and P-waste residue and composite

Microstructure from SEM of the developed composite P-waste and SS leaching residue

The morphology of the composite developed using leaching residue from P-waste and SS was studied using SEM and the obtained images are presented in Figure 3. The composite which contained 50% SS and 50% P by weight exhibited a structure embedded with spherical and needle-like soft particles. The 60SS: 40P composite by weight showed a compact dense particle embedded with a spherical particle. The 40SS: 60P composite by weight exhibited dispersed particles. The EDS micrographs in Figure 4 showed that the composite is composed of Si, Ca, Fe, Al, and P bonded to O. The Ca: Si and Al: Ca ratios are used as an indication for the formation of hydration products responsible for strength development (Mkhonto et al., 2024). The Ca: Si = 0.5 and Al: Ca = 0.4 are responsible for the reported strength gain. The EDS mapping in Figure 5 confirmed the embedded with a spherical particle on 60SS: 40P composite as Fe which confirms the Fe contained the P-waste residue. The even distribution of the Fe and P shows the reaction between the two elements. The is a high concentration and evenly distribution of Si and Al with traces of Ca, which confirms the reaction of cementitious material C-A-S-H exhibiting the strength reported (Eu et al., 2025).

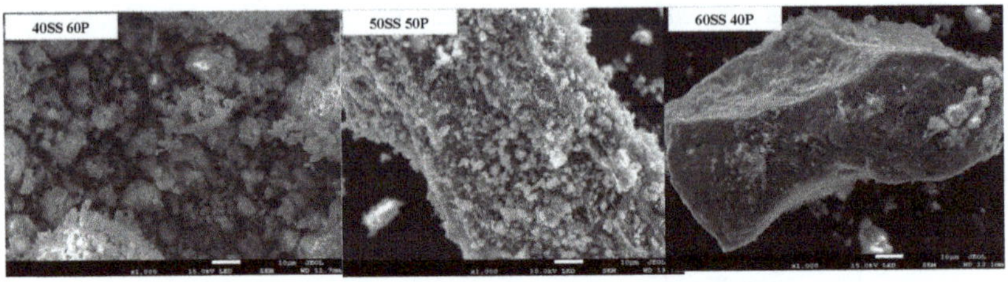

Figure 3 Microstructure of SS and P-waste residue composite

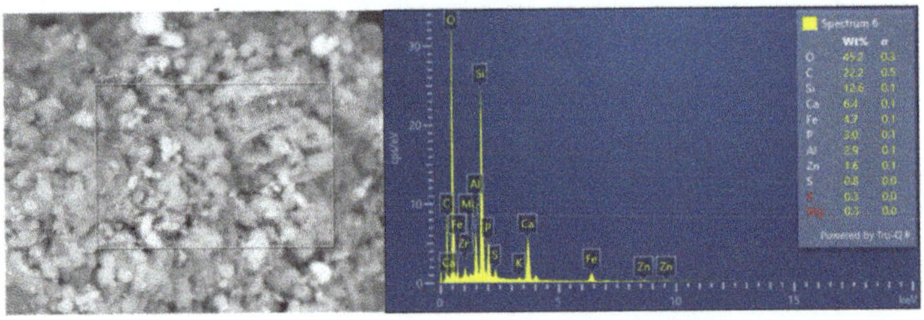

Figure 4 EDS for 60SS 40P composite

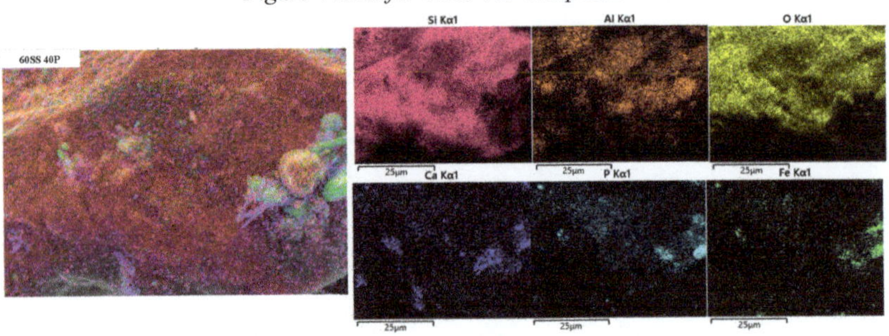

Figure 5 Elemental mapping of 60SS 40P composite

FTIR analysis of the developed composite P and SS leaching residue

The FTIR spectrum of the waste residue composite is presented in Figure 6. The presence of broad absorption bands between 3400-3435 is characterized by the stretching of hydroxyl vibration of water molecules (Maarouf et al., 2021). P-H stretching vibration was absorbed at a wavelength between 2343-2356cm-1, showing the continuous reactivity of phosphorus reaming in both residues (Pu et al., 2021). The vibrational bands reported in 2076 and 2070cm-1 were assigned to the stretching of C-O (Gupta et al., 2016). The absorption band at 1619, 1634, and 1640cm-1 for the different combinations of the residue is assigned to the bending vibration of the H-O-H bonds (Yusuf, 2023). C-H bonds were identified at absorption bands 1399, 1400, and 1384cm-1, these bonds are the ones responsible for the strength development reported for the composite residue (Yang et al., 2018). The vibration band at wavelength 1027-1030cm-1 is ascribed to the stretching of Si-O (Santos et al., 2021). The bands at 779cm-1, and 780cm-1 correspond to Si-O and Al-O bonds (Horgnies et al., 2013). The vibration absorption band at 657cm-1, 675cm-1, 579cm-1,

595cm-1, and 467cm-1 corresponds to S-O stretching (Yusuf, 2023). All the bands representing hydrated phases for a cementitious material were detected with FTIR, this presents the composite of the residue as a cementitious material.

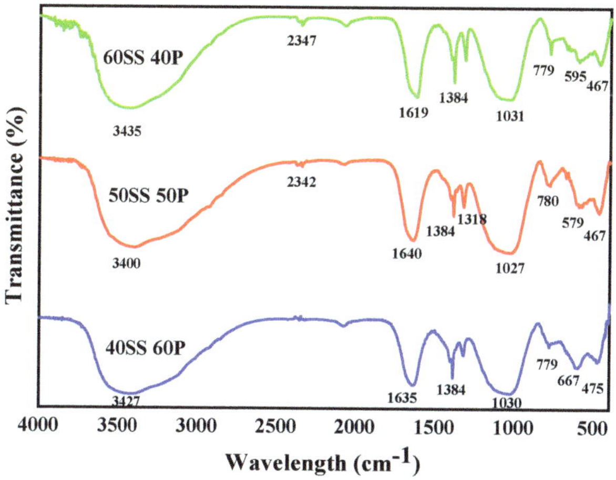

Figure 6 FTIR analysis of SS and P residue composites

TCLP developed composite from P-waste and SS leaching residue

The optimum 60SS 40P composite was evaluated for potential environmental impact using the USEPA toxicity leaching characteristics procedure limits to determine its leachability and prevent environmental pollution. The obtained concentration in comparison with the USEPA allowable concentration is presented in Table 2. The obtained heavy metals concentration for 60SS 40P is below the threshold for USEPA allowable leachable concentrations, with Hg at 0. These results showed that 60SS 40P is safe to be utilized as it will not leach out and contaminate/impact the environment.

Table 2 TCLP metal concentration of 60SS 40P

Metals (mg/L)	As	Ba	Cd	Cr	Pb	Hg	Se
60SS 40P	0.67	0.03	0.01	0.009	0.004	0	0.11
USEPA (1992) limits	5	100	1	5	5	0.2	1

Conclusions

The study showed the feasibility of developing SCM from SS and P. The blended SS and P formed cementitious phases, including C-A-F-S-H, C-S-H, and iron phosphate hydrates without the addition of OPC. The optimum 60SS:40P ratio achieved UCS of 0.89 MPa and exhibited heavy metal leachability below USEPA limit, confirming environmental safety. The developed composite is suitable for application as backfilling and non-load-bearing blocks. The valorization of SS and P contributes to circularity and minimizes emissions from cement production while advancing SDG 12 (responsible consumption) and SDG 13 (climate action).

References

[1] Chang, Z., Long, G., Xie, Y. and Zhou, J.L., 2022. Chemical effect of sewage sludge ash on early-age hydration of cement used as supplementary cementitious material. *Construction and Building Materials*, *322*, p.126116. https://doi.org/10.1016/j.conbuildmat.2021.126116

[2] Dos Santos, V.H.J.M., Pontin, D., Ponzi, G.G.D., e Stepanha, A.S.D.G., Martel, R.B., Schütz, M.K., Einloft, S.M.O. and Dalla Vecchia, F., 2021. Application of Fourier Transform infrared spectroscopy (FTIR) coupled with multivariate regression for calcium carbonate (CaCO3) quantification in cement. *Construction and Building Materials*, *313*, 125413. https://doi.org/10.1016/j.conbuildmat.2021.125413

[3] El Fadili, H., Ait-Khouia, Y., Benzaazoua, M. and Taha, Y., 2025. Symbiotic effects of clayey phosphate sludge and sugar lime on the engineering properties of binary and ternary alkali-activated materials: Harvesting the utilization of industrial by-products. *Construction and Building Materials*, *482*, p.141657. https://doi.org/10.1016/j.conbuildmat.2025.141657

[4] Eu, H., Kim, G., Son, M., Sasui, S., Lee, Y., Choi, H., Kang, S. and Nam, J., 2024. Alkali-Silica Reaction and Residual Mechanical Properties of High-Strength Mortar Containing Waste Glass Fine Aggregate and Supplementary Cementitious Materials. *International Journal of Concrete Structures and Materials*, *18*(1), p.69. https://doi.org/10.1186/s40069-024-00711-x

[5] Goj, P., Wajda, A., Błachowski, A. and Stoch, P., 2021. A new iron-phosphate compound (Fe7P11O38) obtained by pyrophosphate stoichiometric glass devitrification. *Scientific Reports*, *11*(1), 22957. https://doi.org/10.1038/s41598-021-02471-0

[6] Guo, X., Yuan, S., Xu, Y. and Qian, G., 2022. Effects of phosphorus and iron on the composition and property of Portland cement clinker utilized incinerated sewage sludge ash. *Construction and Building Materials*, *341*, p.127754. https://doi.org/10.1016/j.conbuildmat.2022.127754

[7] Gupta, R., Bhardwaj, P., Mishra, D., Mudgal, M., Chouhan, R.K., Prasad, M. and Amritphale, S.S., 2017. Evolution of advanced geopolymeric cementitious material via a novel process. *Advances in Cement Research*, *29*(3), 125-134. https://doi.org/10.1680/jadcr.16.00113

[8] Horgnies, M., Chen, J.J. and Bouillon, C., 2013. Overview of the use of Fourier transform infrared spectroscopy to study cementitious materials. *WIT Trans. Eng. Sci*, *77*, 251-262.

[9] Jhatial, A.A., Nováková, I. and Gjerløw, E., 2023. A review on emerging cementitious materials, reactivity evaluation and treatment methods. *Buildings*, *13*(2), p.526. https://doi.org/10.3390/buildings13020526

[10] Karoui, O., Andrejkovičová, S., Pato, P., Patinha, C., Perná, I., Řimnáčová, D., Hajjaji, W., Ascensão, G., Rocha, F. and Mlayah, A., 2024. Alkali-activated geopolymers based on calcined phosphate sludges and metakaolin. *Environmental Science and Pollution Research*, *31*(32), pp.45138-45161. https://doi.org/10.1007/s11356-024-34025-y

[11] Maarouf, F.Z., Saoiabi, S., Azzaoui, K., Chrika, C., Khalil, H., Elkaouni, S., Lhimr, S., Boubker, O., Hammouti, B. and Jodeh, S., 2021. Statistical optimization of amorphous iron phosphate: inorganic sol–gel synthesis-sodium potential insertion. *BMC chemistry*, *15*, 1-14. https://doi.org/10.1186/s13065-021-00774-x

[12] Marandi, N. and Shirzad, S., 2025. Sustainable cement and concrete technologies: A review of materials and processes for carbon reduction. *Innovative Infrastructure Solutions*, *10*(9), p.408. https://doi.org/10.1007/s41062-025-02213-5

[13]Mkhonto, E., Mashifana, T. and Sithole, N.T., 2024. Synthesis of Acid-Based Fly Ash Geopolymer with Pure Phosphoric Acid Compare with P-Leachate Activator. *Key Engineering Materials, 980*, pp.109-115. https://doi.org/10.4028/p-ZXKyg9

[14]Mojapelo, K.S., Kupolati, W.K., Burger, E.A., Ndambuki, J.M., Snyman, J., Achi, C.G. and Quadri, A.I., 2025. Durability and Environmental Impact of Wastewater Sludge Ash as a Cement Replacement in Concrete: Challenges and Future Directions. *Materials Circular Economy*, *7*(1), pp.1-19. https://doi.org/10.1007/s42824-025-00172-x

[15]Muliawan, J. and Astutiningsih, S., 2018. Preparation and characterization of phosphate-sludge kaolin mixture for ceramics bricks. *International Journal of Technology*, *9*(2), pp.317-324. https://doi.org/10.14716/ijtech.v9i2.1119

[16]Nassiri, S., Butt, A.A., Zarei, A., Roy, S., Filani, I., Pandit, G.A., Mateos, A., Haider, M.M. and Harvey, J.T., 2025. Opportunities for supplementary cementitious materials from natural sources and industrial byproducts: literature insights and supply assessment. *Buildings*, *15*(17), p.3099. https://doi.org/10.3390/buildings15173099

[17]Navarro-Blasco, I., Duran, A., Perez-Nicolas, M., Fernandez, J.M., Sirera, R. and Alvarez, J.I., 2015. A safer disposal of hazardous phosphate coating sludge by formation of an amorphous calcium phosphate matrix. *Journal of Environmental Management*, *159*, pp.288-300. https://doi.org/10.1016/j.jenvman.2015.05.029

[18]Ngayakamo, B.H., Bello, A. and Onwualu, A.P., 2020. Development of eco-friendly fired clay bricks incorporated with granite and eggshell wastes. *Environmental Challenges*, *1*, 100006. https://doi.org/10.1016/j.envc.2020.100006

[19]Pu, S., Zhu, Z., Song, W., Huo, W. and Zhang, J., 2021. Mechanical and microscopic properties of fly ash phosphoric acid-based geopolymer paste: A comprehensive study. *Construction and Building Materials*, *299*, 123947. https://doi.org/10.1016/j.conbuildmat.2021.123947

[20]Świerczek, L., Cieślik, B.M. and Konieczka, P., 2018. The potential of raw sewage sludge in construction industry–a review. *Journal of cleaner production*, *200*, 342-356. https://doi.org/10.1016/j.jclepro.2018.07.188

[21]Xia, Y., Liu, Y., Wang, L., Song, Z., Sun, C., Zhao, Y., Lu, S. and Yan, J., 2023. Value-added recycling of sludge and sludge ash into low-carbon construction materials: current status and perspectives. *Low-carbon Materials and Green Construction*, *1*(1), p.23. https://doi.org/10.1007/s44242-023-00023-5

[22]Yang, H., Che, Y. and Leng, F., 2018. High volume fly ash mortar containing nano-calcium carbonate as a sustainable cementitious material: microstructure and strength development. *Scientific reports*, *8*(1), 16439. https://doi.org/10.1038/s41598-018-34851-4

[23]Yusuf, M.O., 2023. Bond characterization in cementitious material binders using Fourier-transform infrared spectroscopy. *Applied Sciences, 13(5), 3353. https://doi.org/10.3390/app13053353*

[24]Zhang, Y., Nie, S. and Liu, C., 2025. Phosphorus slags: Structural insights, dissolution behavior, and potential as sustainable supplementary cementitious materials. *Chemical Engineering Journal*, *507*, p.160463. https://doi.org/10.1016/j.cej.2025.160463

Materials Research Proceedings 63 (2026) 168-174 https://doi.org/10.21741/9781644904053-19

Suitability of Basic Oxygen Furnace Slag as a Coarse Aggregate in Concrete Made with Supplementary Cementitious Materials

Tsematse TSEMATSE[1,a], Lewis PARSONS[1,b*] and Sunday NWAUBANI[2,c]

[1]University of the Witwatersrand, 1 Jan Smuts Avenue, Johannesburg, South Africa

[2]African University of Science and Technology, Kilometre 10, Airport Road, Abuja, Nigeria

[a]tsematsethabo@gmail.com, [b]lewis.parsons@wits.ac.za, [c]snwaubani@aust.edu.ng

Keywords: Concrete, Aggregate, Supplementary Cementitious Material, Pozzolanic Reaction, Basic Oxygen Furnace Slag

Abstract. Basic Oxygen Furnace Slag (BOFS), a by-product of steelmaking, offers an environmentally friendly alternative to natural coarse aggregates (NCA) in concrete. In South Africa alone, over a million tons of BOFS is produced annually, however, the low utilisation rate of the by-product often leads to growing landfills of the material. The low utilisation rate is primarily due to the free lime content of BOFS, which causes volume instability of the BOFS due to the larger calcium hydroxide crystal which forms during the hydration process and thus limits the potential applications for use. This study explores whether the incorporation of supplementary cementitious materials (SCMs), namely ground granulated blast furnace slag (GGBS) and Metakaolin (MK), may improve the volume stability of concrete made with BOFS as an aggregate but reacting with the BOFS hydration products. Concrete mixes were prepared with 0%, 50%, and 100% volumetric replacement of NCA by BOFS. Furthermore, ordinary portland cement was partially replaced with 60% GGBS in some mixes and 15% MK in others. The compressive strength test was used to assess potential reactivity between BOFS and SCMs, while prism expansion tests evaluated the impact on volume stability and the concrete microstructure was assessed by microcrack analysis method. Results show that BOFS aggregates react with SCMs, due to the difference in the rate of compressive strength gain of the BOFS mixes, relative to those with NCA. Furthermore, including SCMs improved the overall volume stability of concrete made with BOFS, though microcracks were more prevalent. This suggests that while macro-scale stability improved, microstructural issues remain.

Introduction and Background

The world produced 1.8 million tonnes of steel in 2022, and experts expect this figure to grow further due to anticipated global economic growth [1]. South Africa produces between 80 to 100 kilograms of steel per capita, which is the same as India, the second largest steel producer in the world [2]. The steel production processes mainly uses two steel converters, namely, the electric arc furnace or the basic oxygen furnace (BOF) to produce over 3,500 different grades of steel [3]. The EAF route uses scrap steel as feedstock, whereas the BOF process uses a mixture of scrap steel and molten pig iron in the furnace. Research reports that 70% of global steel producers use the BOF, whereas in South Africa, about 60% of steel producers utilise the BOF route [4]. The BOF steel production route results in a solid BOF slag (BOFS) by-product, which accounts for 100 to 200 kg of slag production per ton of steel produced [2]. This translates to approximately 131 to 262 million tons of BOFS produced around the world, every year. Globally, utilisation rates of BOFS remain low, and waste slag landfill sites continue to exist and grow, as the steel production rate increases around the world [5]. Several current uses of BOFS include agriculture, as a soil pH neutraliser, in the cement production kiln to generate tricalcium aluminoferrite, and in the construction sector as backfill, sub-grade, and pavement material [5]. Low utilisation rates of

BOFS result in both an economical and environmental challenge to the steel producer. The economic challenge of landfilling BOFS is the opportunity cost of not considering the by-product as a commercial product. Additionally, the environmental challenge lies in the large land space occupied by these landfills, which exacerbates the environmental footprint of steel plants due to the potential leaching of soluble elements into both surface and underground water resources [6, 7].

Past studies show that the construction sector could use steel slag as an aggregate. However, a mandatory natural aging process of 3 to 18 months limits its immediate use. The aging process helps to hydrate or carbonate the reactive uncombined CaO in the BOFS, which improves its volume stability [8]. Yet researchers such as Mahiex *et al.* [9] found that the BOFS stored in stockpiles during aging continues to remain reactive due to heterogeneous exposure to the environment throughout the stockpile. This unreliability further limits the use of BOFS. Despite these negative aspects, researchers have considered using the BOFS as an aggregate to compare it structural performance, relative to natural aggregates. The use of the by-product exhibits enhanced performance benefits when used as an aggregate in concrete where researchers such as Sharma *et al.* and Zago *et al.* [10, 11], have found that as the replacement percentages of natural aggregates with BOFS in concrete increase, the mechanical strength and abrasion resistance of the concrete improves. This positive aspect makes explorations of novel knowledge related to use of the by-product as an aggregate in concrete a significant research path.

The current study aims to enhance the understanding of using pozzolana and latent cement replacement in concrete made with BOFS. It investigates whether the expansion caused by the hydration of free CaO to form $Ca(OH)_2$ on the surface of the BOFS aggregate can be reduced through pozzolanic reactions. These reactions may consume the $Ca(OH)_2$ crystals as they form, producing calcium silicate hydrates, which could improve the mechanical performance of concrete. An understanding of such a mechanism, will unlock more applications of the BOFS, which is otherwise landfilled.

Material Characterisation;

Physical characterisation of the BOFS: Table 1 provides a comparative overview of the specific density, 24-hour water absorption, and Aggregate Crushing Value (ACV) of the aggregates evaluated. The specific densities were determined using a helium gas pycnometer, the water absorption values were determined using ASTM C127-15 and the ACV was determined using SANS 3001-AG10:2012 as a guideline. As shown in the table, the BOFS exhibits a slightly higher specific density relative to the natural aggregate (NA). This difference was considered during the concrete mix design phase, where volumetric replacements of NA with BOFS and it was also ensured that the particle size distribution (PSD) curve of the BOFS aggregates matched that of the 22 mm Andesite PSD, which conformed to the grading limits specified by BS 822:1992.

Table 1: Physical properties of the coarse aggregates

	22mm Andesite Stone	*BOFS*
Water Absorption (%)	0.38	1.45
Mean Specific Denisty (kg/m³)	2893.1	3274.5
ACV (%)	5.64	11.17

Similarly, the ACV of BOFS is higher than that of NA, indicating that BOFS aggregates are more susceptible to crushing under compressive loads compared to NA, which suggests that NA possesses greater mechanical strength in this regard. The NA has a lower water absorption value (0.38%) relative to BOFS (1.45%). This difference could affect the effective w:b value of concrete and have an impact on the mechanical performance of concrete [12]. Figure 1 illustrates that the shape of the NA and BOFS coarse aggregates is largely consistent

Figure 1: Visual comparison of the particle shape and roughness of the aggregates

Experimental Procedure

Concrete Mix design: Three binder systems were used, each with a unique blend of binders used. The PC replacement percentages of both MK and GGBS blends were strategically selected to have replacement percentages, which are both 1.5 times higher than the known replacement percentages for optimal strength development i.e. 10% PC replacement with MK and 40% PC replacement with GGBS. This was to ensure that sufficient unreacted SCM particles were present in the paste, thereby ensuring that the limiting reagent would be the BOFS hydration products. Therefore, the first binder blend, as a control, was 100% PC, the second binder blend was a 15% replacement of PC with MK, and the third was a 60% replacement of PC with GGBS. The high percentage replacements are selected to ensure that there are unreacted SCM particles in the concrete paste, which could react with the BOFS hydration products.

In addition to the various binder blends, two different coarse aggregates were used in the current study: NA and BOFS. For each binder blend, a concrete mix with 100% NA, and a 100% BOFS mix existed, with a unique nomenclature as per Table 2 below. The mixes are uniquely identified with the binder system used (PC = 100% PC, MK=15% PC replacement with MK, GGBS= 60% PC replacement with GGBS) and their coarse aggregate content/blend (NA = 100% NA, and BOFS -100 = 100% BOFS).

Table 2: Mix Designs used in the current study

MIXES	Mix Proportions [kg/m^3]							Total Density [kg/m^3]
	PC	MK	GGBS	WATER	FINE AGGREGATE	NATURAL COARSE AGGREGATE	BOFS	
NA / PC	388.9	-	-	175	950.5	1073.3	0	2587.7
NA / MK	330.6	58.3	-	175	950.5	1073.3	0	2587.7
NA / GGBS	155.6	-	233.3	175	950.5	1073.3	0	2587.7
BOFS - 100 / PC	388.9	-	-	175	950.5	0	1370.3	2884.7
BOFS - 100 / MK	330.6	58.3	-	175	950.5	0	1370.3	2884.7
BOFS - 100 / GGBS	155.6	-	233.3	175	950.5	0	1370.3	2884.7

Testing Regime: The current study used concrete compressive strength as an indicative testing regime to determine whether a reaction exists between the hydration products of BOFS and the SCM's used in the study. Testing was conducted at 3, 7 and 28-days of curing in a water (22 ± 2 °C), where SANS 5863 was used as a guideline for this. To assess whether the potential reaction affected the overall stability of the concrete, the concrete prism expansion test was conducted, as per the guidance of ASTM C157.

Results and Discussions

Compressive strength test: The results from the compressive strength tests are presented in Figure 2, 3 and 4. For the PC-binder mixes (Figure 2), the strength of the BOFS mix is higher than that of the NA mix at 3- and 7-days, where at 28-days the strength is only around 1% higher than the NA mix.

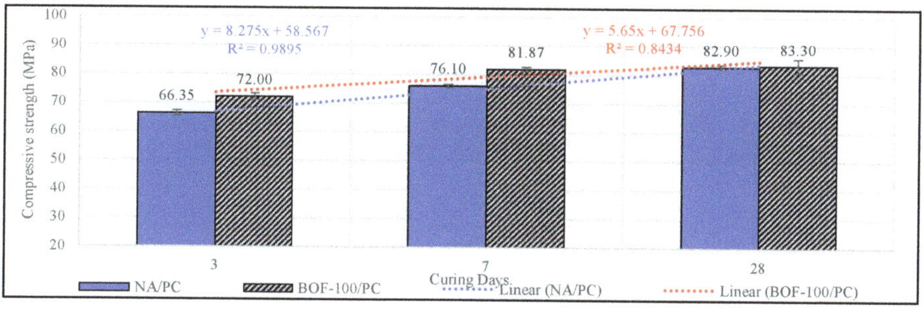

Figure 2: Average compressive strength results of PC-binder concrete

The compressive strength of the concrete made with BOFS and NA mixes in the GGBS-binder system is both 30 MPa at 3-days, but the strength of the BOFS mix is 7% higher, with a 95% confidence interval, than the NA mix at 7-days and 17% higher at 28-days. This is indicative of a faster strength gain of the BOFS mixes relative to the NA mixes, which is indicated by the steeper strength trendline of the BOFS mix, with a GGBS-binder system, shown in Figure 3.

As GGBS is a latent cement replacement, an alkaline environment is required to activate of the amorphous siliceous and aluminous content of the GGBS, resulting in the formation of the expected C-A-S-H formations. This could explain why the enhanced strength benefits of BOFS are not immediately prevalent (3-days), but as the relatively slow BOFS hydration occurs and produces additional CH in the system and thus a more alkaline environment, the GGBS is activated at a faster rate than that in the NA mixes.

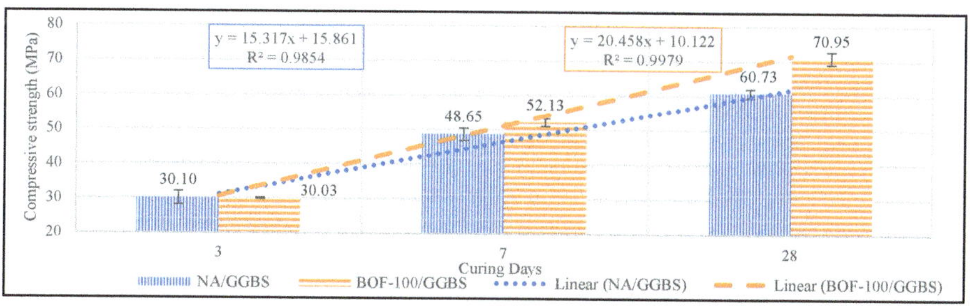

Figure 3: Average compressive strength results of GGBS-binder concrete

Similarly, in the MK-binder system (Figure 4) the early day strength (3- and 7-day) of the BOFS mix is lower than that of the NA mix, however from 28-days the strength of the BOFS mix is about 3% higher than the NA mix. MK is a pozzolan which reacts with the formed CH to produce additional C-S-H and C-A-S-H amorphous gels in the concrete. The difference is that once the

BOFS hydrates, the additional CH introduced into the system would then need to react with the unreacted MK in the paste to form additional amorphous gels, which improve the strength of the concrete. This delayed reaction and subsequent late increase in strength is observed by the steeper gradient (14.667 > 9.8667) of the compressive strength trendline of the BOFS mix, relative to the NA mix.

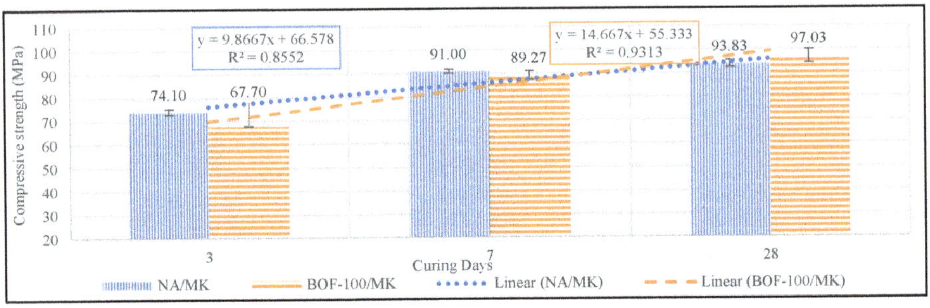

Figure 4: Average compressive strength results of MK-binder concrete

Prism expansion test: The results from the 28-day cured concrete prism expansion test are presented in Figure 5. The difference in the relative length expansion of the concrete prisms at 28-days between the mixes with NA and those with BOFS, is indicative of increased volume instability across all the respective binder systems used when BOFS is used. The difference between the relative length expansion of the BOFS and NA concrete mixes is 0.05 for the PC-binder mixes, 0.04 for the MK-binder mixes and 0.06 for the GGBS-binder mixes. This implies the following regarding the volume stability (VS) of the various binder systems:

$$VS_{BOFS\ in\ MK\text{-}Binder\ system} > VS_{BOFS\ in\ PC\text{-}Binder\ system} > VS_{BOFS\ in\ GGBS\text{-}Binder\ system}$$

The above indicates that the use of an MK-binder system improves the volume stability of concrete made with BOFS, relative to a conventional PC-binder system. This observation is indicative of a potential pozzolanic reaction, which occurs between the CH crystal as it forms during BOFS hydration and the unreacted MK particles in the paste, which limits further expansion.

Additionally, these results suggest that the use of a GGBS-binder system does not improve the volume stability of concrete made with BOFS. This is potentially due to the lower compressive strength of the GGBS-binder system concrete at 28 days, relative to the PC-binder system (shown in Figure 3). The lower compressive strength means that the hardened concrete paste around the BOFS aggregate can counteract less of the internal forces from BOFS expansion than the PC and MK-binder systems. This is believed to result in less restrictive actions by the hardened concrete paste, and thus easier movement of the surrounding concrete, leading to more concrete expansion relative to the PC-binder system.

Materials Research Forum LLC
https://doi.org/10.21741/9781644904053-19

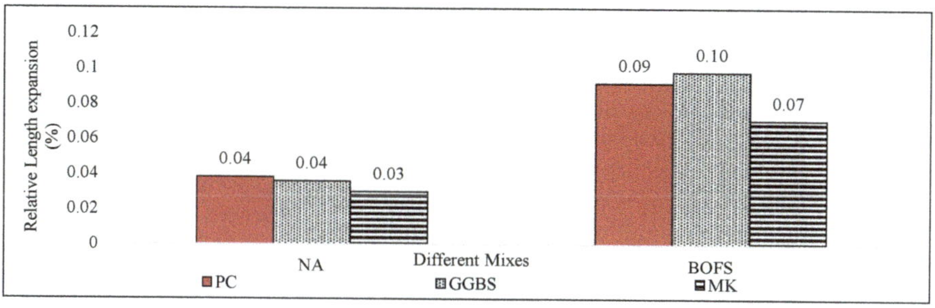

Figure 5: 28- day concrete prism length expansion results

Conclusions

This study demonstrates that BOFS can be effectively used as a coarse aggregate in concrete, particularly when combined with SCMs. The results confirm that pozzolanic and latent hydraulic reactions occur between BOFS hydration products and SCMs, enhancing compressive strength over time. While all mixes with BOFS showed higher volume expansion relative to natural aggregate mixes, the use of MK improved volume stability, indicating potential mitigation of expansion through pozzolanic reactions. These findings support the broader utilisation of BOFS in concrete applications, advancing sustainability in steel by-product management and concrete technology.

References

[1] Bronk & Company. "Green transformation Needs clear orientation – forecasting the steel demand in 2030". Accessed via: https://bronk-company.com/en/2023/01/forecast-steel-demand-2030/ on the 27th of May 2024.

[2] World steel association, "Steelmaking process", undated, accessed via: https://worldsteel.org/about-steel/steelmaking-process/ on the 2nd of March 2024.

[3] Dixon, V. (2023). "6 Steps of the Steel Manufacturing Process". Accessed via: https://dixonvalve.com/en/news-and-events/news/6-steps-steel-manufacturing-process on the 27th of May 2024.

[4] SAISI. (2022). "What is steel and how is it made?", South African Institute of Steel and Iron .Accessed via: https://www.saisi.org/news/what-is-steel-and-how-is-it-made/#:~:text=In%202021%2C%20just%20under%2060,the%20BF%2DBOF%20production%20route. On the 22nd of may 2024.

[5] Topcem cement. (2024)."Main Ingredients in Cement and Their Functions", Topcam cement, 2024, accessed via: https://topcem.in/main-ingredients-in-cement-and-their-functions/#:~:text=Iron%20Oxide%3A%20The%20amount%20of,tricalcium%20aluminoferrite%20at%20high%20temperatures.

[6] Gomes, J. F. P., Pinto, C. G. (2006). "Leaching of heavy metals from steelmaking slags". Accessed via:https://revistademetalurgia.revistas.csic.es/index.php/revistademetalurgia/article/download/39/38 accessed on the 20th of February 2025.

[7] Roadcap, G.S., Kelly, W.R. (1979). "Shallow ground-water chemistry in the Lake Calumet Area, Chicago, Illinois, National Symposium on Water Quality". American Water Resources Association, Herndon, Virginia, 1994.

[8] Trinh S., "Effect of weather aging on volume expansion properties of steel slag", Journal of Science and Transport Technology, issue 1, 2022, Accessed via : https://www.jstt.vn/index.php/en/article/view/96 on the 25th of march 2024.

[9] Mahieux P., Aubert J., Escadeillas G., "Utilization of weathered basic oxygen furnace slag in the production of hydraulic road binders", Construction and Building Materials, Volume 23, Issue 2, 2009, Accessed via https://www.sciencedirect.com/science/article/pii/S0950061808000706 on the 5th of March 2024.

[10] Sharma, G., Goyal, R., Gupta, A., and Sharma D. (2015). "Beneficial Effects of Steel Slag on Concrete," International Journal of Recent Research Aspects. 02(02):51-57, https://www.ijrra.net/Vol2issue2/IJRRA02-02-12.pdf

[11] Zago, S.C.; Vernilli, F.; Cascudo, O. (2023), "The Reuse of Basic Oxygen Furnace Slag as Concrete Aggregate to Achieve Sustainable Development: Characteristics and Limitations". Buildings 2023.13:1193. Accessed via https://doi.org/10.3390/buildings13051193 on the 14th of July 2023.

[12] Sun, J., Luo, S., Wang, Y., Dong, Q.,Zhang Z. (2023)."Pre-treatment of steel slag and its applicability in asphalt mixtures for sustainable pavements". accessed via: https://www.sciencedirect.com/science/article/pii/S138589472305533X#:~:text=Steel%20slag%20is%20a%20potential,slag%20can%20facilitate%20sustainable%20developmen t.

[13] Nugmanova, A., Shon, C.-S., Kim, J. R., Rossi, C. O. (2023). "Characterizing Chronologically Aged Basic Oxygen Furnace Slags as Aggregates and Their Use in Asphalt Concrete Mix as Filler". Applied Sciences. 13(18): 101. Accesed via : https://doi.org/10.3390/app131810126 on the 21st of February 2025.

[14]Venkatraman, S., Choudhary, R., Indurkar, A., Chatterjee, A., Abraham, J., Ostrovskiy, S., Senatov F., Locs J., Swamiappan S.(2021) ."Conversion of Biowaste into Larnite by Sol-Gel Combustion Route for Biomedical Applications".2021. Chemistry Select. Accessed via: https://www.researchgate.net/publication/358084494_Conversion_of_Biowaste_into_Larnite_by_Sol-Gel_Combustion_Route_for_Biomedical_Applications on the 22nd of February 2025.

[15] Yuan Q., Liu Z., Zheng K., Ma C., Chapter 3 - Portland cement concrete, Civil Engineering Materials, Elsevier, 2021, Pages 59-204, Accessed Via:

(https://www.sciencedirect.com/science/article/pii/B9780128228654000039) on the 26th of June 2024.

[16] Tural H., Ozarisoy B., Derogar S., Ince C., "Investigating the governing factors influencing the pozzolanic activity through a database approach for the development of sustainable cementitious materials", Construction and Building Materials, Volume 411, 2024, Accessed via (https://www.sciencedirect.com/science/article/pii/S0950061823039715) on the 12th of September 2024.

[17] Shi, C. (2004). "Steel slag—its production, processing, characteristics, and cementitious properties." Journal of Materials in Civil Engineering, 16(3):230-236.

Materials Research Proceedings 63 (2026) 175-185 https://doi.org/10.21741/9781644904053-20

Performance of Pre-Carbonated Basic Oxygen Furnace Slag as a Replacement for Natural Coarse Aggregates in Concrete

Tsematse TSEMATSE[1,a], Lewis PARSONS[1,b*] and Sunday NWAUBANI[2,c]

[1]University of the Witwatersrand, 1 Jan Smuts Avenue, Johannesburg, South Africa

[2]African University of Science and Technology, 10 Umaru Musa Yar'Adua Road, Abuja, Nigeria

[a]tsematsethabo@gmail.com, [b]lewis.parsons@wits.ac.za, [c]snwaubani@aust.edu.ng

Keywords: Carbon dioxide, Durability, Concrete, Aggregate, Basic Oxygen Furnace Slag

Abstract. Basic Oxygen Furnace Slag (BOFS) is a significant by-product of the steel-making process, where over 250 million tonnes is produced annually around the world. However, the utilisation rate of the steelmaking by-product remains low due to the expansive hydration products formed by the BOFS. To support the potential high-volume use of the by-product as coarse aggregate in concrete, this study aimed to investigate the use of pre-carbonation over 7 days to pre-react the constituents that cause expansive behaviour. For this purpose, five concrete mixes were designed with a volumetric replacement of the Natural Coarse Aggregates (NCA) with the BOFS and the pre-carbonated BOFS (BOFC) at 0, 50 and 100% replacement percentages. X-ray diffraction (XRD) and water absorption analyses were conducted on both types of BOFS aggregates to support the compressive strength, prism expansion, and oxygen permeability tests, which were also performed in the study. The results from the compressive strength test showed a 10.7% and 1.4% higher 56–day compressive strength result for the concrete made with BOFC and BOFS, respectively, compared to the natural aggregates. Similarly, the oxygen permeability test showed that the concrete made with 50% BOFC was 2% less permeable than the untreated BOFS and at 28-days of curing, exactly the same compared to NCA mixes.. Lastly, the prism expansion test indicated that pre-carbonating the BOFS improved the volumetric stability of concrete made with BOFS over the 84-day curing regime. Therefore, the above results suggest that pre-carbonation can increase the utilisation rate of BOFS as a coarse aggregate in concrete.

Introduction

Economic development closely aligns with industrial growth, with steel usage being a key indicator [1]. Countries in the Brazil-Russia-India-China-South Africa (BRICS) forum account for 68% of global steel production [2], with South Africa producing 80–100 kg of steel per capita, which is comparable to India, the world's second-largest producer [3]. Roughly 70% of global steel is produced via the Basic Oxygen Furnace (BOF) process, which generates 100–200 kg of Basic Oxygen Furnace Slag (BOFS) per ton of steel [3]. In this process, oxygen is blown into molten scrap and pig iron to oxidise impurities, followed by limestone addition. The limestone decomposes to calcium oxide (CaO), which combines with impurities, forming liquid BOFS [4]. Though BOFS is used in agriculture, cement kilns, and road construction [4, 5], a significant portion of the more than 250 million tons of BOFS produced annually in the world is not utilised, where potential large-scale utilisation lies in the construction sector [4]. The predominant concern over the utilisation of the by-product as a construction material, specifically an aggregate due to its coarse nature, is the volume instability of the BOFS during hydration. The unreacted CaO remaining in the BOFS after steelmaking hydrates into calcium hydroxide [$Ca(OH)_2$], which has lower density relative to CaO and thus a larger volume for the same mass, which results in expansion of the BOFS in the process [6, 7].

Various treatments have been explored to mitigate BOFS's expansion, minimising concrete cracking and spalling [8]. Sun *et al.* [9] evaluated pre-treatment methods, such as soaking in water, acetic acid, and silane coatings, which all reduced expansion. Silane proved most effective (0.5% expansion after 10 days) but is costly and industrially complex. Alkaline activation with sodium silicate improves the cementitious properties of BOFS by forming calcium silicate hydrate (C-S-H) gels [10], but low grindability of the BOFS raises energy costs. As BOFS has a vitreous structure, using it as aggregate instead of binder may offer better scalability, particularly given concrete's 60 to 80% aggregate content [11] which would promote high volume usage.

Early carbonation is another promising method, converting the accessible unreacted CaO to solid $CaCO_3$ in the pores and on the surface of the BOFS, thus limiting hydration and subsequent expansion [12]. While this method has been tested under ideal lab conditions (i.e. high CO_2 concentration and temperature), minimal research exists on ambient conditions. Additionally, limited studies assess "worst-case" BOFS (unweathered and untreated), highlighting the need for more practical assessments across variable feedstock. As such, this study investigates early carbonation of unweathered BOFS under ambient conditions (room temperature, atmospheric pressure, 20% CO_2, and 65% humidity) before its use as coarse aggregate as a means of investigating the adequacy of pre-carbonation as a solution to improving the volume stability of BOFS in concrete.

Pre-carbonation process

The following section describes the pre-carbonation process along with the procedure followed and the findings from the treatment process.

The principle: Mineral carbonation is a method of carbon sequestration in which carbon dioxide readily reacts with natural or synthetic mineral phases, such as CaO and MgO, to form various carbonates, effectively permanently trapping CO_2 in stable mineral phases [13]. This carbon sequestration method is considered permanent since it would require high energy levels through thermal decomposition to separate the CO_3^{2-}ions. Furthermore, the resulting products are known to be stable insoluble carbonates [13]. Given that a major limitation to the utilisation of BOFS is the presence of free CaO, in the early carbonation approach, the CaO and MgO in the solid BOFS feedstock readily react with the gaseous CO_2 to form the respective insoluble carbonates on the aggregate's surface and internal pores for further use in concrete. In the case where CaO has already hydrated to form $Ca(OH)_2$, the carbonation reaction still occurs without any major changes to the environmental conditions. The early carbonation approach aims to pre-react this compound, forming a stable and insoluble carbonate on the aggregate's surface and internal pores for further use in concrete.

It was reported by Zhao *et al.* [14], that the reaction of CO_2 with $Ca(OH)_2$ actually occurs at faster rates relative to the reaction of CaO and CO_2 where it was found that the former, also experienced higher conversion degree to $CaCO_3$. Furthermore it was found by Dostie *et al.* [15], that the $Ca(OH)_2$ was consumed highly to produce $CaCO_3$ in high humidity environments (60-95%) as it was found that the dissolution of $Ca(OH)_2$ and the precipitation of $CaCO_3$ are greater at higher RH where the most optimal was found to be 80%. The reactions which are anticipated to occur during the carbonation process are the following Eq. 1 – Eq. 3, where the source of the CaO is the BOFS feedstock.

$$CaO_{(s)} + CO_{2\,(g)} \rightarrow CaCO_{3\,(s)} \quad (1)$$

$$CaO_{(s)} + H_2O_{(g)} \rightarrow Ca(OH)_{2\,(s)} \quad (2)$$

$$Ca(OH)_{2\,(s)} + CO_{2\,(g)} \rightarrow CaCO_{3\,(s)} + H_2O_{(l)} \quad (3)$$

The Process: In a system where there is sufficient CO_2 gas available, reactions in Eq. 1 – Eq. 3 above are expected to occur within the system until the CaO which is accessible to fluids in the BOFS has been consumed. The various parameters which can be altered to make the process of CaO consumption occur faster are well investigated and have been reported to be: the concentration and pressure of the CO_2 gas in the system, the air pressure of the system temperature and relative humidity of the system and the addition of any admixtures to the system [16]. In the current study, a focus on selecting parameters which would require the least external inputs, thus improving the economic feasibility of the treatment process. As such, the process parameters which were selected in the current study are: Air with a 20% concentration of CO_2 gas in the chamber, relative humidity of 65 ± 2 %, room temperature of around 18°C, and atmospheric pressure.

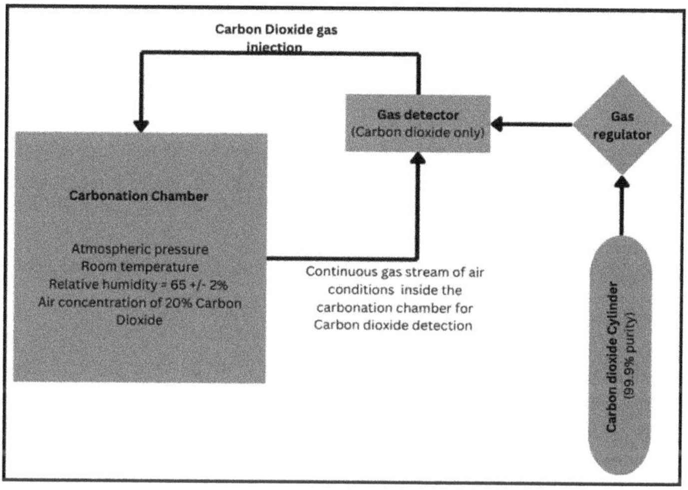

Figure 1: Schematic Diagram of the carbonation chamber

The carbonation process was carried out in a chamber which was kept closed, where all openings were covered with petroleum jelly to adequately seal the openings of the chamber. An air circulation fan was used to ensure uniform air conditions throughout the chamber. Figure 1 above shows a schematic diagram of the system used.

The Result: Using the selected process parameters, as mentioned above, a pre-investigation was required to establish the adequate period which the BOFS would have to remain in the chamber to ensure that the accessible CaO has reacted as per Eq. 1 – Eq. 3. Using mass as an indication of ongoing reactivity, a BOFS sample of 1472.4 g was placed in a 4.75mm aperture weighed sieve, then placed into the carbonation chamber. The sieve was used to ensure adequate gas circulation around the specimen. This weighed sieve was then removed from the chamber and weighed every 24-hours, with the change in mass recorded, where this change in mass is assumed to be due to the formation of $CaCO_3$ from the reaction of either the CaO or $Ca(OH)_2$, as per Eq. 1 or Eq. 3. This mass reading is shown in Figure 2, where the most reactivity occurs within the first four days of exposure in the chamber. After four days of exposure, the change in mass becomes negligible, where it can be assumed that the accessible CaO in the BOFS has been converted to $CaCO_3$ at this period of treatment.

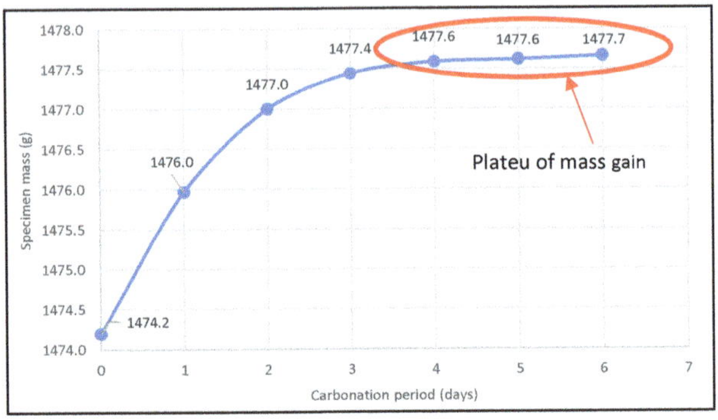

Figure 2: Results of carbonation pre-investigation

A period which was selected based on this set of results was a treatment period of 7 days. This is on the basis that after 5-days the gain in mass was negligible, but to account for any variations in treatment on a larger volume of BOFS, a 7-day treatment period was selected for certainty.

Characterisation of the material used in the study

The natural aggregate used in the study is a 22 mm Andesite stone and crusher sand from the Eikenhof quarry and the BOFS was from the Vanderbjilpark workings steel plant. The binder used was CEM I 52,5R Portland Cement (PC).

Physical properties of the aggregates used in the study: Table 1 below compares the specific density, water absorption and Aggregate Crushing Value (ACV) of the aggregates used in the study. The specific densities were determined using a helium gas pycnometer, the water absorption values were determined using ASTM C127-15 and the ACV was determined using SANS 3001-AG10:2012 as a guideline. Table 1 shows that the specific densities of the BOFS and BOFC are slightly higher than that of the NA. This difference will be factored into the study when considering the concrete mix design, where the replacements of NA are carried out on volumetric quantities. During this process, it was ensured that the particle size distribution curves of the replacing BOFS, matched that of the replaced NA. Additionally, it was observed that the ACV of the BOFS and BOFC is higher than that of the NA, which means that the NA is more resistant to crushing under a compressive load than the BOFS and BOFC. SANS 3001-AG10:2012 was used as a guide to determining the ACV.

Table 1: Physical properties of the aggregates used in the study

	Natural Coarse Aggregate	BOFS coarse aggregate	BOFC coarse aggregate
Specific Density [kg/m^3]	2893.1	3274.5	3274.5
Water Absorption [%]	0.38	1.45	1.08
Aggregate Crushing Value (ACV) [%]	5.64	11.17	11.17

Additionally, the NA exhibited the lowest water absorption, followed by the BOFC, while the untreated BOFS showed the highest water absorption at 1.45%. The difference between the BOFS and BOFC is attributed to the early carbonation process which resulted in the formation of calcite in the material's pores. This effect effectively clogs the internal pores of the BOFS and makes the aggregate less porous, thus less water is trapped internally, leading to lower water absorption [17].

Figure 3: Visual representation of the coarse aggregates used in the study

As seen in Figure 3, the surface roughness and aggregate shape and of the different coarse aggregates used is highly comparable. It should be noted that the early carbonation process does not change the visual aspects of the BOFS aggregates.

Crystal composition of the BOFS: An analysis of the crystal composition of the BOFS and BOFC is used to provide evidence of treatment adequacy and indicate whether any additional reactions occurred during the treatment process. This analysis was carried out using an X-Ray Diffraction (XRD) analysis. The equipment used to obtain the diffraction pattern was the D2 Phaser diffractometer where each sample was exposed to incoming X-rays for 10 minutes using a pre-programmed setting, where peak identification was done using the Diffrac Eva database, and 2Theta locations confirmed from literature [18, 19]. Two different BOFS (Named BOFS A and BOFS B) and two different BOFC (named BOFC A and BOFC B) single 22 mm aggregates were randomly selected and pulverised to powder form, such that XRD analysis can be undertaken, with the diffractograms shown in Figure 4 below. Two random aggregates were used to investigate whether there is a distinct difference in the mineralogy of the BOFS batch of aggregates used in the study, as the production of BOFS is not regulated, and a difference could exist. The same approach was considered for the BOFC aggregate batch. The diffractograms for BOFS A and B were plotted to identify any differentiation in the observed peaks, and the same approach was taken for the BOFC A and B diffractograms, though on a different vertical axis.

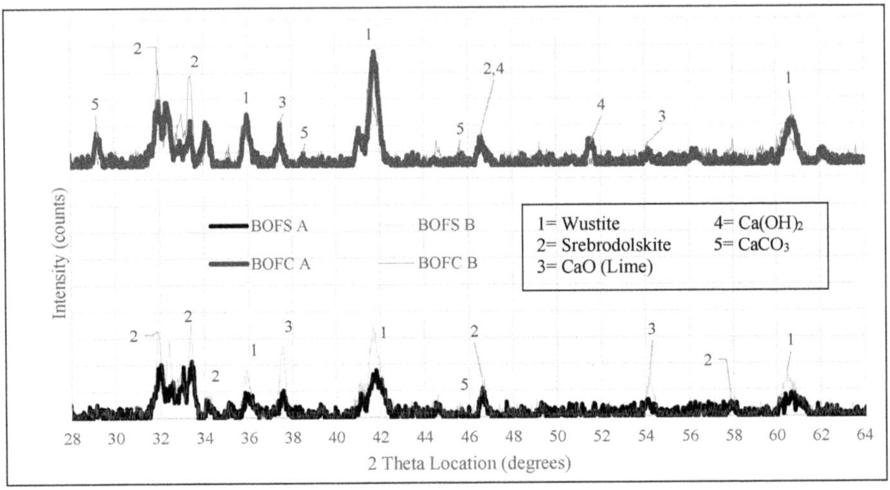

Figure 4: XRD for BOFS and BOFC

The combined diffractograms showed that despite the differences in the recorded peak intensities for the various crystals in both samples of both BOFS and BOFC the prevailing dominant phases in both BOFS and BOFC include Wustite (FeO) and Srebrodolskite ($Ca_2Fe^{3+}{}_2O_5$), which appear quite similar. A notable difference between BOFS and BOFC is the presence of calcite and portlandite peaks in the BOFC samples, alongside a significant reduction in the intensity of the CaO peaks relative to the BOFS samples. The CaO peak is not expected to disappear entirely after early carbonation, as CO_2 gas can only react with the free CaO content that is accessible to fluids in an unpulverized aggregate. However, not all lime pockets are accessible to fluids, as some are located within the internal matrix of the BOFS [20]. Therefore, due to the pulverisation of the BOFC aggregate following treatment for XRD analysis, the free CaO that remained inaccessible to fluids will be released, resulting in an expected CaO peak. It can be appreciated from Figure 4, that the reactions which occur during the treatment process, are those identified by Equation 1 to 3.

Experimental Procedure

Concrete Mix design: A single PC-binder system was used in the current study, where the fine aggregate quantity and water content were kept constant throughout all the mixes. The natural coarse aggregate was replaced, by volume, at 50% and 100% with either BOFS or BOFC. The 100% replacement percentage were selected to focus the study on the complete replacement and 50% was used to confirm any trends as an midway replacement. Each mix design used in the study had a unique blend of coarse aggregate content, where blends either consisted of NA and BOFS aggregates or NA and BOFC aggregates. Table 2 below shows the various unique mix designs used in the current study. The nomenclature used to identify the mixes includes the BOFS or BOFC name and the number after the dash represents the percentage which the NA has been replaced by in that mix i.e. BOFS – 50 / PC, means that 50% of the coarse aggregates BOFS and 50% is the NA.

Materials Research Forum LLC
https://doi.org/10.21741/9781644904053-20

Table 2: Concrete mix Designs used in the current study

MIXES	Mix Proportions [kg/m³]						Total Density [kg/m³]
	PC	WATER	FINE AGGREGATE	NATURAL COARSE AGGREGATE	BOFS	BOFC	
NA / PC	388.9	175	950.5	1073.3	0	0	2587.7
BOFS - 50 / PC	388.9	175	950.5	536.6	685.2	0	2736.2
BOFC - 50 / PC	388.9	175	950.5	536.6	0	685.2	2736.2
BOFS - 100 / PC	388.9	175	950.5	0	1370.3	0	2884.7
BOFC - 100 / PC	388.9	175	950.5	0	0	1370.3	2884.7

Testing Regime: The current study used concrete compressive strength to investigate the effect which precarbonation of BOFS coarse aggregates has on the performance of concrete. Testing was conducted at 3, 7, 28, 56 and 90-days of curing in a water (22 ± 2 °C), where SANS 5863 was used as a guideline for this. To assess the effect of precarbonation of the BOFS on the volume stability of the concrete, the concrete prism expansion test was conducted, as per the guidance of ASTM C157. Additionally, to assess the effect of precarbonation of the BOFS on the gas permeability of concrete, the Oxygen Permeability Index (OPI) was determined for the concrete mixes, as guided by the Durability Index Manual and SANS 3001-CO3-1:2015.

Results and Discussion

Compressive strength test results: Figure 5 shows that the average compressive strength of concrete made with either BOFS or BOFC is higher than that of concrete made with NA (at 56-days, 100% BOFS mix is 1.5% stronger and BOFC mix is 5% stronger than NA mix). Furthermore, these results indicate that the use of early carbonation, to produce BOFC, results in a slightly stronger concrete at all curing days, relative to untreated BOFS. Given that all the physical properties of the materials which were characterised were similar, except for the water absorption and the ACV, it is believed that these properties could be the reason for the difference. However, a higher ACV value for the BOFS or BOFC, wouldn't positively affect the compressive strength of concrete, and if there had to be an effect, it would result in a decrease in the strength of concrete due to the aggregate cracking before the paste [21]. This leaves the different water absorption values as the physical property of the aggregates which could be the reason for the difference in concrete strength. Given that the NA has a lower water absorption and results in a lower compressive strength and the inverse is the case for BOFS, a relationship could be drawn between the water absorption and the strength. However, the BOFS has a higher water absorption relative to the BOFC, yet BOFC results in a higher concrete strength. This introduces the possibility of another property of the BOFS and BOFC aggregates also affecting the concrete strength, such as the volumetric expansion.

Another observed difference when characterising the BOFS and BOFC aggregates, are the peaks of $Ca(OH)_2$ and $CaCO_3$ in the BOFC aggregates which formed during the treatment process where the hydratable CaO was pre-reacted.

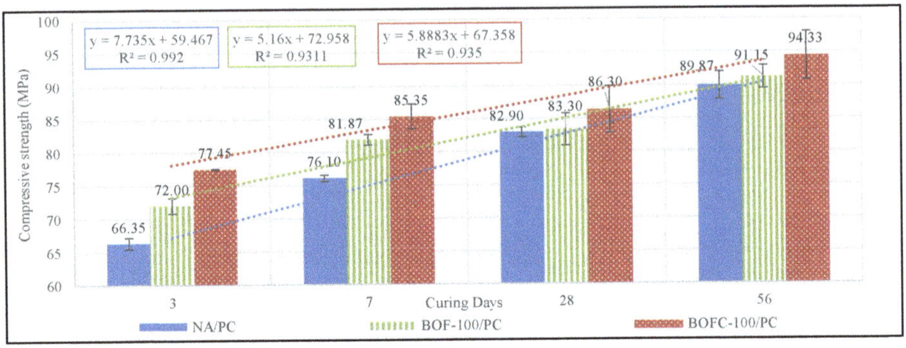

Figure 5: Average compressive strength of concrete mixes

The CaO in the BOFS is known for resulting in volumetric instability of the BOFS during hydration, due to the crystal formation of the hydration product [$Ca(OH)_2$] which has a larger crystal, relative to CaO. This expansive reaction causes a tensile force on the surrounding concrete paste, which induces additional cracking of the concrete. This cracking is believed to make crack propagation easier under loading, and as such, result in a lower bearing capacity of the concrete [22]. This phenomenon could explain why the relationship between the water absorption value and compressive strength of the concrete made with the BOFS and BOFC was unclear, as the BOFS is more volumetrically unstable, as seen in Figure 6, which would induce more internal cracking, and thus negatively affect the compressive strength of concrete. Figure 6 shows the results from the concrete prism expansion test, where it is seen that at 100% replacement of NA with BOFS, a 177% higher length expansion is recorded for the concrete . Furthermore, the opposite is recorded where a 71% reduction in length expansion is recorded in the concrete mixes where 100% NA is replaced with BOFC (improved volume stability). This indicates the effect of the pre-carbonation process on improving the volume stability of concrete made with the steelmaking by-product.

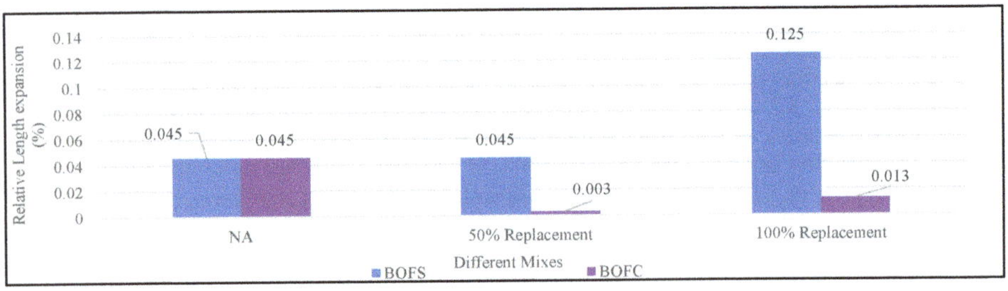

Figure 6: Results from the concrete prism expansion test

Additionally, another positive effect of the pre-carbonation process on the BOFS is the improved resistance to oxygen permeability of the resultant concrete. This is seen in Figure 7 where the OPI of concrete made with BOFS decreases more than that made with BOFC, which is represented by less steep ($-0.1018 < -0.1579$) trend line of the BOFC concrete mixes, relative to the BOFS mixes. Additionally, the OPI values for the 50% BOFS mix is 2.88% lower than of the 50% BOFC mix. This lower OPI occurs at all replacement percentages of NA with BOFS. The inclusion of both BOFS or BOFC however makes the concrete more permeable by gases as the OPI values are generally lower than those of the NA concrete mixes, however this can not be said with a 95% confidence, as the error bars of the NA and BOFC concrete mixes overlap at both 50% and 100% replacements. The 50% replacement of NA with BOFC has an identical OPI value to the NA mix. The superior performance of the BOFC, relative to the BOFS is potentially due to the $CaCO_3$ formation within the pores of the BOFS after the pre-carbonation process [23]. A look back at Figure 4, shows that the BOFC aggregates had a $CaCO_3$ peak after the carbonation, which supports this statement. Additionally, the lower water absorption rate which was determined for the BOFC aggregates, relative to the BOFS is also indicative of the $CaCO_3$ formation occurring within the pores of the aggregates which clogged the pores and reduced the permeability of gases.

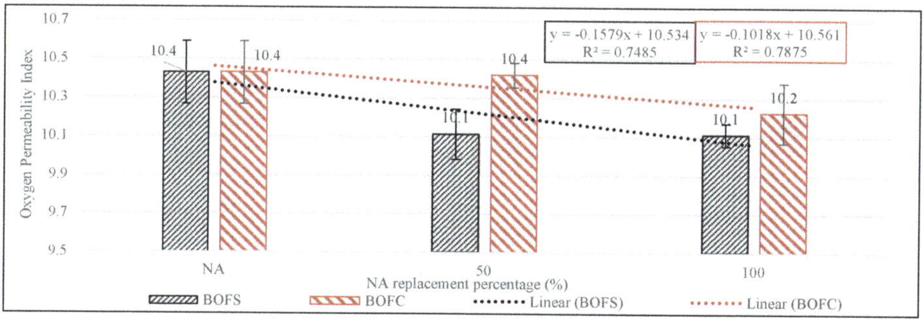

Figure 7: Average OPI test results of the concrete mixes

Conclusion

This study demonstrates that carbonation of BOFS under ambient conditions significantly enhances its performance as a coarse aggregate in concrete. The treatment effectively reduced the water absorption of BOFS and volume instability of the concrete, while improving compressive strength and gas impermeability. The formation of stable $CaCO_3$ within the BOFS microstructure played a key role in these improvements. These findings suggest that carbonation is a practical and scalable approach to increasing the utilisation of BOFS in concrete applications, supporting both sustainability and circular economy objectives in the construction and steel industries. The BOFC-100/PC showed superior compressive strength and volume stability, relative to all other concrete mixes and the oxygen permeability performace was only around 1% less than the NA mix, thus sugesting very insignificant differences in performance. As such the BOFC-100/PC would be the favourable mix for further investigations.

References

[1] Banerjee, S. (2016). "Steel Industry: Here's What Is The Most Significant Element Of Development". Financial Express. Accessed Via: Https://Www.Financialexpress.Com/Market/Steel-Industry-Heres-What-Is-The-Most-Significant-Element-Of-Development/369488/ On The 27th Of March 2024.

[2] Conte, N. (2023). "Ranked: The World's Biggest Steel Producers, By Country". Accessed Via: Https://Www.Visualcapitalist.Com/Biggest-Steel-Producers- Country/#Google_Vignette Accessed On The 20th Of February 2025.

[3]World Steel Association, (2021), "Steel Industry Co-Products Fact Sheet", Accessed Via: Https://Worldsteel.Org/Wp-Content/Uploads/Fact-Sheet-Steel-Industry-Co-Products.Pdf

[4] Ruth, M. (2004). "Basic Oxygen Furnace Production, Encyclopedia Of Energy". Accessed On The 31st Of May 2024 Via: Https://Www.Sciencedirect.Com/Referencework/9780121764807/Encyclopedia-Of- Energy

[5] Topcem Cement. (2024).“Main Ingredients In Cement And Their Functions”, Topcam Cement, 2024, Accessed Via: Https://Topcem.In/Main-Ingredients-In-Cement-And-Their-Functions/#:~:Text=Iron%20oxide%3a%20the%20amount%20of,Tricalcium%20aluminoferrite%20at%20high%20temperatures.

[6] Mei, X., Zhao, Q., Han, C., Wang, Z., Min, Y., Liu, C., Saxén, H., Zevenhoven, R. (2024).“Crystallization Behavior Of Bof Slag During The Cooling Process Towards Leaching For Co2 Sequestration”. Process Safety And Environmental Protection. 191: 2663-2673. Available Via : (Https://Www.Sciencedirect.Com/Science/Article/Pii/S0957582024012680) On The 2nd Of March 2025.

[7] Zago, S.C.; Vernilli, F.; Cascudo, O. (2023), “The Reuse Of Basic Oxygen Furnace Slag As Concrete Aggregate To Achieve Sustainable Development: Characteristics And Limitations”. Buildings 2023.13:1193.

[8] Chen, J., Wei, S. (2016). "Engineering Properties And Performance Of Asphalt Mixtures Incorporating Steel Slag". Accessed Via: Https://Www.Sciencedirect.Com/Science/Article/Abs/Pii/S0950061816316415?Via%3dihub On The 3rd Of April 2024.

[9] Sun, J., Luo, S., Wang, Y., Dong, Q.,Zhang Z. (2023)."Pre-Treatment Of Steel Slag And Its Applicability In Asphalt Mixtures For Sustainable Pavements”. Accessed Via: Https://Www.Sciencedirect.Com/Science/Article/Pii/S138589472305533x#:~:Text=Steel%20slag%20is%20a%20potential,Slag%20can%20facilitate%20sustainable%20developmen T

[10] Zhang, Y. (2023). Journal Of Cleaner Production.

[11] Caltrans. (2013). "Concrete Technology Manual". Accessed Via: Https://Dot.Ca.Gov/-/Media/Dot-Media/Programs/Engineering/Documents/Structureconstruction/Ctm/Sc-Ctm-Chpt3-A11y.Pdf On The 16th Of April 2024.

[12] Liu P., Mo L., Zhang Z., (2023) „Effects Of Carbonation Degree On The Hydration Reactivity Of Steel Slag In Cement-Based Materials“, Construction And Building Materials, Volume 370, 2023, Accessed Via : (Https://Www.Sciencedirect.Com/Science/Article/Pii/S0950061823003641 On The 7th Of September 2024)

[13] Lee, H., Kim, T. W., Kim, S. H., Lin, Y.-W., Li, C.-T., Choi, Y., & Choi, C. (2023). “Carbon Dioxide Capture And Product Characteristics Using Steel Slag In A Mineral Carbonation Plant”. Processes, 11(6), 1676. Https://Doi.Org/10.3390/Pr11061676

[14] Zhao, Q., Liu, C., Mei, X., Saxén, H., & Zevenhoven, R. (2025). “Research Progress Of Steel Slag-Based Carbon Sequestration”. Fundamental Research, 5(1), 282-287. Https://Doi.Org/10.1016/J.Fmre.2022.09.023

[15] Dostie L., Rausis K., Power I., (2024), “Passive Direct Air Capture Using Calcium Oxide Powder: The Importance Of Water Vapor”, Journal Of Cleaner Production, Volume 457, 2024, Accessed Via: Https://Www.Sciencedirect.Com/Science/Article/Pii/S0959652624018420 On The 5th Of Oxtober 2024.

[16] Nielsen, P., Boone, M.A., Horckmans, L. (2020). "Accelerated Carbonation Of Steel Slag Monoliths At Low Co2 Pressure–Microstructure And Strength Development[J]". J. Co2 Util. 36 (2020) 124–134.

[17] Mustafayeva, A., Moon, S.-W., Satyanaga, A., Kim, J. (2024). "Enhancing Mechanical Properties Of Expansive Soil Through Bof Slag Stabilization: A Sustainable Alternative To Conventional Methods". Minerals. 14(11): 1145. Accessed Via: Https://Doi.Org/10.3390/Min14111145 On The 15th Of February 2025.

[18] Nugmanova, A., Shon, C.-S., Kim, J. R., Rossi, C. O. (2023). "Characterizing Chronologically Aged Basic Oxygen Furnace Slags As Aggregates And Their Use In Asphalt Concrete Mix As Filler". Applied Sciences. 13(18): 101. Accesed Via : Https://Doi.Org/10.3390/App131810126 On The 21st Of February 2025

[19] Venkatraman, S., Choudhary, R., Indurkar, A., Chatterjee, A., Abraham, J., Ostrovskiy, S., Senatov F., Locs J., Swamiappan S.(2021) ."Conversion Of Biowaste Into Larnite By Sol-Gel Combustion Route For Biomedical Applications".2021. Chemistry Select. Accessed Via: Https://Www.Researchgate.Net/Publication/358084494_Conversion_Of_Biowaste_Into_Larnite_By_Sol-Gel_Combustion_Route_For_Biomedical_Applications On The 22nd Of February 2025.

[20] Ko M., Chen Y., Jiang J., (2015), "Accelerated Carbonation Of Basic Oxygen Furnace Slag And The Effects On Its Mechanical Properties", Construction And Building Materials, Volume 98, Accessed Via: Https://Www.Sciencedirect.Com/Science/Article/Pii/S0950061815302695 On Th3 11th Of April 2024.

[21] Gora J., Piasta W., (2020), "Impact Of Mechanical Resistance Of Aggregate On Properties Of Concrete ", Case Studies In Construction Materials, Volume 13, Accesssed Via : Https://Www.Sciencedirect.Com/Science/Article/Pii/S2214509520301108 On The 15th Of February 2025.

[22] Miura T., Sato K., Nakamura H., (2021), " Influence Of Primary Cracks On Static And Fatigue Compressive Behaviour Of Concrete Under Water", Construction And Building Materials, Volume 305, Accessed Via : Https://Www.Sciencedirect.Com/Science/Article/Abs/Pii/S0950061821025095 On The 25th Of February 2025.

[23] Mustafayeva, A., Moon, S.-W., Satyanaga, A., Kim, J. (2024). "Enhancing Mechanical Properties Of Expansive Soil Through Bof Slag Stabilization: A Sustainable Alternative To Conventional Methods". Minerals. 14(11): 1145. Accessed Via: Https://Doi.Org/10.3390/Min14111145 On The 15th Of February 2025.

Long Term Compressive Strength of Concrete made with Ceramic Waste as a Partial Substitute for Cement and Fine Aggregates

Omphemetse ZIMBILI[1,a*], Ramadhan SALIM[2,b], Julius NDAMBUKI[3,c], Bolanle Deborah IKOTUN[4,d]

[1]Department of Civil and Environmental Engineering and Building Science, University of South Africa, Pretoria, South Africa

[2]Retired Professor, Pretoria, South Africa

[3]Department of Civil Engineering, Tshwane University of Technology, Pretoria, South Africa

[4]Department of Civil and Environmental Engineering and Building Science, University of South Africa, Pretoria, South Africa

[a]zimbiao@unisa.ac.za, [b]wanjalars@gmail.com, [c]NdambukiJM@tut.ac.za, [d]ikotunbd@unisa.ac.za

Keywords: Ceramic, Compressive, Pozzolanic, Validation, Waste

Abstract. In response to environmental concerns from cement production and natural aggregate extraction, this study investigates ceramic waste (CW) as a partial substitute for cement and fine aggregates in concrete. With cement production contributing up to 8% of global CO_2 emissions, this research promotes the use of ceramic tile waste to support sustainable construction and the circular economy. It focuses particularly on the long-term mechanical performance of CW-modified concrete—an area with limited prior exploration. The study followed a two-stage experimental approach: an Initial Experiment (IE) conducted and completed ~9 years ago, and a more recent Validation Experiment (VE) to confirm and extend the findings. The optimized mix, labelled 16CC50CS, used 16% ceramic waste powder (CWP) in place of cement and 50% crushed ceramic sand (CS) instead of natural sand. Concrete specimens were tested for compressive strength over extended curing periods - up to 270 days in the VE. Results showed notable long-term strength improvements in CW-based mixes. Although the control mix showed better early strength in the IE, the 16CC50CS mix achieved 54.5 MPa vs. 44.4 MPa at 28 days, and in the VE, 69.7 MPa vs. 51.8 MPa at 270 days. Strength gains were linked to continued pozzolanic reactions, improved particle packing, and denser microstructure. It was concluded that ceramic waste, when properly processed and proportioned, enhances strength and durability while lowering the environmental impact. The study also confirmed reproducibility across different materials and test settings. The findings supported the practical use of CW in conventional concrete and call for further exploration in high-performance applications and under diverse conditions.

Introduction

The global construction industry significantly impacts the environment, primarily due to cement production and the extensive use of natural aggregates. Cement manufacturing is responsible for approximately 8% of worldwide CO_2 emissions, emphasizing the need for sustainable alternatives [1]. Incorporating waste materials like ceramics helps mitigate these environmental issues while also improving certain properties of concrete.

Ceramic waste (CW) is predominantly composed of silica and alumina, materials known for their pozzolanic properties. These properties enable CW to interact chemically with calcium hydroxide during cement hydration, forming additional calcium silicate hydrate (C-S-H), which improves the microstructure and mechanical properties of concrete over time [2]. The high abrasion resistance and durability of CW also make it suitable as a partial substitute for fine aggregates. This material also improves the microstructure of concrete by reducing void spaces and improving interfacial bonding, ultimately leading to higher compressive strength [3]

The compressive strength of concrete is a fundamental measure of its load-bearing capacity and durability. Studies have revealed that replacing 20–30% of fine aggregates with CW improves long-term compressive strength by optimizing particle packing and minimizing porosity [4]. Additionally, substituting 10–20% of cement with ceramic waste powder (CWP) enhances mechanical properties due to its pozzolanic reactivity, resulting in better bonding within the concrete matrix [5]. Similarly, substituting up to 20% of cement with CWP results in an optimized mix design that achieves comparable or improved long-term strength [6].

Research also highlights the role of CW in improving concrete's durability under harsh environmental conditions, such as resistance to freeze-thaw cycles and chloride ingress. These properties are critical for the longevity of structures in adverse climates [7]. In self-compacting concrete (SCC), CWP improves both early-age and long-term compressive strength. The performance gains are attributed to the refined particle size distribution and pozzolanic reactivity, which densify the cementitious matrix[1]. Furthermore, SCC incorporating CW has shown enhanced workability and reduced shrinkage, making it ideal for use in complex architectural designs [8]

The adoption of CW in concrete production supports the circular economy by repurposing waste into a valuable resource. This practice reduces the volume of CW destined for landfills and minimizes the extraction of natural aggregates [3]. Moreover, it significantly lowers the carbon footprint of cement production [9]. Despite its advantages, challenges remain regarding the variability of CW composition and the need for standardized processing techniques. Current research efforts focus on exploring the long-term performance of concrete made from optimized mixes of partial substitutions of CW as both cement and fine aggregates. Future studies should also evaluate the suitability of CW in advanced applications such as high-performance concrete and its effects on properties like thermal resistance and dimensional stability [7]

Methodology

This study investigated the long-term compressive strength of concrete incorporating CW as a partial replacement for both cement and fine aggregates. The research focused on comparing the performance of a control mix with a modified mix containing CW as substitute for cement (16%) and fine aggregates (50%), designated as 16CC50CS. This study is trying to replicate or confirm an initial experiment (IE) result, which was conducted previously, and further explore the long-term compressive strength results with this validation experiment (VE). The experimental program included material selection, concrete mix design, specimen preparation, and compressive strength testing at various curing ages.

Materials Selection (IE and VE). Ordinary Portland Cement (OPC) CEM I, sourced from PPC, South Africa, was used as the primary binder in both the control and test mixes.

Aggregates: During the IE, 19 mm Coarse aggregates of the Norite group from Raumix Quarry in Rosslyn, Pretoria North, South Africa and fine aggregates obtained from crushed Meta Quartzite Group's quarry in Baviaanspoort, Pretoria, South Africa were used. During the VE, All natural aggregates were obtained from the AfriSam Jukskei Quarry in Midrand, South Africa. A detailed particle size distribution analysis was performed on the crusher sand (fine aggregates), and a similar particle size distribution was created using CW, to ensure consistency during substitution. A similar concept was used during the IE. The coarse aggregates conformed to the South African National Standard (SANS), SANS 1083:2013 [10].

Ceramic Waste (CW): Ceramic wall tile wastes were sourced from the SAMCA plant in Babelegi, Hammanskraal, South Africa for both the IE and VE. The CW was processed into two fractions:

- *Ceramic Waste Powder (CWP – also referred to as Ceramic Cement (CC)):* Fine ceramic waste passing through a 75-micron sieve was used to replace 16% of OPC in the test mix.
- *Ceramic Sand (CS):* Crushed ceramic waste was processed to match the particle size distribution of natural crusher sand and was used to replace 50% of the fine aggregate in the test mix.

The mentioned proportions of 16% CWP as cement, and 50% CS as fine aggregate were optimised during the IE (termed 16CC50CS).

Methods. Concrete Mix Design and Preparation. The Cement & Concrete Institute (C&CI) of South Africa provides commonly used mix design methods, heavily influenced by the ACI Standard 211.1 (1991) [11]. This research utilized the C&CI method, which is based on four key assumptions [12]. The compressive strength of fully compacted concrete, when cured under standard conditions, is primarily determined by the water-to-cement ratio (w:c) and the type of cementitious material used.

- The water required per unit volume to achieve the desired workability remains relatively constant, largely depending on aggregate and cement properties, maximum aggregate size, and workability range. This assumption is generally valid for strengths between 20 and 30 MPa, but adjustments may be needed outside this range.
- Each concrete mix has an optimum coarse aggregate content, influenced by aggregate size, shape, compacted bulk density, fineness modulus of sand, and required consistency.
- The volume of compacted concrete is the sum of the absolute volumes of its components—cement, aggregates, water, and air—with absolute volume calculated using mass and relative density (D) as per Equation 1 below:

$$\text{Absolute volume, m}^3 = \frac{\text{Mass of material,kg}}{D \times 1000} \tag{1}$$

A controlled mix design was developed during the IE and employed for both the control and 16CC50CS specimens (Table 1 and 2). The control mix contained OPC and natural aggregates, while the test mix incorporated CW substitutions. All materials were batched by weight to maintain accuracy and consistency.

Table 1: Control Mix design - per 10 000 cm3 (1 m3 / 100)

Cast	Control	Avg Slump	140 mm
Total no of Cubes	21 Cubes (VE)	Cast Time	12:40
Ingredients	Calc. Weight (kg)	Actual Weight (kg)	
Sand	7,105	7,105	
Coarse Aggregate	11,195	11,195	
Cement	4,64	4,64	
Water	2,784	2,784	

Table 2: 16CC50CS Mix design - per 10 000 cm3 (1 m3 / 100)

Cast	16CC50CS	Avg Slump	30 mm
No of Cubes (Per Mix)	40 Cubes (VE)	Cast Time	13h55
Ingredients	Calc. Weight (kg)	Actual Weight (kg)	
Natural Sand	3,553	3,553	
Ceramic Sand	3,552	3,552	
Coarse Aggregate	11,195	11,195	
Ceramic Cement	0,742	0,742	
Cement	3,898	3,898	
Water	2,784	2,784	

The same mix design was employed during the VE. Concrete was mixed in a drum mixer under standard laboratory conditions. The mixing process involved the sequential addition of dry materials, followed by water and thorough mixing to achieve uniform consistency. Slump tests were conducted according to SANS 5862-1:2006 [13] for both mixes to assess workability.

Specimen Casting and Curing (VE). A total of 61 concrete cube specimens (100 × 100 x 100 mm) were cast, comprising 20 control specimens and 41 16CC50CS specimens. The fresh concrete was poured into standard cube moulds in three layers, with each layer compacted using a steel rod to eliminate entrapped air. The moulds were vibrated to ensure proper consolidation. After casting, the specimens were covered with plastic sheets and left undisturbed for 24 hours. They were then demoulded and transferred to a curing tank, where they were submerged in water at a controlled temperature of ±22°C until the designated testing ages.

Compressive Strength Testing (VE). Compressive strength tests were conducted at 3, 7, 28, 56, 92, and 270 days to evaluate strength development over time. Testing was performed in accordance with SANS 5863:2006 [14] using a calibrated compression testing machine shown in figure 1. Each specimen was placed centrally in the testing apparatus, and a compressive load was applied at a constant rate until failure. The maximum load recorded was used to calculate compressive strength. It is important to note that the IE was conducted as an exploratory test, with compressive strength measured only at 3 and 28 days. During the IE, the research objectives differed from those of the VE, which focused on long-term strength development. The results obtained from the test specimens were compared against the control mix to assess the impact of CW substitution on long-term compressive strength development.

Figure 1: Compressive strength testing machine that was used

Results and Discussion

The particle size distribution results for the crusher sands used in both the IE and VE are shown in figure 2 below.

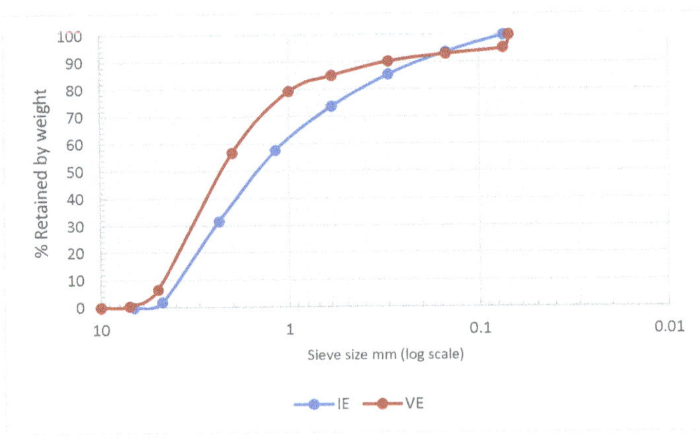

Figure 2: Granulometric curve for fine aggregates (IE vs VE)

The particle size distribution of fine aggregates significantly influenced concrete's workability, strength, and durability. It was important to note that the sieve sizes used in the experiments were not entirely identical, as the SABS standard sieve sizes were modified over time. The particle size distribution data for fine aggregates from both the IE and the VE exhibited a high degree of consistency, indicating reliable and reproducible results. Both experiments followed similar trends, with most of the material passing through finer sieves (0.3 mm and 0.15 mm) and minimal retention at coarser sizes. The cumulative percentage passing values closely aligned, particularly at key sieve sizes such as 0.6 mm (85.18% for VE and 73.75% for IE) and 0.15 mm (93.06% for

VE and 93.78% for IE), confirming the predominance of fine particles. A notable difference in the comparison was the absence of material retained on the pan in the IE, which suggested the presence of even finer particles in the VE. Additionally, in both the VE and IE, the 16CC50CS mixes exhibited virtually no slump, indicating stiff mixtures due to the absence of admixtures. These results demonstrated strong agreement between the datasets, ensuring accuracy and consistency in comparison.

Discussion of Concrete Compressive Strength Results. Figures 2 and 3 below show the compressive strength results of the IE.

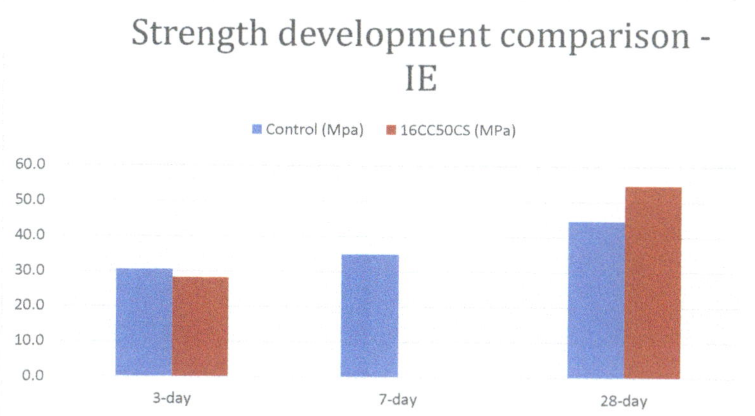

Figure 3: Strength development comparison - IE

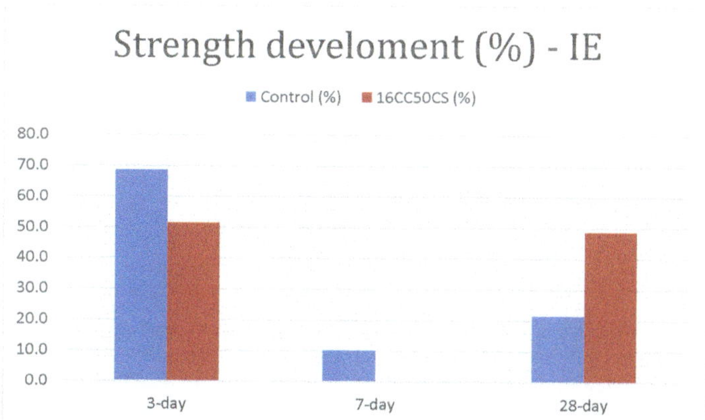

Figure 4: % Strength Development (in stages, up to 100% strength) - IE

During the IE, the control concrete exhibited a higher early-age strength than the 16CC50CS mix at 3 days (30.4 MPa vs. 28.0 MPa). However, by 28 days, 16CC50CS concrete surpassed the control with a compressive strength of 54.5 MPa compared to 44.4 MPa.

The results for the VE are shown in figure 4 and 5.

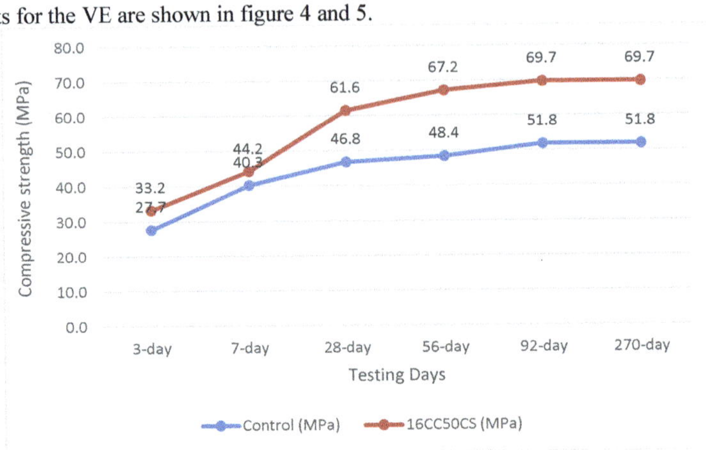

Figure 5: Compressive strength comparison – VE

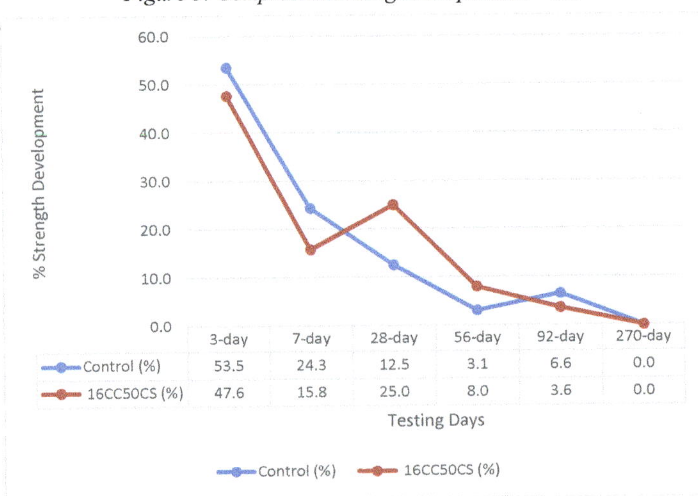

Figure 6: % Strength Development (in stages, up to 100% strength) - VE

The VE results further validated these trends, showing an even more pronounced difference in strength between the two mixes. At 3 days, the 16CC50CS mix already exceeded the control mix (33.2 MPa vs. 27.7 MPa), reversing the early-age trend seen in the IE. This could be attributed to the VE mix having more fines in the fine aggregates content. This advantage persisted throughout all testing ages, with the 16CC50CS mix achieving significantly higher compressive strength at 28 days (61.6 MPa vs. 46.8 MPa) and continuing to outperform the control up to 270 days (69.7 MPa vs. 51.8 MPa). The percentage strength development data further highlighted the long-term benefits of CW substitution, with 16CC50CS concrete demonstrating a higher rate of strength gain between 28 and 56 days (25.0% vs. 12.5% for the control), indicating continued pozzolanic activity and improved particle packing.

Conclusions

Both experiments showed that 16CC50CS concrete exhibited superior long-term strength development compared to control concrete, despite a slightly slower initial gain in the IE. The VE results reinforced the findings of the IE, showing consistent strength improvements at all ages. The enhanced performance of the 16CC50CS mix can be attributed to the pozzolanic reactions of CW, which contribute to a denser microstructure and increased long-term strength. These results indicated that CW could serve as an effective sustainable alternative to conventional cement and fine aggregates, enhancing both mechanical performance and environmental sustainability.

Acknowledgments

The authors would like to express sincere gratitude to the Ceramic Industries SAMCA Wall Tiles Factory in Babelegi, Hammanskraal, for their generous donation of ceramic wall tile waste, which was critical to the successful execution of this study. Appreciation is also extended to the staff at the Civil Engineering Laboratory at the University of South Africa (UNISA) for their support and cooperation throughout the experimental phase. A special thanks is due to Ms. Palesa Tshetlanyane, Technologist at the UNISA laboratory, whose technical assistance and commitment were invaluable to the progress and completion of the research.

References

[1] A. S. Tarek, E.-D. A. Salah, and T. M. Reda, "Effect of High-Volume Ceramic Waste Powder as Partial Cement Replacement on Fresh and Compressive Strength of Self-Compacting Concrete," *Journal of Materials in Civil Engineering*, vol. 31, no. 2, p. 04018374, Feb. 2019. https://doi.org/10.1061/(ASCE)MT.1943-5533.0002588

[2] M. Samadi *et al.*, "Waste ceramic as low cost and eco-friendly materials in the production of sustainable mortars," *J Clean Prod*, vol. 266, p. 121825, Sep. 2020. https://doi.org/10.1016/J.JCLEPRO.2020.121825

[3] A. Sivakumar, S. Srividhya, V. Sathiyamoorthy, M. Seenivasan, and M. R. Subbarayan, "Impact of waste ceramic tiles as partial replacement of fine and coarse aggregate in concrete," *Mater Today Proc*, vol. 61, pp. 224–231, Jan. 2022. https://doi.org/10.1016/J.MATPR.2021.08.142

[4] S. A. Zareei, F. Ameri, P. Shoaei, and N. Bahrami, "Recycled ceramic waste high strength concrete containing wollastonite particles and micro-silica: A comprehensive experimental study," *Constr Build Mater*, vol. 201, pp. 11–32, Mar. 2019. https://doi.org/10.1016/J.CONBUILDMAT.2018.12.161

[5] S. Ahmad, A. Anwar, B. S. Mohammed, M. B. A. Wahab, and S. A. Ahmad, "Strength behavior of concrete by partial replacement of fine aggregate with ceramic powder," *International Journal of Recent Technology and Engineering*, vol. 8, no. 2, pp. 5712–5718, Jul. 2019. https://doi.org/10.35940/ijrte.B3366.078219

[6] L. Gautam, J. Kumar Jain, T. Alomayri, N. Meena, and P. Kalla, "Performance evaluation of self-compacting concrete comprising ceramic waste powder as fine aggregate," *Mater Today Proc*, vol. 61, pp. 204–211, Jan. 2022. https://doi.org/10.1016/J.MATPR.2021.08.063

[7] M. Amin, B. A. Tayeh, and I. S. Agwa, "Effect of using mineral admixtures and ceramic wastes as coarse aggregates on properties of ultrahigh-performance concrete," *J Clean Prod*, vol. 273, p. 123073, 2020. https://doi.org/https://doi.org/10.1016/j.jclepro.2020.123073

[8] S. Ray, M. Haque, S. A. Soumic, A. F. Mita, M. M. Rahman, and B. B. Tanmoy, "Use of ceramic wastes as aggregates in concrete production: A review," *Journal of Building Engineering*, vol. 43, p. 102567, Nov. 2021. https://doi.org/10.1016/J.JOBE.2021.102567

[9] P. Jain, R. Gupta, and S. Chaudhary, "A literature review on the effect of using ceramic waste as supplementary cementitious material in cement composites on workability and compressive strength," *Mater Today Proc*, vol. 65, pp. 871–876, Jan. 2022. https://doi.org/10.1016/J.MATPR.2022.03.453

[10] SANS 1083, Aggregates from natural resources – Aggregates for concrete, South African Bureau of Standards, Pretoria, 2013

[11] ACI 211.1, Standard Practice for Selecting Proportions for Normal, Heavyweight, and Mass Concrete, ACI Committee 211, Farmington Hills, MI: American Concrete Institute, 1991.

[12] B. Addis and J. Goodman, "Concrete mix design," *Fulton's concrete technology*, pp. 219–228, 2009.

[13] SANS 5862-1, Concrete Tests - Consistence of freshly mixed concrete - Slump test, South Africa Bureau of Standards, Pretoria, 2006

[14] SANS 5863, Concrete tests - Compressive strength of hardened concrete, South African Bureau of Standards, Pretoria, 2006

Materials Research Proceedings 63 (2026) 195-202 https://doi.org/10.21741/9781644904053-22

Crushed Burnt Clay Bricks as Partial Replacement of Sand in Portland Cement Concrete – A Sustainable Construction Approach

Getachew M ASHEBIR[1*], Walied A ELSAIGH[1]

[1]Department of Civil and Environmental Engineering and Building Science, UNISA, South Africa

21002932@mylife.unisa.ac.za

Keywords: Recycled Fine Aggregates, Waste Clay Bricks, Tensile and Compressive Strength, Drying Shrinkage, Sustainable Construction, Portland Cement Concrete

Abstract: The rapid pace of urbanization and infrastructure expansion has triggered a notable escalation in the generation of Construction and Demolition (C&D) waste. This investigation seeks to assess the performance of recycled fine aggregates from Crushed Burnt Clay Brick (CBCB) as a viable sustainable partial substitute for sand in the production of Portland cement concrete. Waste clay bricks were obtained from a land fill near Johannesburg, South Africa and were crushed. The fine particles that pass sieve 5 mm was utilized in this study. Concrete mixes with water- to- cement ratio of 0.55 was created to the control, and with an increase in percentage of CBCB were adjusted to 0.61, 0.66, 0.71, and 0.77. CBCB content of 0, 25, 50, 75, and 100% by mass was used to replace the crusher sand in the mixes. Standard tests were conducted to evaluate the workability, cube compressive strength, tensile strength, and drying shrinkage. While the samples for tensile strength were tested only after 28 days of a standard water curing, the samples for compressive strength were cured for 3, 7, and 28 days before testing. The results showed that up to 75% crusher sand substitution is possible for both tensile and compressive strengths. Also, an increase in CBCB content led to greater shrinkage. Consequently, incorporation of CBCB fines to replace natural sand in Portland cement concrete is viable. Such use can mitigate environmental impacts associated with sand mining and waste burnt clay bricks disposal.

1. Introduction

The rapid expansion of the construction industry has led to a heightened demand for resource materials. Today, structures such as buildings, bridges, roads, and railways are being constructed with increasing frequency, propelled by the growth in population and urbanization trends. This rise has necessitated the building of high-rise structures and the replacement of older, smaller buildings [1]. Concrete, a widely used material in the construction field across the globe, is expected to continue its prevalence [2]. It is a composite made of water, cement, and aggregates. These aggregates, which include sand and gravel, make up approximately 60% to 75% of the concrete's total volume and significantly affect both its fresh and hardened properties, mixture composition, and cost-effectiveness [3]. Portland cement concrete contains approximately 33% to 40% fine aggregates [4]. The demand for aggregates including sand is expected to double over the next 20-30 years [5]. Consequently, the extensive utilization of fine aggregates within the construction sector exerts a considerable impact on the natural environment, particularly influencing riverbeds, which serve as the primary sources of natural sand for industrial purposes. Substantial quantities of construction and demolition (C&D) waste are produced annually during the activities of constructing, renovating, and demolishing buildings and infrastructure. It has been documented that the global annual generation of C&D waste surpasses 10 billion tons [6]. Developing nations encounter notable challenges in waste management stemming from insufficient practices [7]. Also, the pace of construction is rising due to rapid urbanization, industrialization, and economic growth in the developing countries [8]. Research estimates that, over the coming 50 years, bricks will continue to be the second most important building material, following concrete [9]. The disposal of numerous waste clay bricks in South Africa, resulting from construction and demolition debris, construction leftover and bricks' production waste, primarily involves landfilling. Nonetheless, these materials present an opportunity to be utilized in concrete manufacturing, serving as a partial substitute for natural aggregates.

The construction industry faces challenges due to depleting natural resources and the accumulation of C&D wastes, which frequently ends at landfills. Utilizing C&D wastes into concrete as partial replacements of natural aggregates in concrete is vital for reducing their environmental impact and preserving natural resources for future use [10]. To ensure sustainability, a key priority should be exploring every possible means to minimize waste from construction sites [2, 10].

Sustainable development encompasses energy conservation, environmental stewardship, and safeguarding nonrenewable resources. According to various findings, as noted in [11], the use of CBCB helps in lowering the reliance on natural resources and decreases landfill waste, thereby promoting environmental sustainability. The pozzolanic activity of crushed clay bricks improves the durability and lifespan of concrete, which in turn diminishes the ecological impact of construction ventures [12]. Various researchers have explored and advocated for the use of CBCB aggregate in concrete as a partial substitute. The study by Fadia [13] found that utilization of recycled aggregates from crushed clay reduces compressive and tensile strengths of concrete as compared to conventional concrete. In another study [14], substituting recycled clay brick aggregate (0, 50, or 100%) for fine aggregate increased the porosity and pore volume of the concrete; however, the chloride migration coefficient decreased because of the recycled clay brick's pozzolanic activity.

In the study conducted by [15] researchers investigated the impact of substituting fine aggregate with crushed clay brick debris at proportions of 5, 10, 15, and 20%. It was found that a 10% substitution led to optimal compressive strength compared to the control mixture. Additionally, research suggests that up to 50% of natural aggregates can be replaced with recycled material without negatively affecting mechanical properties such as tensile strength and water absorption [16]. The water absorption rate of recycled clay brick is expected to be higher than the natural fine aggregate. Incorporating recycled CBCB into concrete affects its shrinkage behavior, as substituting more recycled aggregates for natural ones enhances drying shrinkage [17]. The use of recycled crushed brick as a sand replacement in concrete offers both advantages and disadvantages, particularly in terms of mechanical properties, durability, and practical feasibility. It was found that incorporating of CBCB in place of natural fine aggregate during production of Portland cement concrete, reduced slump by roughly 71% at 50% replacement and 90% at full replacement, producing noticeably dry and difficult to work concrete [18]. While some results relating to the use of CBCB as a sand substitute in concrete exist, the physical and chemical properties differ worldwide, pointing out to the need for additional research involving local materials.

This study explored the feasibility of employing recycled crushed brick waste as a substitute for crusher sand in Portland cement concrete. The experimental work included X-ray fluorescence (XRF) and sieve analysis on CBCB prior to the use of the material. Concrete mixes containing CBCB of 0, 25, 50, 75, and 100% by mass of the fine aggregate were created. The effect of CBCB content on workability, compressive strength, direct tensile strength, and drying shrinkage of concrete were then assessed.

2. Materials and Experimental Design

Recycled clay bricks were sourced from a landfill (Latitude: -26.237°, Longitude: 27.943°) close to Johannesburg, South Africa, and were crushed. The fines that passed through a 5 mm sieve were used in this study. The chemical composition and particle size distribution were established for the Crushed Burnt Clay Bricks (CBCB). Control concrete mix (i.e., contains crusher sand) and mixes including CBCB with 25%, 50%, 75% and 100% by mass replacement of the natural sand were created. The mixes were used to cast standard cubes of 100 x 100 x 100 mm and drying shrinkage beams of 100 x100 x300 mm. A total of seventy-five samples were prepared to evaluate the compressive strength according to SANS 5863-3:2006 [19], tensile strength as per SANS 6253:2006 [20] and drying shrinkage in accordance with the procedure described in SABS Method 1085:2021 [21]. Slump tests were conducted according to SANS 5862:2006 [22]. The physical properties of the coarse aggregate, crusher sand and CBCB were assessed and compared. Table 1 shows the constituent materials for the control mix and CBCB percentage replacements for one cubic meter.

***Table 1**: Mix composition of concrete cubes.*

Replacement Level	Cement (kg/m^3)	Designed (w/c)	Adjusted (w/c)	Water (1) (kg/m^3)	Crusher Sand (kg/m^3)	CBCB (kg/m^3)	C-Aggregate (2) (kg/m^3)
0%(Control)	340	0.55	0.55	187.00	760	0.00	1113
25%	340	0.55	0.61	207.40	570	190	1113
50%	340	0.55	0.66	224.40	380	380	1113
75%	340	0.55	0.71	241.40	190	570	1113
100%	340	0.55	0.77	261.80	0.00	760	1113

([1]) For mixtures containing CBCB, water is added into the mix until the workability value approaches that of the control mix.
([2]) Granite stones with a maximum size of 22 mm.

The cube and beam specimens were subjected to a standard water curing. Compressive strength tests were performed at 3, 7, and 28 days while cube split tensile strength tests were conducted at 28 days. The drying shrinkage measured using a digital extensometer. Specimens were oven-dried at 55 ± 5°C and cooled before measurement. Shrinkage readings were taken every 48 hours.

3. Results and Discussion

3.1 Chemical Composition of CBCB

The chemical composition of the crushed burnt clay bricks obtained using X-ray fluorescence spectrometer (XRF) technique revealed that the sum of SiO_2, Al_2O_3, & Fe_2O_3 is 89.83% and SO_3 content is 0.06%, which is less than 5.0% by mass. This qualifies the CBCB as class (F) pozzolan according to criteria presented in ASTM C- 618-19 [23].

One can speculate that in the presence of water, CBCB fines (< 0.075 mm) on the surface of the sand particles can chemically react with the calcium hydroxide produced during cement hydration, leading to the formation of additional calcium silicate hydrate (C-S-H) and, consequently, expected to enhanced strength over time

3.2 Particle size distribution

Figure 1 shows the Particle Size Distribution (PSD) for both the crusher sand and the CBCB sand. The results show that CBCB has a slightly more fines the crusher sand. The presence of finer materials increases the surface area and eventually the water demand.

Materials Research Forum LLC
https://doi.org/10.21741/9781644904053-22

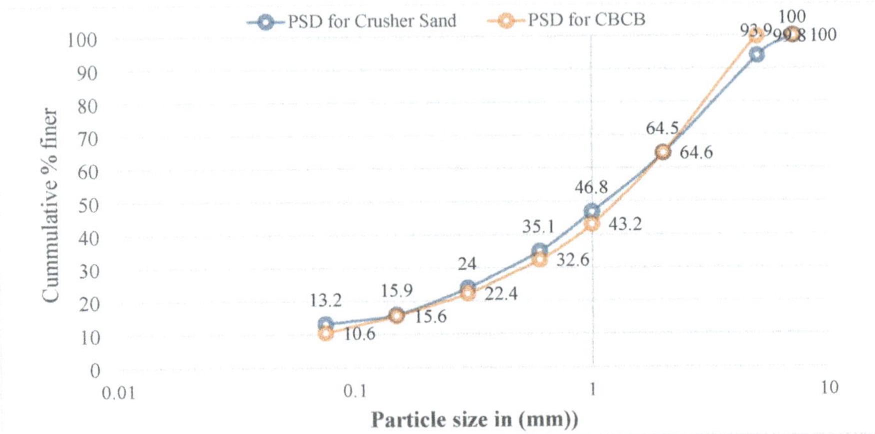

Figure 1: Particle size distribution for crusher sand and CBCB.

3.3 Physical properties of crusher sand and CBCB

The physical property test results of coarse aggregate, recycled CBCB fine aggregate and crusher sand were evaluated as per the relevant SANS standard, and the results have been summarized in Table 2. Crusher sand demonstrates a superior density, reduced silt content, and significantly lower absorption compared to CBCB. Nevertheless, both materials exhibit similar fineness moduli, suggesting analogous particle size distributions.

Table 2: Physical test results of aggregates.

Type of Test	Coarse Aggregate	Crushed Sand	CBCB	Standard Applied
Silt content (%)	-	4.8	12	SANS 1083
Fineness modulus (%)	7.2	3.2	3.22	SANS 1083:2017
Moisture content (%)	0.6	1.4	6.04	SANS 3001-GR20:2010
Loose Bulk Density (kg/m^3)	1,390	1,770	1,110	SANS 5845:2006
Compacted Bulk Density (kg/m^3)	1,500	1,880	1,286	SANS 3001-AG23:2014
Apparent Density (kg/m^3)	2,664	2,651	2,162	SANS 3001-AG20/AG21
Absorption capacity (%)	0.6	1.6	13.8	SANS 5843

3.4 Workability test

Figure 2 indicates that the slump value ranges between 68 and 86 mm. The slump value increases with the increase in the water-to-cement ratio for each mixture. Intuitively, the surface area and absorption capacity of the CBCB have played a role in terms of consuming the extra water (i.e., higher water-to cement ratio) added for the higher replacement percentage. Hence, to maintain workability, the amount of water requirement for CBCB mixtures should be increased as the percentage of replacement of CBCB increased.

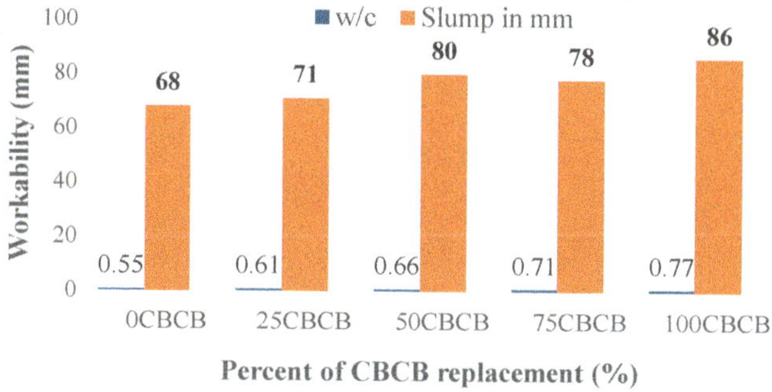

***Figure 2**: Slump test results with different w/c and CBCB percent replacement.*

3.5 Compressive Strength

The compressive strength test results are presented in Figure 3 below.

	0CBCB	25CBCB	50CBCB	75CBCB	100CBCB
Series I (3 days curing)	34.77	34.01	33.56	30.20	30.14
Series I (7 days curing)	43.08	43.03	43.59	37.97	36.75
Series I (28 days curing)	56.13	54.35	56.72	53.65	49.29

Percent of replacement (%)

***Figure 3:** Compressive strength variation of with CBCB percentage and curing periods.*

Figure 3 details the compressive strength of concrete cubes at intervals of 3, 7, and 28 days, using varying CBCB replacement. The control mix (i.e., 0 CBCB), exhibits a strength increase from 34.77 MPa at 3 days to 56.13 MPa at 28 days. Remarkably, the 50CBCB mix achieves a slightly higher 28-day strength (56.72 MPa) compared to the control (56.13 MPa), suggesting a beneficial replacement threshold where CBCB enhances strength, owing to filler effects or improved packing. However, further increase in CBCB percentage lowers strength. There is a negligible decline in compressive strength for the mixes 75CBCB, but greater strength reduction is noticeable as CBCB replacement increases to 100%. In addition, the strength growth for mixes containing CBCB is comparable to that of the control mix. Additional investigation is needed beyond 28 days to determine if CBCB fines are reactive and can enhance long-term strength.

3.6 Direct Tensile Strength
The tensile strength test results at age of 28 days are presented in the Figure 4 below.

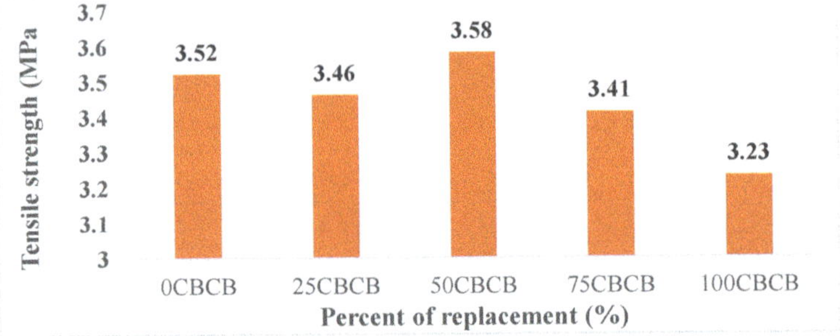

Figure 4: *Tensile strength variation of with CBCB percentage.*

The control mixture, 0CBCB, has a tensile strength of 3.52 MPa Except for the 50CBCB mix, the tensile strength reduces as the CBCB replacement increases. Nonetheless, the decrease in tensile strength is minimal with a replacement of up to 75%, after which the reduction in tensile strength is comparatively significant.

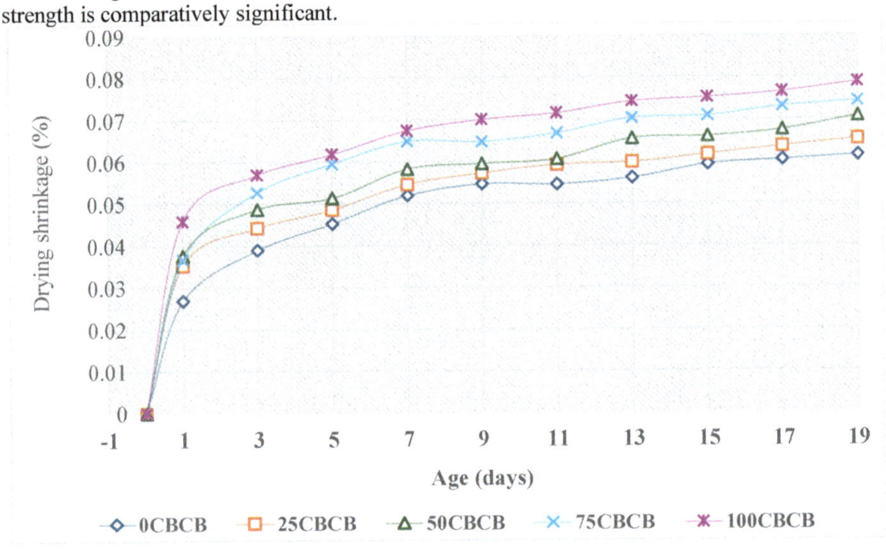

Figure 5: *Accelerated drying shrinkage strain test results.*

3.7 Accelerated Drying Shrinkage Tests
Figure 5 shows the drying results obtained after 19 days. The result reveals that increasing the CBCB fines replacement increases drying shrinkage. The control mix (0CBCB) displays the minimal shrinkage strain, while the 100CBCB mix demonstrates the maximal shrinkage strain. An initial considerable rise in shrinkage observed, followed by a deceleration in the shrinkage rate across all mixes. The high-water absorption of the CBCB fines influences the mix's free water content, thereby impacting drying shrinkage strain. The increase in drying shrinkage associated with higher CBCB contents can be attributed to the absorption capacity of burnt clay bricks

4. Conclusion

In this study, the effect of Crushed Burnt Clay Bricks (CBCB) used as a crusher sand replacement on physical and mechanical properties of concrete was assessed. Based on the experimental results, the following conclusions drafted:

- Crushed Burnt Clay Bricks (CBCB) had high water absorption capacity and higher silt content than crusher sand, which affect the fresh and hardened states of concrete.
- The research indicates that CBCB is highly effective as a partial replacement for fine aggregates (crusher sands) in concrete, though only up to the feasible limits.
- Up to 75% crusher sand can be replaced with CBCB with negligible reduction in compressive and tensile strengths of concrete. However, with excessive CBCB replacement both compressive and tensile strengths decreased.
- The drying shrinkage increased with increasing CBCB replacement percentage. This attributes to the high-water absorption capacity.

From this research, the construction, and the demolition CBCB wastes could replace the crusher sand up to 75%, and this could reduce environmental impacts due to waste disposal and sand mining.

References

[1] N. Kisku, H. Joshi, M. Ansari, S. K. Panda, S. Nayak, and S. C. Dutta, "A critical review and assessment for usage of recycled aggregate as sustainable construction material," *Constr. Build. Mater.*, vol. 131, pp. 721–740, 2017. https://doi.org/10.1016/j.conbuildmat.2016.11.029

[2] M. Saidi, B. Safi, M. F. Samar, and F. Ait-Medjber, "Recycling of Aggregates from Construction Demolition Wastes in Concrete: Study of Physical and Mechanical Properties," *Int. J. Civil, Environ. Struct. Constr. Archit. Eng.*, vol. 8, no. 12, pp. 1236–1240, 2014.

[3] M. L. KOSMATKA, Steven H.; WILSON, *Design and Control of Concrete Mixtures – The Guide to Applications, Methods and Materials.* 2011.

[4] N. K. Pranita and D. . Gadhe, "Use of Recycled Aggregate in Concrete," *Int. J. Eng. Res. Technol.*, vol. 2, no. 2, pp. 1–8, 2013.

[5] R. S. Ingalkar and S. M. Harle, "Replacement of Natural Sand by Crushed Sand in the Concrete," *Landsc. Archit. Reg. Plan.*, vol. 2, no. 1, pp. 13–22, 2017. https://doi.org/10.11648/j.larp.20170201.12

[6] H. Wu, J. Zuo, G. Zillante, J. Wang, and H. Yuan, "Status quo and future directions of construction and demolition waste research : A critical review.

[7] A. Akhtar and A. K. Sarmah, "Construction and demolition waste generation and properties of recycled aggregate concrete: A global perspective," *J. Clean. Prod.*, vol. 186, pp. 262–281, 2018. https://doi.org/10.1016/j.jclepro.2018.03.085

[8] W. Z. Taffese, "Suitability Investigation of Recycled Concrete Aggregates for Concrete Production: An Experimental Case Study," *Adv. Civ. Eng.*, vol. 2018, 2018. https://doi.org/10.1155/2018/8368351

[9] A. G. Office, *Scoping Study to Investigate Measures for Improving the Environmental Sustainability of Building Materials*, no. December. 2006.

[10] S. P. Chandar, N. Thakur, C. L. K. Reddy, and T. S. P. Reddy, "Experimental Investigation on Partial Replacement of Fine Aggregates by Demolished Concrete in the Production of Normal Concrete," vol. 8, no. 4, pp. 386–393, 2017.

[11] J. Andal, M. Shehata, and P. Zacarias, "Properties of concrete containing recycled concrete aggregate of preserved quality," *Constr. Build. Mater.*, vol. 125, pp. 842–855, 2016. https://doi.org/10.1016/j.conbuildmat.2016.08.110

[12] U. Kalsum, "Recycling of Clay Based Demolition Wastes for the Production of Concrete Block," vol. 2008, no. Icenv, pp. 1–5, 2008.

[13] D. Wu, W. Chu, L. Wang, W. Wang, and H. Wang, "Preliminary Feasibility Investigation on Reutilization of Recycled Crushed Clay Bricks from Construction and Demolition Waste for Cement-Stabilized Macadam," 2022.

[14] S. K. Fadia, "Use of Crushed Bricks as Coarse Aggregate in Concrete.," *Tikrit J. Eng Sci.*, vol. 16(3), pp. 64–69, 2009. https://doi.org/doi:10.25130/tjes.16.3.06

[15] J. Dang, J. Zhao, S. Dai, and S. Zhao, "Durability and microstructural properties of concrete with recycled brick as fine aggregates," *Constr. Build. Mater.*, vol. 262, p. 120032, 2020. https://doi.org/10.1016/j.conbuildmat.2020.120032

[16] R. Veerakumar and R. Saravanakumar, "A Detailed Study on Partial Replacement of Fine Aggregate with Brick Debries," vol. 9, no. 3, pp. 1028–1036, 2018.

[17] M. C. Llanes, "Construction and demolition waste as recycled aggregate for environmentally friendly concrete paving," pp. 9826–9840, 2022.

[18] J. M. Gómez-Soberón, J. C. Gómez-Soberón, and L. A. Gómez-Soberón, "Shrinkage and creep of recycled concrete interpreted by the porosity of their aggregate," *Vol. 5*, pp. 563–576, 2002.

[19] SANS 5863-3: 2006: Concrete Testing: Making and Curing of Test Specimens. In: South African Bureau of Standards (ed.), Pretoria, South Africa.

[20] SANS (South African National Standard) 2006a. *SABS Method 6253:2006 Concrete Tests - Tensile Splitting Strength of Concrete.* Pretoria

[21] SABS Test Method, 1085:2021: Initial Drying Shrinkage and Wetting Expansion of Concrete (2021).

[22] South African National Standards SANS 5862: 2006. SABS; Pretoria: 2006. Consistence of Freshly Mixed Concrete – Slump Test.

[23] ASTM C618-23e. 2023. Standard Specification for Coal Ash and Raw or Calcined Natural Pozzolan for Use in Concrete.

Advances in Cement and Concrete Research - ICACCR 2025
Materials Research Proceedings 63 (2026) 203-210

Materials Research Forum LLC
https://doi.org/10.21741/9781644904053-23

Valorization of Corn Cob Ash as a Sustainable Cement Substitute for Low-Carbon Concrete Block Production

Uche Emmanuel EDIKE[1,2*]

[1]Department of Civil Engineering Science, Faculty of Engineering and the Built Environment, University of Johannesburg, South Africa

[2]Department of Building, Faculty of Environmental Sciences, Southern Delta University, Ozoro, Nigeria

uchee@uj.ac.za, edikeuch84@gmail.com

Keywords: Cement Composition, Constituents of Pozzolans, Concrete Block, Global Warming Potential, Pozzolans; Sustainability

Abstract. The construction industry is one of the most vicious contributors to global GHG emissions, and the sector has a pivotal role to play in mitigating environmental degradation. This study examines the use of corn cob ash (CCA) as a partial substitute for cement in the production of concrete blocks. The study addresses the urgent need to identify and adopt sustainable alternative materials in the construction industry to mitigate greenhouse gas (GHG) emissions and promote environmental sustainability. CCA was used to partially replace OPC at 0%, 5%, 10%, 15%, and 20% by mass. The EDXRF, mechanical properties, embodied energy, global warming potential, and sustainability index of the composite materials were evaluated for each mix. The study found that the optimum mix of 15% CCA led to a 14.43% reduction in GHG emissions compared to the control sample. The study reveals a complex relationship between compressive strength and carbon footprint in concrete composites. The research provides empirical evidence supporting the valorization of CCA as a viable pozzolan in concrete block production. The adoption of CCA as a partial cement replacement leads to lower embodied energy, enhanced waste management benefits, and a reduced carbon footprint for masonry units.

1. Introduction

The construction industry is a significant contributor to global greenhouse gas (GHG) emissions, with substantial implications for environmental sustainability, resource efficiency, and climate change [1]. The GHG emissions challenge is particularly acute in developing countries like Nigeria, where the urgent demand for infrastructural development intersects with the imperative to decarbonize. In a multi-regional input-output (MRIO) analysis of 44 countries, [2] established that the built environment accounts for approximately 39% of global GHG emissions, with notable intensity in countries reliant on concrete and steel production. To mitigate GHG emissions, [3] suggests the adoption of green construction practices and robust policy frameworks.

Pozzolanic materials are either naturally occurring or artificially produced from agricultural or industrial waste. Natural pozzolanic materials include volcanic ash, diatomaceous earth, and calcined clays [4]. Pozzolans and pozzolanas improve the mechanical performance of concrete and enhance its durability properties [5], thereby reducing the environmental impact of cementitious composites and promoting more sustainable construction practices.

Despite the numerous benefits of CCA incorporation in concrete, CCA applications in cementitious systems are not without limitations. Excessive replacement levels (>15–20%) can impair workability, slow setting times, and reduce early-age strength due to insufficient calcium availability [6]. The variability in CCA's chemical composition due to source and combustion conditions poses a challenge to standardization [7], and it may not allow direct engineering application of corn cob ash from any region without thorough investigation. The research optimizes the application of CCA in block production and provides insights into the sustainability potentials of CCA in block production.

2. Materials and experimental procedures

The study utilized CCA, OPC, natural river sand, and water. The natural river sand used for producing the blocks has 80% of the particle size of less than 4.75 mm. The corn cob, CCA and concrete blocks produced are presented in Fig 1. Compressive and flexural strength tests were carried out on six samples of blocks produced with 0%, 5%, 10%, 15%, and 20% of corn cob ash replacement of ordinary Portland cement (OPC) by mass.

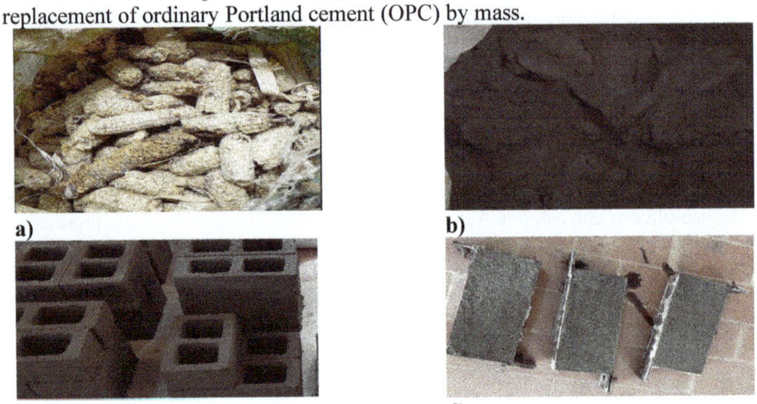

a) b) c) d)

Fig 1: Corn cob ash production and samples prepare for testing: *a) Corn cob; b) Corn cob ash after calcination; c) CCA and OPC concrete blocks; d) CCA and OPC concrete prisms for flexural test*

Embodied energy assessment, Global Warming Impact and Sustainability index

The energy intensity and embodied Carbon emission of the various materials employed in the production process of the concrete block are presented in Table 1, while material haulage distances are shown in Table 2. The Embodied energy of corn cob ash was determined using Equation (1)

$$EE = (1.20)\sum_{i=1}^{n}(m_w \times EE_f) \tag{1}$$

Where EE= Embodied energy (MJ/kg); m_w = mass of material (kg); EE_f = coefficient of embodied energy.

Table 1: Embodied energy and GHG emissions carbon equivalent coefficients

Materials	**Embodied energy coefficient, EEf (MJ/kg)**	**Embodied carbon equivalent, GWPf ($kgCO_2e/kg$)**	**Source**
Ordinary Portland Cement	5.85; 5.50	0.95	[8], [9]
Sharp sand	0.081	0.0048; 0.0051	[8], [10]
Water	0.01	-	[8]
Transportation, truck by road	2.4 MJ/tkm	0.15 $kgCO_2e/tkm$	[8]

Table 2: Material haulage distance

Material	Place of collection	Transportation distance (km)	Coordinates of location
OPC	Ibese	62.4	6.9528° N, 3.0388° E
River sand	Ota	7.5	6.6926° N, 3.2366° E
Corn cob	Ota	10	6.6888° N, 3.2265° E
Water	Bells University of Tech.	0	6.6859° N, 3.1711° E

Global warming impact is an environmental effect indicator that evaluates the combined impact of different GHG emissions that have the potential to alter the mean temperature of the atmosphere [11]. The global warming impact of utilizing corncob ash was assessed using the Inventory of Carbon and Energy (ICE) following [12]. The global warming potential (GWP) was computed as a function of the mass of material and the coefficient of the greenhouse gas emission in $KgCO_2$ eq/kg using Equation 3. The corn cob ash carbon equivalent coefficient was computed using Equation 2 [8], [10]. The carbon equivalent coefficient of corn cob ash is 5.88×10^{-4} $kgCO_2e/kg$. The embodied Carbon equivalent for sand and cement is 0.0051, and 0.95 $kgCO_2e/kg$, respectively [8].

$$EC = 0.021 \times EE \tag{2}$$

$$GWP = (1.15)\sum_{i=1}^{n}(m_w \times GWP_f) \tag{3}$$

Where EC is the embodied carbon equivalent coefficient in $kgCO_2e/kg$; EE is the embodied energy in MJ/kg; m_w is mass of material in kg; GWP_f is material embodied carbon equivalent coefficient in $kgCO_2e/kg$.

The sustainability index is the measure of the environmental impact of the concrete in relation to the characteristic compressive strength [13], [14]. Sustainability Index is expressed in two forms as the characteristic compressive strength compared to kg of Carbon dioxide equivalent emissions as presented in Equations 4 and 5 [13][14], [15], [16].

$$SI = \frac{f_{cu}}{GWP} \tag{4}$$

$$SI = \frac{GWP}{f_{cu}} \tag{5}$$

Where SI in the Sustainability Index ($N/mm^2.kgCO2e$ or $MPa/kgCO_2e$), f_{cu} is the characteristic compressive strength at 28 days (N/mm^2 or MPa); GWP is global warming potential ($kgCO_2e$).

3. Results and Discussions

Chemical characteristics of corn cob ash (CCA)

Energy Dispersive X-ray Fluorescence (EDXRF) analysis was performed to investigate the elemental and oxide constituents of corn cob ash (CCA) and Ordinary Portland Cement (OPC). The EDXRF spectrum of CCA indicated Zirconium Zr, and Strontium Sr, with a peak atom amount reaching 22000 cps/mA. Details of the principal constituent elemental oxides of the CCA and OPC measured in % weight as obtained in the EDXRF spectrum are shown in Table 3.

Table 3: Chemical analysis of CCA and OPC

Constituents	SYMBOLS	OPC (%)	CCA (%)
Calcium Oxide	CaO	49.60	2.98
Strontium Oxide	SrO	9.70	9.70
Silica	SiO_2	13.13	47.29
Aluminium Oxide	Al_2O_3	2.57	13.70
Iron Oxide	Fe_2O_3	0.64	2.32
Magnesia	MgO	1.01	5.04
Sulphur Trioxide	SO_3	0.88	0.84
Phosphorus Pentoxide	P_2O_5	0.47	4.88
Potassium Oxide	K_2O	0.21	7.80

The Calcium Oxide (CaO) content in CCA and OPC is approximately 2.98% and 49.6%, respectively. The low CaO content in CCA suggests that CCA may not substantially contribute to early strength development. Thus, in an OPC and CCA binder blend, the OPC is responsible for the early development of strength.

Effect of corn cob ash on compressive and flexural strength of concrete blocks

The results of the compressive and flexural strength tests are presented in Fig. 2. From the figure, the compressive strength of all the concrete block samples increased with curing age, with the highest value for each sample at the 28-day test. This finding aligns with the results of [17], [18], which indicate that pozzolans enhance the compressive strength of concrete. The flexural strength test results show a steady decrease in flexural strength with an increase in CCA content (Fig. 4b).

Embodied energy, Global Warming Impact and Sustainability index

The result of the embodied energy of corn cob ash is presented in Table 4, while Table 5 and Fig. 3 show the embodied energy of the control concrete. The embodied energy is less than the 1.35 MJ/kg obtained for corn cob ash by Abubakar et al. [10] . This result is similar to the 0.97 MJ/kg provided for 1:5 mortar by Hammond and Jones [8] in the Inventory of Carbon and Energy (ICE). The result in Fig. 4 indicates a steady reduction in global warming potential (GWP) with the increase in CCA content reaching 19.23% at the 20% replacement of OPC.

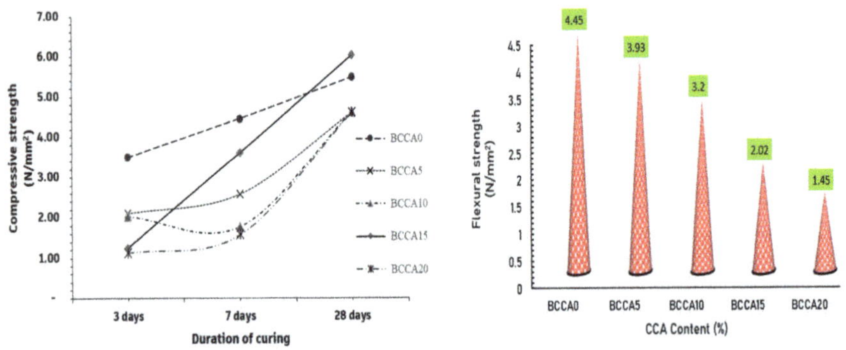

a) b)

Fig. 2: Effect of corn cob ash on the strengths of concrete blocks: a) compressive strength; b) flexural strength

Table 4: Embodied energy of Corn cob ash

Materials	Mass (kg)	Energy Coefficient (MJ/kg)	Transportation distance	Transportation coefficient (MJ/tkm)	Energy sum (MJ)	20% waste (MJ)	Total
Corn cob	20	0.0	10	2.4	0.48	0.096	0.576
Thus, energy coefficient of Corn cob ash per kg							0.0288 MJ/kg

Table 5: Embodied energy of control concrete 0% corncob ash

Materials	Mass (kg)	Energy Coefficient (MJ/kg)	Transportation distance (km)	Transportation coefficient (MJ/tkm)	Energy sum (MJ)	15% waste (MJ)	Total (MJ)
Corn cob	0	0.0	0	2.4	0	0	0
OPC	1.5	5.5	62.4	2.4	8.47	1.27	9.74
River sand	9	0.081	7.5	2.4	0.891	0.134	1.023
Water	0.75	0.01	0	0	0.0075	0.0011	0.0086
Total	11.25				9.366	1.405	10.77
Thus, energy coefficient of concrete mix per kg (MJ/kg)							0.957

Sustainability Index serves as a pivotal metric in advancing eco-efficient concrete technologies. The Sustainability Index, SI, was assessed using the ISO 14064-1 [12]. The result of the analysis of sustainability Index is shown in Fig. 5.

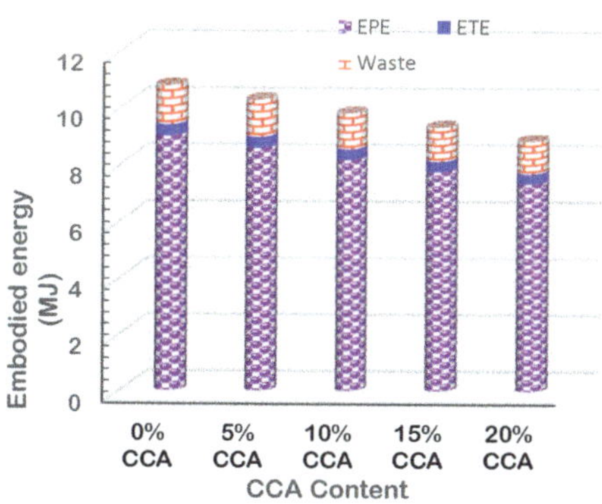

(a) Embodied energy constituents (MJ)

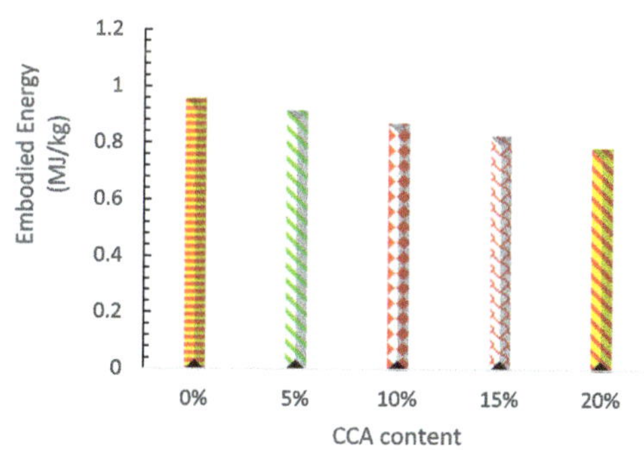

(b) Embodied energy per unit mass

Fig. 3: Impact of corn cob ash on embodied energy of concrete

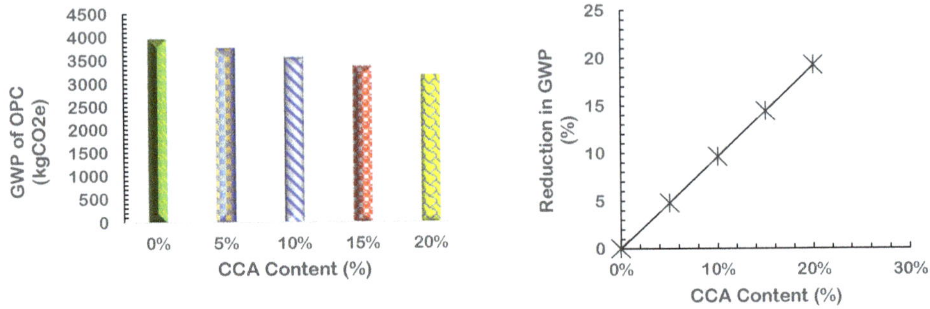

(a) Global warming potential (b) Percentage reduction in GWP

Fig. 4: Impact of CCA on global warming potential of concrete

The results highlight the complex relationship between compressive strength and the global warming potential (GWP) of concrete. The substitution of OPC with CCA initially led to a decline in the Sustainability Index at 5% and 10% replacement levels, indicating a reduction in carbon efficiency. However, at 15% replacement, the Sustainability Index improved significantly by 28.61%, suggesting that moderate inclusion of CCA can optimize both mechanical and environmental performance.

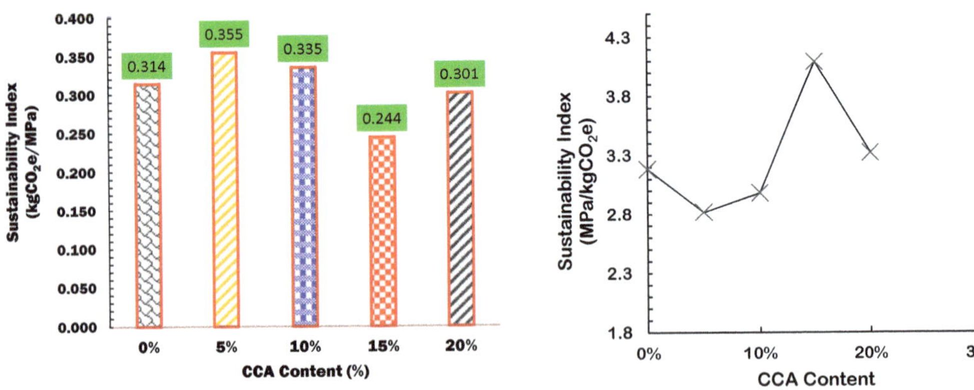

(a) GHG emission per unit MPa concrete (b) SI effect on strength gains per unit GHG

Fig. 5: Impact of CCA content on Sustainability Index.

Conclusion

This study explored the use of Corn Cob Ash (CCA) as a replacement for Ordinary Portland Cement (OPC) in concrete blocks. Chemical analysis revealed that CCA meets the requisite pozzolanic standards, with a significant content of silicon dioxide. CCA-based concrete exhibited initial strength retardation but showed enhanced strength at 28 days, with a 15% substitution, driven by secondary hydration resulting from CCA's pozzolanic properties. Additionally, increasing CCA content resulted in reduced energy consumption and a 14.43% decrease in global warming potential (GWP) at optimal levels. While lower substitution levels indicated a decline in sustainability, a 15% replacement improved the sustainability index by 28.61%. The study emphasizes that utilizing agricultural waste, such as CCA, can effectively reduce the carbon footprint of concrete while maintaining its structural integrity, thereby promoting sustainable concrete design through innovative material utilization.

References

[1] S. Junaid, A. H. Khan, and P. Alam, "Cement replacement using pozzolanic materials: a sustainable approach," in *Recent Developments and Innovations in the Sustainable Production of Concrete*, Elsevier, 2025, pp. 185–214. https://doi.org/10.1016/B978-0-443-23895-6.00008-X

[2] R. H. Crawford, "Greenhouse Gas Emissions of Global Construction Industries," *IOP Conf. Ser.: Mater. Sci. Eng.*, vol. 1218, no. 1, p. 012047, Jan. 2022. https://doi.org/10.1088/1757-899X/1218/1/012047

[3] M. Tampubolon, "The Impact of CO2 Emissions in Indonesia's Construction Industry: A Human Development Perspective," Jan. 22, 2025, *Environmental and Earth Sciences*. https://doi.org/10.20944/preprints202501.1668.v1

[4] B. C. Olaiya, M. M. Lawan, K. A. Olonade, and O. O. Segun, "An overview of the use and process for enhancing the pozzolanic performance of industrial and agricultural wastes in concrete," *Discov Appl Sci*, vol. 7, no. 3, p. 164, Feb. 2025. https://doi.org/10.1007/s42452-025-06586-1

[5] M. Amin, Y. Elsakhawy, K. Abu El-Hassan, and B. A. Abdelsalam, "Effects of pozzolanic treatment method of aggregate on recycled aggregate high strength concrete properties," *Journal of Building Engineering*, vol. 108, p. 112898, Aug. 2025. https://doi.org/10.1016/j.jobe.2025.112898

[6] K. A. Mujedu, S. A. Adebara, and I. O. Lamidi, "The Use of Corn Cob Ash and Saw Dust Ash as Cement Replacement in Concrete Works," *The International Journal of Engineering and Science*, vol. 3, no. 4, pp. 22–28, 2014.

[7] J. Ahmad, M. M. Arbili, H. Alabduljabbar, and A. F. Deifalla, "Concrete made with partially substitution corn cob ash: A review," *Case Studies in Construction Materials*, vol. 18, p. e02100, July 2023. https://doi.org/10.1016/j.cscm.2023.e02100

[8] G. Hammond, C. Jones, F. Lowrie, and P. Tse, *Embodied carbon: the Inventory of Carbon and Energy (ICE)*. Bracknell: BSRIA, 2011.

[9] B. V. Venkatarama Reddy and K. S. Jagadish, "Embodied energy of common and alternative building materials and technologies," *Energy and Buildings*, vol. 35, no. 2, pp. 129–137, Feb. 2003. https://doi.org/10.1016/S0378-7788(01)00141-4

[10] A. Abubakar, A. Mohammed, and D. Samson, "Assessment of Embodied Energy and Carbon IV Oxide Emission of Concrete Containing Corncob Ash," *IJSGE*, vol. 10, no. 2, p. 76, 2021. https://doi.org/10.11648/j.ijrse.20211002.15

[11] C. Ma, G. Long, Y. Shi, and Y. Xie, "Preparation of cleaner one-part geopolymer by investigating different types of commercial sodium metasilicate in China," *Journal of Cleaner Production*, vol. 201, pp. 636–647, Nov. 2018. https://doi.org/10.1016/j.jclepro.2018.08.060

[12] ISO 14064-1, *Greenhouse Gases—Part 1: Specification with Guidance at the Organization Level for Quantification and Reporting of Greenhouse Gas Emissions and Removals*. Geneva, Switzerland: International Organization for Standardization, 2018.

[13] N. H. Auliya, M. A. Caronge, and M. W. Tjaronge, "Compressive strength and sustainability indicators of concrete manufactured with recycled refractory brick furnace waste," *Case Studies in Chemical and Environmental Engineering*, vol. 11, p. 101177, June 2025. https://doi.org/10.1016/j.cscee.2025.101177

[14] K. Bavithra and R. Mohana, "Sustainable development of durable and novel nano GGBS impregnated eco-friendly green concrete using micro structural characterization and techno-economic sustainability analysis," *Construction and Building Materials*, vol. 489, p. 140565, Apr. 2025. https://doi.org/10.1016/j.conbuildmat.2025.140565

[15] S. Oyebisi *et al.*, "Sustainability assessment of geopolymer concrete synthesized by slag and corncob ash," *Case Studies in Construction Materials*, vol. 17, p. e01665, Dec. 2022. https://doi.org/10.1016/j.cscm.2022.e01665

[16] H. Ulusu, H. Yılmaz Aruntaş, A. Burcu Gültekin, M. Dayı, M. Çavuş, and G. Kaplan, "Mechanical, durability and microstructural characteristics of Portland pozzolan cement (PPC) produced with high volume pumice: Green, cleaner and sustainable cement development," *Construction and Building Materials*, vol. 378, p. 131070, May 2023. https://doi.org/10.1016/j.conbuildmat.2023.131070

[17] A. A. Abdalla *et al.*, "Microstructure, chemical compositions, and soft computing models to evaluate the influence of silicon dioxide and calcium oxide on the compressive strength of cement mortar modified with cement kiln dust," *Construction and Building Materials*, vol. 341, p. 127668, July 2022. https://doi.org/10.1016/j.conbuildmat.2022.127668

[18] U. E. Edike, O. A. Adenuga, D. U. Idusuyi, and A. A. Oke, "Effect of curing media on the performance of clay bricks partially stabilised with pulverised cow bone ash," *JEDT*, vol. 23, no. 2, pp. 442–457, Feb. 2025. https://doi.org/10.1108/JEDT-11-2022-0578

Advances in Cement and Concrete Research - ICACCR 2025 Materials Research Forum LLC
Materials Research Proceedings 63 (2026) 211-218 https://doi.org/10.21741/9781644904053-24

Mechanical Behavior of Low-Carbon Concrete Produced with Wood Ash as a Sustainable Cement Replacement

Uche Emmanuel EDIKE[1,2 *], Mark Omeiza ONIPE[2], Oluwafemi Peter ADENIYI[3], and Adepoju Muhammed MUNTASIR[3]

[1]Department of Civil Engineering Science, Faculty of Engineering and the Built Environment, University of Johannesburg, South Africa

[2]Department of Building, Faculty of Environmental Sciences, Southern Delta University, Ozoro, Nigeria

[3]Department of Building Technology, College of Environmental Sciences, Bells University of Technology, Ota, Nigeria

uchee@uj.ac.za, edikeuch84@gmail.com

Keywords: Supplementary Cementitious Material, Portland Cement Concrete, Setting Time, Wood Ash, Workability

Abstract. This study examines the mechanical behavior of low-carbon concrete produced using a binder blend of cement and wood ash (WAS) to promote the application of supplementary cementitious materials (SCMs) and enhance environmental sustainability. WAS blended concrete was produced by replacing ordinary Portland cement OPC) at varying percentages, 0%, 5%, 10%, 15%, 20%, and 25% of WAS. Some of the index properties of the materials and the mechanical behavior of the WAS-blended concrete were tested to evaluate the performance of the concrete. The particle size distribution curve of the natural sand employed in the study indicated a well-graded but predominantly uniformly distributed fine particles with a low Uniformity Coefficient of 1.37 and a Coefficient of Curvature of 1.04. The fineness analysis revealed that the OPC exhibited a residue of 3.13%, while the WAS recorded a slightly higher value of 3.83%. The study reveals that the initial setting time of WAS-OPC paste steadily increased with the WAS content, while workability declined at higher OPC replacement levels. The optimum mix of 15% wood ash blended concrete achieved 18.0% and 35.9% increase in compressive and flexural strengths at 28 days, respectively. The study suggests that wood ash is a promising SCM when used at controlled dosages between 5% and 15%, promoting the environmental sustainability of the construction industry.

1. Introduction

As the search for alternative and sustainable supplementary cementitious materials (SCMs) continues to gain momentum, it is evident that the global environmental footprint reduction protocol of the construction industry has become a research priority. The high energy demands of the calcination process of limestone in the production of clinker for cement contribute approximately 7 – 8 % of global CO_2 emissions [1], [2]. This has led to increasing regulatory pressure to decarbonize the construction sector and meet the international climate commitments [3]. Consequently, the substitution of cement with SCMs derived from industrial and agricultural wastes is globally gaining attention as an effective strategy to mitigate the environmental footprint of the construction sector [4], [5].

The utilization of WAS in concrete production not only addresses the environmental footprint of the construction sector but also the environmental burden of wood waste disposal, which often involves landfilling or uncontrolled burning, both of which pose risks of groundwater contamination and air pollution [6], [7], [8]. The integration of WAS into construction materials

represents a sustainable waste management strategy, offering a high-value end-use application that aligns with circular economy principles [9], [10]. However, the performance of WAS in concrete is not yet fully standardized in widely adopted cement specifications like ASTM C618 [11] or EN 450-1 [12] as the optimal replacement percentages, chemical thresholds, and durability benchmarks vary across studies, indicating a lack of consensus in the literature. It is therefore expedient to carry out focused research to provide the empirical evidence needed to support the development of standards and best practices for WAS incorporation in cementitious materials.
This study investigates the effect of WAS on the initial setting time, workability, and mechanical properties of concrete, aiming to develop structurally safe cementitious composites for the construction industry. The study contributes to unlocking the full potential of WAS, mitigating the environmental challenges associated with wood ash disposal, and low-carbon cementitious composites for the future construction industry.

2. Materials and experimental procedures

Wood ash (WAS), ordinary Portland cement (OPC), natural sand, and portable water were used to produce concrete at 0%, 5%, 10%, 15%, 20%, and 25% replacement of cement by mass. The specific gravities of the WAS, river sand, coarse aggregate, and OPC are 2.63, 2.61, 3.48, and 2.69, respectively. The particle size analysis of the natural sand was carried out in line with ASTM D2487-17 (2025), while the wood ash and OPC were tested for fineness and initial setting time in accordance with ASTM C430 [7], ASTM C188 [8], and ASTM C191 [9], respectively. The WAS-blended concrete was tested for workability, compressive strength, and flexural strength in accordance with ASTM C143/C143M [10], ASTM C39/C39M [11], and ASTM C293/C293M [12], respectively. The compressive and flexural strength tests were conducted at 7, 14, and 28 days. The WAS sample and slump test are presented in Fig. 1.

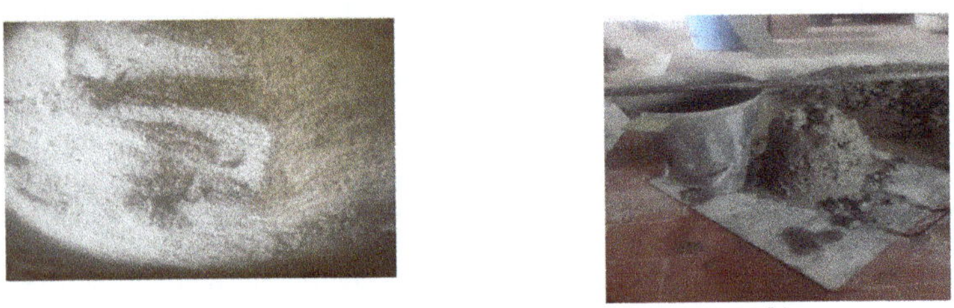

a) b)

Fig 1: (a) Wood ash and (b) Slump test

3. Results and discussions

Particle size analysis

The natural sand employed in the present study is characterized by fine sand and very fine sand as the median particle diameter (D_{50}) is approximately 0.165 mm (see Fig. 2). The particle size distribution curve of the natural sand also indicates a well-graded but predominantly uniformly distributed fine particles, as shown by the low Uniformity Coefficient, 1.37 and Coefficient of Curvature, 1.04 of the sand material. The majority of the particles fall within a narrow size range rather than spanning across multiple size fractions. Narrow size distribution and uniformly graded sand are prone to lower packing density, which may increase void content, requiring more binder or fines to achieve cohesion in concrete [20], [21].

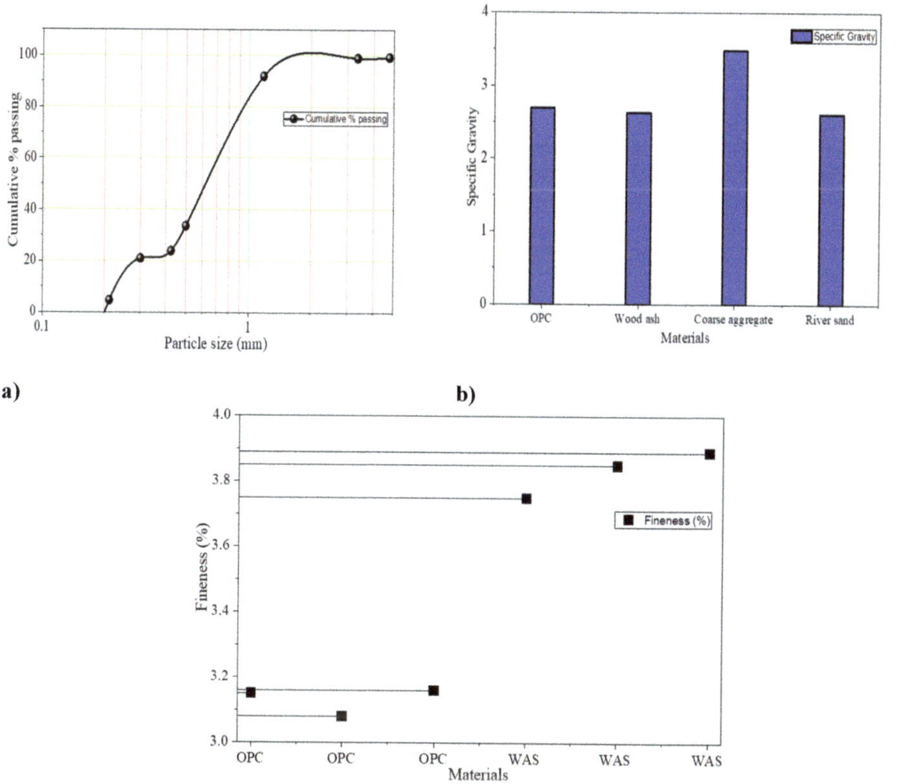

Fig. 2: a) Particle size distribution of natural sand; b) Specific gravity; c) Fineness of OPC and wood ash

Specific gravity

The specific gravity results presented in Fig. 2b revealed distinct density characteristics among the constituent materials used for the concrete production. The specific gravity of both OPC and the natural sand fell within the expected ranges for conventional binders and fine aggregates, indicating material quality consistency. WAS a slightly lower specific gravity of 2.63, a typical trait of biomass-derived pozzolanic materials [22], [23], [24]. The similarity in specific gravity to cement and wood ash suggests that WAS can be incorporated as a supplementary cementitious material or fine filler without inducing significant density variations in concrete mixes. Fig. 2b indicates that the granite coarse aggregate exhibits a significantly higher value of 3.48, indicative of a dense mineral composition that contributes substantially to the concrete's weight and load-bearing performance. The results confirm the compatibility of WAS with conventional concrete constituents.

Fineness

The fineness analysis revealed that the ordinary Portland cement exhibited a residue of 3.13%, whereas the WAS recorded a slightly higher value of 3.83%. The low residue of the WAS indicates that the material is within a range suitable for use as a supplementary cementitious material that could enhance filler effects and improve packing density within the cementitious matrix[25], [26], [27]. Also, the fineness characteristics of the WAS suggest that it can be suitably incorporated into blended cement systems without significant adverse effects on workability or hydration kinetics. However, the coarser nature of the WAS may reduce early-age reactivity, but it is expected to contribute effectively to long-term pozzolanic activity. The slightly coarser nature of the ash can be attributed to incomplete combustion and the presence of partially calcined and carbonaceous particles.

Workability

The inclusion of WAS as a partial replacement for OPC had a significant but nonlinear influence on the workability of the blended concrete, as shown in Fig. 3a. At a low replacement level, 0% the slump was 35 mm, but increased to 45 mm at 5%, indicating a beneficial filler and packing effect and reduced interparticle friction. Modest replacements between 10 to 15% produced slumps similar to the control mix, 38 to 42 mm, while higher OPC replacement of 20–25% caused the slumps to decline to 35 mm and 30 mm, respectively [28], [29], [30]. The reduction in workability at higher WAS contents is attributed to increased water demand arising from the WAS particles and reduced available cementitious paste.

Initial setting time

The result of the initial setting time is displayed in Fig. 3b. The results show a consistent increase in initial setting time of the blended mixtures with the WAS content, reaching 260 minutes at 20% WAS inclusion. The result aligns with Ahmed [30], who states that agricultural by-products, such as rice husk ash retards the hydration process.

Flexural Strength

The result of the flexural strength test of concrete incorporating WAS as a partial replacement for ordinary Portland cement is shown in Fig. 3c. The flexural Strength demonstrated a dependence on both replacement level and curing age. At all ages, the strength increased with WAS content up to 15%, after which it declined. Specifically, at 7 days, the 5- to 15% blends recorded increases from 8.4% to 27.5% relative to the control, while at 28 days, the 10% and 15% mixes achieved increases in flexural strength of 19.8% and 35.9%, respectively. The early improvements are attributed to filler and packing effects, whereas the pronounced long-term gains reflect slow pozzolanic reactions that densify the matrix and the interfacial transition zone [31], [32]. Reductions in flexural strength at 20% replacement are attributed to cement dilution, increased porosity resulting from WAS particles and possible unburnt carbon, and higher water demand, which collectively weaken the matrix and bond.

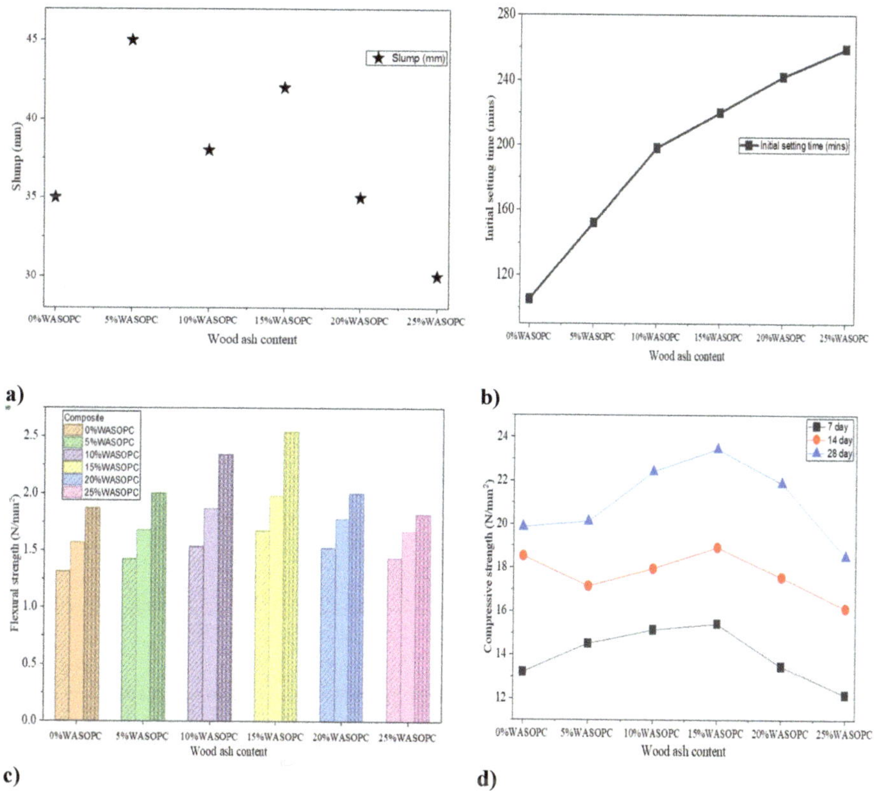

Fig. 3: Effects of wood ash content a) Workability b) Initial setting time; b) Flexural Strength; c) Compressive Strength

Compressive Strength

The compressive strength of concrete containing WAS as a partial replacement for ordinary Portland cement increased with curing age and WAS content up to 15% (see Fig. 3d). Strength increased with curing time for all mixes and improved with WAS addition up to 15% replacement, beyond which the compressive strength declined. At 7 days, mixes with 5–15% WAS ash recorded increases of 9.8–16.7% relative to the control mix, while at 28 days, the 10% and 15% blends achieved 12.8% and 18.0% increases, respectively. The early and sustained improvements at low replacement are attributable to filler and packing effects, as well as the subsequent pozzolanic reaction between ash-derived amorphous silica and portlandite. The reduction in strength at higher replacements of 20 and 25% is ascribed to dilution of cementitious material, increased porosity, potential unburnt carbon, and higher water demand [33], [34], [35], [36], [37]. These results indicate that WAS is a promising supplementary cementitious material when used at controlled dosages of 5-15%. Higher OPC replacement with WAS exceeding 15% may require further mix optimization to prevent strength loss.

4. Conclusion

This study investigated the use of WAS as a replacement for ordinary Portland Cement (OPC) in concrete. The study demonstrates that WAS can improve workability at low replacement levels not exceeding 15%. The particle size distribution curve for the natural sand used in this research showed that it was well-graded, primarily consisting of uniformly distributed fine particles, as indicated by a low Uniformity Coefficient and Coefficient of Curvature. The fineness analysis suggests that the OPC and wood provided good surface area for hydration and pozzolanic reactions, as shown by the low residue contents. The findings indicate that the initial setting time of the WAS-OPC paste increased gradually with higher percentages of WAS content, while the workability decreased at greater OPC replacement percentages. The compressive and flexural strengths of WAS-blended concrete have demonstrated that WAS is an effective supplementary cementitious material for enhancing flexural performance when used at controlled dosages not exceeding 15%. The study observed that WAS is a promising supplementary cementitious material when used at controlled dosages between 5% and 15%. However, higher OPC replacement with WAS beyond 15% may require further mix optimization to avoid performance loss.

References

[1] J. De Brito and R. Kurda, "The past and future of sustainable concrete: A critical review and new strategies on cement-based materials," *Journal of Cleaner Production*, vol. 281, p. 123558, Jan. 2021. https://doi.org/10.1016/j.jclepro.2020.123558.

[2] N. Tripathi, C. D. Hills, R. S. Singh, and J. S. Singh, "Offsetting anthropogenic carbon emissions from biomass waste and mineralised carbon dioxide," *Sci Rep*, vol. 10, no. 1, p. 958, Jan. 2020. https://doi.org/10.1038/s41598-020-57801-5.

[3] K. Scrivener, M. Ben Haha, P. Juilland, and C. Levy, "Research needs for cementitious building materials with focus on Europe," *RILEM Tech Lett*, vol. 7, Apr. 2023. https://doi.org/10.21809/rilemtechlett.2022.165.

[4] K. L. Scrivener, V. M. John, and E. M. Gartner, "Eco-efficient cements: Potential economically viable solutions for a low-CO2 cement-based materials industry," *Cement and Concrete Research*, vol. 114, pp. 2–26, Dec. 2018. https://doi.org/10.1016/j.cemconres.2018.03.015.

[5] R. Siddique and M. I. Khan, *Supplementary Cementing Materials*, vol. 37. in Engineering Materials, vol. 37. Berlin, Heidelberg: Springer Berlin Heidelberg, 2011. doi: 10.1007/978-3-642-17866-5.

[6] F. F. Udoeyo, H. Inyang, D. T. Young, and E. E. Oparadu, "Potential of Wood Waste Ash as an Additive in Concrete," *J. Mater. Civ. Eng.*, vol. 18, no. 4, pp. 605–611, Aug. 2006. https://doi.org/10.1061/(ASCE)0899-1561(2006)18:4(605).

[7] I. Obernberger, F. Biedermann, W. Widmann, and R. Riedl, "Concentrations of inorganic elements in biomass fuels and recovery in the different ash fractions," *Biomass and Bioenergy*, vol. 12, no. 3, pp. 211–224, Jan. 1997. https://doi.org/10.1016/S0961-9534(96)00051-7.

[8] U. E. Edike, O. J. Ameh, H. S. Yohanna, I. C. Osuizugbo, and D. O. Nduka, "Mechanical Properties of Concrete Blocks Incorporating Recycled Waste Plastic," *Mater Circ Econ*, vol. 6, no. 1, p. 10, Dec. 2024. https://doi.org/10.1007/s42824-024-00101-4.

[9] R. Abdulwahab, B. D. Ikotun, A. A. Raheem, E. A. Adetoro, and O. K. Oriti, "Supplementary cementitious influence of cashew leaf ash as cement replacement in the production of mortar," *J Build Rehabil*, vol. 10, no. 2, p. 116, Dec. 2025. https://doi.org/10.1007/s41024-025-00624-6.

[10] I. Ekop, I.-I. N. Etim, E. Ambrose, and U. E. Edike, "Bio-Based Cement Concrete Admixtures for Green Recovery in the Construction Industry: a Critical Survey," *Mater Circ Econ*, vol. 6, no. 1, p. 31, Dec. 2024. https://doi.org/10.1007/s42824-024-00113-0.

[11] ASTM C618, *Specification for Coal Fly Ash and Raw or Calcined Natural Pozzolan for Use in Concrete*, 2023. doi: 10.1520/C0618-22.

[12] BS EN 450-1, *Fly ash for concrete Definition, specifications and conformity criteria*, Confirmed. 2012.

[13] ASTM D2487-17, "Standard Practice for Classification of Soils for Engineering Purposes (Unified Soil Classification System)." ASTM International, 2025.

[14] ASTM C188, "Standard Test Method for Density of Hydraulic Cement." ASTM International, 2025.

[15] ASTM C430, *Test Method for Fineness of Hydraulic Cement by the 45-m (No. 325) Sieve*, 2025. doi: 10.1520/C0430-25.

[16] ASTM C191, *Test Methods for Time of Setting of Hydraulic Cement by Vicat Needle*, 2021. doi: 10.1520/C0191-21.

[17] ASTM C143/C143M, *Test Method for Slump of Hydraulic-Cement Concrete*, 2020. doi: 10.1520/C0143_C0143M-20.

[18] ASTM C39/C39M, *Test Method for Compressive Strength of Cylindrical Concrete Specimens*, 2024. doi: 10.1520/C0039_C0039M-24.

[19] ASTM C78/C78M-22, *Test Method for Flexural Strength of Concrete (Using Simple Beam with Third-Point Loading)*, 2022. doi: 10.1520/C0078_C0078M-22.

[20] K. Moolchandani, "Industrial byproducts in concrete: A state-of-the-art review," *Next Materials*, vol. 8, p. 100593, July 2025. https://doi.org/10.1016/j.nxmate.2025.100593.

[21] U. E. Edike, O. J. Ameh, and M. O. Dada, "Production and optimization of eco-bricks," *Journal of Cleaner Production*, vol. 266, p. 121640, Sept. 2020. https://doi.org/10.1016/j.jclepro.2020.121640.

[22] W. Sun, Q. Sun, and Z. Xu, "Preparation of Biomass Ash Low-Carbon Sustainable Grouting Materials: Optimal Regulation, Diffusion Characteristics, Synergistic Mechanism, and Environmental and Economic Evaluation," 2025, *SSRN*. doi: 10.2139/ssrn.5334779.

[23] U. E. Edike, O. A. Adenuga, D. U. Idusuyi, and A. A. Oke, "Effect of curing media on the performance of clay bricks partially stabilised with pulverised cow bone ash," *JEDT*, vol. 23, no. 2, pp. 442–457, Feb. 2025. https://doi.org/10.1108/JEDT-11-2022-0578.

[24] J. H. S. Rêgo, A. A. Nepomuceno, E. P. Figueiredo, and N. P. Hasparyk, "Microstructure of cement pastes with residual rice husk ash of low amorphous silica content," *Construction and Building Materials*, vol. 80, pp. 56–68, Apr. 2015. https://doi.org/10.1016/j.conbuildmat.2014.12.059.

[25] K. Ganesan, K. Rajagopal, and K. Thangavel, "Evaluation of bagasse ash as supplementary cementitious material," *Cement and Concrete Composites*, vol. 29, no. 6, pp. 515–524, July 2007. https://doi.org/10.1016/j.cemconcomp.2007.03.001.

[26] W. Sun, Q. Sun, and Z. Xu, "Preparation of Biomass Ash Low-Carbon Sustainable Grouting Materials: Optimal Regulation, Diffusion Characteristics, Synergistic Mechanism, and Environmental and Economic Evaluation," 2025, *SSRN*. doi: 10.2139/ssrn.5334779.

[27] M. Salvo *et al.*, "Biomass ash as supplementary cementitious material (SCM)," *Advances in Applied Ceramics*, vol. 114, no. sup1, pp. S3–S10, Sept. 2015. https://doi.org/10.1179/1743676115Y.0000000043.

[28] N. A. Razeman *et al.*, "A Review on The Compressive Strength and Workability of Concrete with Agricultural Waste Ash as Cement Replacement Material," *IOP Conf. Ser.: Earth Environ. Sci.*, vol. 1135, no. 1, p. 012058, Jan. 2023. https://doi.org/10.1088/1755-1315/1135/1/012058.

[29] I. F. Oseke and F. P. Dangiwa, "OPTIMAL CEMENT REPLACEMENT LEVELS IN CONCRETE USING GROUND GRANULATED BLAST FURNACE-SLAG FOR WORKABILITY AND STRENGTH DEVELOPMENT.," *OJES*, vol. 6, no. 1, pp. 1–14, Apr. 2025. https://doi.org/10.52417/ojes.v6i1.786.

[30] A. Ahmed, "Assessing the effects of supplementary cementitious materials on concrete properties:a review," *Discov Civ Eng*, vol. 1, no. 1, p. 145, Dec. 2024. https://doi.org/10.1007/s44290-024-00154-z.

[31] B. N. AL-Kharabsheh, M. M. Arbili, A. Majdi, J. Ahmad, A. F. Deifalla, and A. Hakamy, "A Review on Strength and Durability Properties of Wooden Ash Based Concrete," *Materials*, vol. 15, no. 20, p. 7282, Oct. 2022. https://doi.org/10.3390/ma15207282.

[32] S. Chowdhury, A. Maniar, and O. M. Suganya, "Strength development in concrete with wood ash blended cement and use of soft computing models to predict strength parameters," *Journal of Advanced Research*, vol. 6, no. 6, pp. 907–913, Nov. 2015. https://doi.org/10.1016/j.jare.2014.08.006.

[33] R. Yu *et al.*, "Mechanical properties and microstructural discrepancies of concrete with flotation-modified fly ash from circulating fluidized bed and pulverized coal furnaces," *Construction and Building Materials*, vol. 428, p. 136320, May 2024. https://doi.org/10.1016/j.conbuildmat.2024.136320.

[34] B. F. Aygun, T. Bilir, and M. Uysal, "Coal bottom ash and its applications in cement and concrete technologies: a review," *Discov Civ Eng*, vol. 1, no. 1, p. 86, Sept. 2024. https://doi.org/10.1007/s44290-024-00090-y.

[35] A. Schackow *et al.*, "Coal Bottom Ash as Supplementary Cementitious Material: Optimizing Reactivity Through Particle Grinding," *Sustainability*, vol. 17, no. 11, p. 5031, May 2025. https://doi.org/10.3390/su17115031.

[36] Y. Wang, L. Burris, R. D. Hooton, C. R. Shearer, and P. Suraneni, "Effects of unconventional fly ashes on cementitious paste properties," *Cement and Concrete Composites*, vol. 125, p. 104291, Jan. 2022. https://doi.org/10.1016/j.cemconcomp.2021.104291.

[37] A. K. Saha, "Effect of class F fly ash on the durability properties of concrete," *Sustainable Environment Research*, vol. 28, no. 1, pp. 25–31, Jan. 2018. https://doi.org/10.1016/j.serj.2017.09.001.

Advances in Cement and Concrete Research - ICACCR 2025 Materials Research Forum LLC
Materials Research Proceedings 63 (2026) 219-229 https://doi.org/10.21741/9781644904053-25

Sustainable and High-Performance Composite Materials for Civil Engineering

Vivek BHADIYADRA

Jersey City, NJ, USA 07307

Vivekbhadiyadra11@gmail.com

Keywords: Composite Materials, Sustainability, Fiber-Reinforced Polymers, Textile-Reinforced Mortar, High-Performance Materials

Abstract. Composite materials, one of the fastest-growing materials in civil engineering, offer superior performance, weight-saving potential, and long-term durability over conventional materials. New generation materials like FRPs (Fiber-Reinforced Polymers), TRM (Textile-Reinforced Mortars), and hybrid composites have emerged, which can permit structure optimization and sustainable ecology by being more reliable and environmentally friendly. In this review, the existing peer-reviewed papers are considered to examine the prospects of composites in civil engineering, focusing on their sustainability, environmental impact, and high-performance profile. Emphasis is placed on LCA, mechanical properties, bridges, pavements, seismic upgrade, and retrofitting applications. Challenges, like cost, durability, recyclability, and standardization, are also elaborated, and then future trends of smart composites and green reinforcement are discussed. This paper emphasizes the contribution of composites to sustainable, resilient, and high-performance infrastructures.

1. Introduction

The construction industry is currently suffering from a vast, unprecedented challenge regarding sustainability, as it accounts for nearly half of the world's greenhouse gas emissions, with cement and steel production being the two major contributors to this emission [1,2]. Reinforced concrete is the key component of current fast-paced infrastructure, which relies heavily on steel reinforcement, which is energy-intensive to produce. Not only that, but it is also highly vulnerable to corrosion in some areas, such as marine and aggressive environments [3]. Corrosion-induced degradation is the most dominant reason for the high maintenance expenditures and short life of service of the reinforced concrete structures, which poses severe environmental and economic burdens [4]. In this scenario, composite materials are gaining increased attention as potential alternatives to address durability issues and ecological sustainability.

In the past two decades, significant interest has been drawn toward fiber-reinforced polymers (FRPs), textile-reinforced mortars (TRM), and hybrid composites owing to their high strength-to-weight ratio, corrosion resistance, and versatility to cater to different structural applications [5–7]. With tensile properties several times greater than steel itself (but at only a fraction of the weight), FRPs are lightweight and can be handled and installed much more easily [8,9]. The TRM system, which involves using cementitious mortars reinforced with high-performance textiles, is a particularly interesting option for strengthening masonry and historical structures due to the required compatibility (with the host structure) and breathability [11,13]. Recently, natural fiber composites from sustainable sources, e.g., flax and jute, have been developed as the more eligible candidates, which are in line with the general intentions of the circular economy in construction [14,15,22].

The environmental gains of composites, besides performance benefits, are being justified with life-cycle analysis (LCA). The reduced maintenance demand and longer service life of FRP-reinforced structures result in a pending lower cradle-to-grave carbon footprint when compared to steel, even though the cradle-to-gate carbon footprint of CFRP (Carbon Fiber Reinforced Polymer) production is much higher than that of steel [18,20,29]. This effect has been found in TRM, where the lightness of these systems leads to a reduction of the transport impact, and material compatibility with seawater concrete significantly contributes to the reduction of the embodied energy [16]. This observation is consistent with the global trend towards net-zero carbon infrastructure, demonstrated by concerns expressed by recent RILEM technical committee reports on introducing non-metallic reinforcements into the next generation of design codes [28].

The use of composite materials in civil engineering is also a cause for cost issues, recyclability concerns, and the non-availability of uniform design codes worldwide. However, substantial developments in the area have supported their usage [18]. However, these obstacles are being overcome by further development of fiber technology, resin chemistry, and production methods. The development of thermoplastic FRPs with easier recyclability, the use of hybrids for improving the natural fiber durability, and smart composites with self-sensing and self-healing properties are the most promising lines of future application [19,31,32].

This review summarizes the recent progress in sustainable and high-performance composites for civil engineering. The interest is in classifying the composite types, sustainability considerations via LCA, and their use in concrete bridges, pavements, seismic retrofitting, and conservation of cultural heritage. In so doing, the paper seeks to offer a critical review of the state-of-the-science and articulation of the barriers and opportunities that need to be resolved for composites to be integrated as core materials in the sustainable infrastructure of the future.

2. Composite Materials in Civil Engineering

2.1 Fiber-Reinforced Polymers (FRPs). Fiber-reinforced polymers (FRPs) became one of the most extensively studied and used materials in civil engineering due to their excellent mechanical properties and ability to withstand adverse environmental conditions. They are fabricated by integrating high-strength fibers (glass, carbon, or basalt) into a polymeric matrix. The resultant composite combines the tensile strength of the fiber and the binding and load-carrying of the matrix [5,6].

Among these materials, CFRP is known for its very high tensile strength, which may exceed 2000 MPa, and modulus values between 150 and 200 GPa [7]. This makes them perfect for high-performance structural applications such as bridge decks and prestressed tendons requiring stiffness, strength, and durability. GFRPs, although less stiff (modulus ~40–60 GPa), are economical (for large-scale reinforcement in concrete elements) when compared with CFSs [8]. Basalt FRPs (BFRPs) made from natural volcanic rocks are considered alternative, eco-friendly materials. It has been reported that the tensile strength of BFRP bars is about 1200 MPa and exhibits better chemical resistance in an alkaline environment than that of GFRP [10,28].

The tensile stress–strain curve in FRPs is linear until failure, whereas steel yields. The constitutive law can be formulated as follows:

$$\sigma = \frac{F}{A}, \quad E = \frac{\sigma}{\varepsilon} \tag{1}$$

Where σ\sigma is tension stress, F is applied force, A is cross-sectional area, ε is strain, and E is elasticity. Without the yielding of flexural members reinforced with FRP, civil engineers have had to reorient design philosophies, which now can be based on ultimate limit state design principles and safety factors that are fine-tuned to the unique behavior of FRP reinforced structures [9,28].

Recent research has emphasized not only the mechanical behavior of FRPs but also their effect on sustainability. De Felice [18,20,29] showed that CFRP strengthening, with higher cradle-to-gate emissions, led to lifecycle costs and carbon impacts reduction when maintenance needs were considered. Similarly, Papanicolaou et al. [28] revealed better durability of FRP reinforced concrete in chloride conditions by comparison to traditional steel materials used in marine infrastructure.

2.2 Textile-Reinforced Mortars (TRM). Textile-reinforced mortars (TRMs) are composite systems consisting of an open-mesh textile, such as carbon, glass, basalt, or other fibers, embedded in a mortar made up of a cementitious matrix. In contrast to resin-bounded FRPs, the matrix of the TRMs is inorganic, which gives it better Type equation here.compatibility with masonry and increases the level of fire resistance [14,15]. This renders TRMs particularly beneficial in the retrofitting of and rehabilitation of historic structures, given the importance of the breathability and the chemical compatibility of the strengthening material.

The bond between fabric and mortar is one of the dominant parameters affecting the performance of TRMs, which can be described in the form of bond stress:

$$\tau_b = \frac{F}{\pi d l} \quad (2)$$

Where F is the pull-out force, d is the equivalent diameter of the textile yarn, and l is the embedment length. Laboratory work has demonstrated that TRM does not lose bond strength under wet-dry cycling and freeze-thawing, which would indicate the durability of TRM [16].

Significant enhancement in both the ductility and shear capacity has been achieved by the use of TRM in seismic retrofit of URM walls. Suriani et al. [17] reported that the lateral load capacity of masonry walls strengthened with carbon TRM could be up to three times higher, and Hafner et al. [18] demonstrated significant enhancement of energy dissipation under cyclic loading. Zeng [19] also have reported a sustainable alternative, the textile-reinforced alkali-activated mortars (TRAAM), where the mortar matrix is composed of industrial by-products. Such developments show a promising direction for higher performance and environmentally friendly TRM systems.

2.3 Natural and Hybrid Composites. The use of natural fibers such as flax, hemp, jute, and coir for reinforcement is still under investigation as they are non-toxic, relatively inexpensive, and biodegradable [20]. Their mechanical properties are much inferior to those of their synthetic counterparts, and tensile strengths are usually in a range of 200–500 MPa, and elastic moduli are 20–30 GPa [21]. Alkali soaking and silane coupling agent surface treatments have been proven effective in increasing fiber–matrix adhesion and reducing water absorption, leading to increased durability [22].

The so-called "hybrid composites" are considered a new generation material in which natural fibers are reinforced along with glass or carbon fibers in the polymer matrix to attain sustainability, along with other mechanical characteristics. Tatar et al. [23] reported water resistance and accelerated aging performance for flax–glass hybrid FRPs that were increased by 40% and by 40%, respectively, compared with those of pure natural fiber composites. Similarly, they have good potential within the 3D-print end cementitious system, to provide structural integrity and sustainable capability [24]: polymer fiber–reinforced mortars with natural fiber.

The use of natural fiber resources also enhances sustainability indicators. Life-cycle analysis showed much lower embodied carbon in natural-fiber composites, which is typically lower than 1 kgCO_2/kg material, indicating that natural-fiber composites are good low-carbon materials for construction [25]. While durability still remains the primary concern for hybrids, hybridization and protective coatings are being further developed, which is likely to enhance their potential usage in mainstream civil engineering [26].

3. Sustainability of Composite Materials

Construction materials' performance relative to the environment is evaluated more and more with LCA in a comprehensive fashion, which takes into account raw material extraction, processing, usage, maintenance, and end-of-life measures. For civil engineering-based composites, LCA offers insight into whether efficiency gains of the kind described actually offer gains in terms of sustainability LCA [1, 12, 14].

3.1 Life-Cycle Assessment (LCA) Framework. The total environmental burden of a material system can be expressed as:

$$LCA_{total} = \sum_{i=1}^{n} (m_i \times EF_i) \tag{3}$$

Where m_i is the mass of input and EF_i is its associated emission factor. For reinforcement materials, the key contributors include resin production, fiber manufacturing, transportation, and energy demand during fabrication [1,17]. Despite the CFRP cradle-to-gate GHG(Greenhouse Gas) emissions being greater than steel as a result of the energy-intensive fabrication of carbon fiber, reduced maintenance and extended service life can result in lower cradle-to-grave life cycle emissions [2,13].

Sbahieh et al. [29] investigated the climate change impacts of GFRP, CFRP, sand-coated GFRP (SGFRP), and steel bars (diameter 12 mm and length 1 m). As illustrated in Fig. 1, the emissions at the production stage were the largest for CFRP (approximately 3.3 kg CO 2 eq), followed by 1.7 kg CO 2 eq for steel, and 0.6–1.1 kg CO 2 eq for glass-based composites. Switching to service life and the frequency of maintenance, the gross impacts over 50 years of the timeframe strongly favored FRP reinforcement, in particular when used in a corrosive environment.

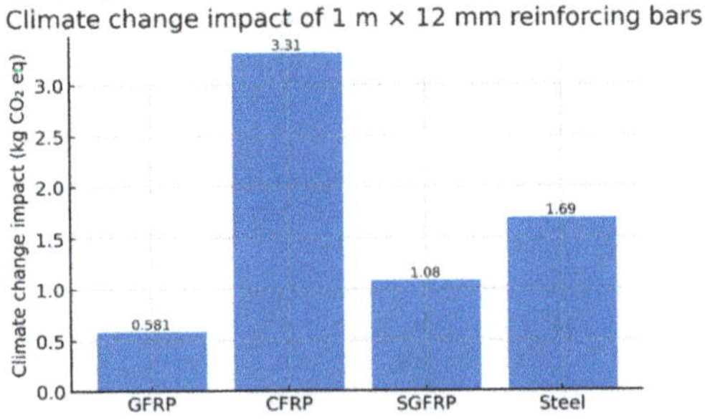

***Fig. 1**: Climate change impact (kg CO_2 eq) 1 m of reinforcing bars (diameter 12 mm) for GFRP, CFRP, sand-coated GFRP, and steel [29].*

3.2 Structural-Level Assessments. Differences in embodied emissions become more apparent at the structural level. Freshwater concrete beams with steel wraps resulted in an overall impact of 417 kg CO_2 eq, while in seawater-mixed concrete and using GFRP, this value dropped to 322 kg CO_2 eq, and dropped even more to ~401 kg CO_2 eq for CFRP under the same conditions [29] and this reduction is specific to the results reported in that study. This emphasizes how material selection, as well as companion technologies (e.g., seawater concrete), can interact to amplify the reduction of carbon footprints.

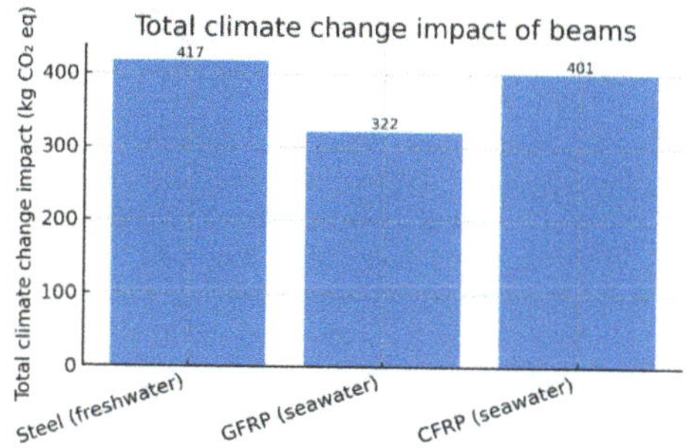

***Fig 2**: Total climate change impact (kg CO_2 eq) for beams with steel (freshwater), GFRP (seawater), and CFRP (seawater) reinforcement [29].*

These results highlight the value of systems-level analysis: even if the compound has less favorable raw material inputs for composites use, the net advantage from using it as part of an optimized structural system can still be positive in a sustainability context. Comparable findings have been reported for FRP bridge decks, enabling foundation requirements to be reduced and indirectly lower amounts of emissions [18].

3.3 Embodied Carbon and Comparative Metrics. Recent works have synthesized values of embodied carbon reinforcement. CFRP usually represents 3.0–3.5 kg CO_2 eq/kg, GFRP is 1.2–1.5, and BFRP is 1.0–1.3. In contrast, the figure is roughly 1.8–2.0 for steel, but at greater lifetime maintenance costs [17,19]. Natural fiber composites, such as flax-FRP, are usually found in scans with less than 1.0 kg CO_2 eq/kg, as these materials originate from renewable sources [20].

Table 1: Embedded Specification of CO2 Reinforcement Materials.

Material	Embodied CO_2 [kg CO_2 eq/kg]	Source(s)
Steel	1.8–2.0	[17,19]
GFRP	1.2–1.5	[1,17]
CFRP	3.0–3.5	[1,2]
BFRP	1.0–1.3	[10,17]
Flax-FRP	<1.0	[20,23]

3.4 Recycling and End-of-Life Scenarios. Recycling and end-of-life for composites remains one of the most pressing sustainability issues. Thermosetting matrices that predominate in FRPs are not recyclable. End-of-life composites, when mechanically grounded, generate filler that is often down-cycled [21]. Pyrolysis has been found to reclaim carbon fibers with some strength preservation, but its high energy cost prevents it from industrial acceptance [22].

Thermoplastic-derived FRPs with potential are further thermoplastic materials with the possibility of reheating and other hybrid systems that utilize biodegradable natural fiber with synthetic resin. Recent studies on flax-glass hybrids indicated enhanced durability and still partial biodegradability, providing a route to a more circular usage [23].

3.5 Durability as a Sustainability Factor. Durability directly influences lifecycle sustainability. Long-term investigations show that the tensile strength of GFRP bars in alkaline conditions deteriorates by 20–30% for a 15-year period, whereas the degradation of CFRP is significantly lower [13]. TRM systems were also found to have minimal degradation upon accelerated aging, which included exposure to chlorides and freeze–thaw [14,16]. Not only is the equation more in favor of composites taking into account durability in sustainability assessments, but also, a lower frequency of maintenance becomes the starting point for less emissions by default.

Table 2: Selected composite systems durability results

Composite Type	Durability Factor Tested	Performance Summary	Source(s)
GFRP bars	Alkali resistance	20–30% strength loss after 15 years	[13,16]
CFRP bars	Chloride ingress	<5% strength loss, excellent retention	[13]
TRM systems	Freeze–thaw cycles	Stable bond and tensile performance	[14,16]
Flax-FRP	Water absorption	High unless surface-treated	[22,23]

4. High-Performance Applications of Composites in Civil Engineering

Civil engineering applications for composites are driven not only by sustainability, but also by the need for high performance in the presence of complex loading, aggressive environments, and long design lives. Fiber-reinforced polymer (FRP) composites, textile-reinforced mortars (TRM), and hybrid systems are being widely used for structural strengthening, seismic performance enhancement, and new construction [3,7,15]. With unparalleled high tensile strength, lightweight, and corrosion resistance, you can design slimmer, durable, and more efficient systems free from the imposed limitations of steel-reinforced concrete.

4.1 Structural Strengthening and Retrofitting. One of the first and most common uses of FRPs has been in the rehabilitation of reinforced concrete (RC) beams and columns. External bonding of CFRP sheets leads to a significant increase in both the flexural and shear strengths by taking advantage of the high tensile capacity of the carbon fibers [8]. The increment in ultimate flexural load can then be defined as:

$$\Delta P = L A_f f_{fu} \varepsilon_{fe} \tag{4}$$

where A_f is the cross-sectional area of FRP reinforcement, f_{fu} is the ultimate tensile strength of the fiber, ε_{fe} is the effective strain developed, and L is the bonded length [9]. According to experimental studies, the increase in the flexural strength ranges from 30 to 80% in relation to the type of FRP and the method of anchorage of the strips [10].

TRM has emerged as a promising alternative for masonry and heritage structures, where compatibility with the support and reversibility of the intervention are paramount. The carbon-TRM and basalt-TRM systems become compatible with vapor transmission and offer shear resistance, which is applicable to the masonry of historic structures [12].

4.2 Seismic Applications. In seismic retrofitting, composites provide enhanced ductility and confinement. Jacketing of RC columns with CFRP wraps increases the ultimate axial strain and ductility ratio by up to 2.5 times compared to unconfined specimens [11]. The stress–strain response of confined concrete can be described by the well-established Mander model, modified to account for FRP confinement pressure:

$$f'_{cc} = f'_c\left(2.254\sqrt{1 + 7.94\frac{f_l}{f'_c}} - 2\frac{f_l}{f'_c} - 1.254\right) \tag{5}$$

where fc′ is the unconfined concrete strength, and fl is the effective lateral pressure provided by the FRP [11,13]. Representative stress–strain curves for unconfined concrete and CFRP-confined concrete, showing the enhancement in peak strength and ductility plateau.

4.3 Bridge Decks and Other Components of Structure. The lightweight FRP bridge deck was introduced for additional dead load reduction and service life extension in a corrosive environment [18]. North American case studies show that projected service lives for GFRP girders will exceed 75 years with little or no maintenance relative to steel girders [19]. Pultruded GFRP sections are also employed for modular pedestrian bridges to reduce the construction phase by 50% while maintaining structural performance [21].

In a hybrid concrete-FRP structure, CFRP tendons prestressed in cable-stayed bridges have smaller relaxation losses and better fatigue properties than steel ones [20]. Furthermore, FRP rebar can help to reduce chloride-induced corrosion – a leading cause of premature structural failure, as seen in traditional steel-reinforced marine cribs and walls in this case study [14].

4.4 Fire and High-Temperature Performance. FRPs are subject to high temperatures due to the degradation of polymer matrices despite their mechanical benefits. The glass transition temperature (Tg) of epoxy resin ranges from 60 to 120 °C, which limits the upper temperature of reliable operation [22]. Studies of high-temperature matrices, such as phenolic resins and nano-modified epoxies, have demonstrated enhanced tensile strength retention above 200 °C [23].

Protecting methods are intumescent coatings, sacrificial concrete covers, or hybrid systems with fire-protective fabrics. Recent research has shown that basalt-FRP holds better fire-resistant properties over GFRP and CFRP as it has a greater resistance to fire and a mineral-based structure, which is non-combustible [16].

4.5 Fatigue and Long-Term Loading. Composite fatigue strength is also better than the steel fatigue strength. CFRP tendons have much less creep and stress relaxation when under long-term stress. Unedited long-term fatigue tests performed on CFRP tendons demonstrated endurance limits exceeding 2 million cycles at 60% of ultimate tensile strength [17]. This results in cyclic load control and predictable performance in applications such as bridge cables and wind turbine towers.

5. Challenges, Limitations, and Future Directions

Although composite materials have shown excellent potential for use in civil engineering, their application has been hampered by a number of technical, economic, and legislative restrictions. These challenges need to be addressed to see the shift of composites from a niche to a mainstream structural material.

5.1 Cost and Economic Barriers. The initial cost of FRPs/ TRMs is one of the major disadvantages when compared with the cost of normal steel reinforcement. The CFRP can be 10–15 times more expensive per kg than steel [1]. Even though LCC analysis usually shows that there are economic benefits because of lower maintenance and longer service lives [2], the investments necessary for using FRCP have often scared off parties in financially sensitive projects like infrastructure work for municipalities. Economies of scale and developments in automated manufacturing techniques, including pultrusion, 3D weaving, and resin transfer molding, are driving costs down; however, a wider industrialization will be required Approach [3].

5.2 Recyclability and Environmental Impact. Sustainable, while often referred to as an important benefit, is a more complicated consideration because the recyclability of polymer-based composites is an issue. The majority of FRPs are thermoset, so their recycling calls and crosslinked polymer structure mean they are hard to recycle. Mechanical grinding and pyrolysis are also commercialized methods, but they are energy costly and negatively affect the quality of the fibers [4]. Recent studies have shown potential with thermoplastic FRPs, which are capable of remolding and recycling, although energy for production is higher [5]. Natural fiber composites can be a greener alternative but still suffer from durability and moisture sensitivity issues [6]. Figure 3 shows recycling flow patterns of steel, thermoset FRPs, thermoplastic FRPs, and natural fiber composites in a comparative way.

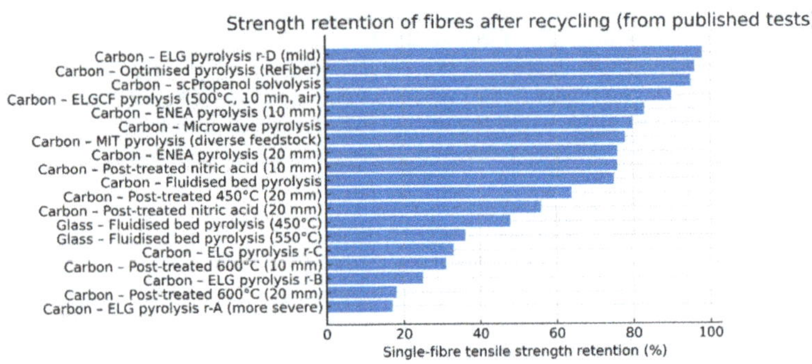

Fig 3: Recycling routes of typical civil engineering reinforcement materials [4–6].

5.3 Standardization and Design Codes. A lot also prevents against the weaknesses of the codes and specifications between the different types of composites. Although ACI 440 in the US and fib Bulletin 40 in Europe give directions towards the utilization of FRP materials, there are still gaps in knowledge, such as on durability models, fire performance, and the long-term behavior [7]. Local variations in codes can be barriers to supranational adoption as well as constraints to managing for global construction companies [8]. Since none of the fire resistance test methods are standardized, this is especially problematic as elevated temperatures may give rise to a catastrophic loss of mechanical properties in polymer composites [9].

5.4 Durability Under Harsh Environments. The long-term behavior of composites under the combined action of some stressors, high humidity, chloride ingress, ultraviolet radiation, and freeze–thaw action, is still being studied. Although FRPs are resistant to corrosion, they are susceptible to microcracking and resin decomposition. Previous studies have shown that the CFRP tends to lose up to 15% of its tensile strength after sustained exposure to pore alkalinity in concrete [10]. Natural fiber composites also undergo fiber swelling and matrix debonding due to repeated wetting–drying cycles [11].

5.5 Emerging Research Directions and Future Outlook. A number of novel approaches are being developed to resolve the above problems:

Biomaterials and natural fibers: Substituting Petro-based epoxies with lignin or soy resins so far has demonstrated their capability to reduce the carbon footprint without significant downsizing of mechanical properties [12]. Hybridization of natural fibers with synthetic fibers is expected to provide greater performance at a cost-effective price with reduced environmental impact.

Smart composites: Integration of self-sensing fiber and conductive fillers into FRP for online structural health monitoring. These multifunctional composites have built-in sensors for strain, cracks, and temperature changes [13].

Self-healing composites: Recent microencapsulation of healing agents in resin matrices has also shown autonomous regeneration of cracks under fatigue loading, thereby prolonging the service life [14].

Additive manufacturing: 3D printing of continuous fiber composites. The 3D printing of continuous fiber composites is a disruptive technology for individualized reinforcement segments, which saves material waste and topology-optimized structures for sectioning components [15].

Performance versus long-term sustainability. Probably, consider the kind of balance to make in the future. Conventional and modern composite systems are compared in Table 3, based on cost, strength, recyclability, and environmental footprint.

Table 3: Performance and sustainability comparison of selected reinforcement materials.

Material Type	Tensile Strength [MPa]	Density [g/cm³]	Recyclability	Relative Cost	Durability in Aggressive Environments
Steel Reinforcement	400–600	7.8	High (re-melting)	Low	Moderate (corrosion-prone)
CFRP	2400–4000	1.6	Low (thermoset)	Very High	Excellent (non-corrosive, UV-sensitive)
GFRP	1000–1600	2.0	Low–Moderate	Moderate	Good (alkaline resistance issues)
Basalt FRP	1500–2000	2.7	Low–Moderate	Moderate	Very Good (better thermal stability)
TRM	800–1500	2.0–2.5	Moderate	Moderate	Excellent (heritage applications)
Natural Fiber Composites	200–800	1.2–1.5	High (biodegradable)	Low–Moderate	Poor–Moderate (moisture sensitivity)
Thermoplastic FRPs	1200–2000	1.6–2.0	High	High	Good (pending long-term data)

Conclusion

Civil engineering is gradually being transformed by composite materials, which provide an attractive combination of high performance, potential for sustainability benefits, and durability. From fiber-reinforced polymers to textile-reinforced mortars and natural fiber composites, these advanced materials have shown the potential to increase life spans, lower maintenance costs, and enable new design concepts. However, despite the benefits, their wider application is still constrained by the high initial investment needed, restricted reprocessing rate, and the lack of standardized design codes.

Composites for civil engineering in the near future. The further development of composites for civil engineering in the future will be focused on the following: a) Use of bio-based resins; b) Use of recyclable thermoplastic systems; and c) Multifunctional smart composites. The development of additive manufacturing and self-healing technology is expected to expand the application range, leading to structures that are not only stronger and lighter but also more adaptable and environmentally friendly.

As the demand for infrastructure continues to expand in size and complexity, composite materials will play a key role in finding creative, sustainable, and resilient solutions. Yet bringing them to bear on the real world will still require researchers, players in the industry, and policymakers to work together to address economic, regulatory, and durability hurdles. By filling these voids, composites can progress from being niche materials to serving as mainstream building blocks of the next generation of infrastructure, enabling a more sustainable built environment.

References

[1] Andrew, R. M. (2019). Global CO_2 emissions from cement production, 1928–2018. Earth System Science Data, 11(4), 1675–1710. https://doi.org/10.5194/essd-11-1675-2019

[2] Habert, G., Miller, S. A., John, V. M., et al. (2020). Environmental impacts and decarbonization strategies in the cement and concrete industries. Nature Reviews Earth & Environment, 1(11), 559–573. https://doi.org/10.1038/s43017-020-0093-3

[3] Bertolini, L. (2013). Corrosion of Steel in Concrete: Prevention, Diagnosis, and Repair (2nd ed.). Wiley.

[4] Page, C. L., & Page, M. M. (2007). Durability of Concrete and Cement Composites. Woodhead Publishing/Elsevier.

[5] Hollaway, L. C. (2010). A review of the current and future utilization of FRP composites in civil infrastructure. Construction and Building Materials, 24(12), 2419–2445. https://doi.org/10.1016/j.conbuildmat.2010.04.062

[6] Nanni, A., De Luca, A., & Zadeh, H. (2014). Mechanics and Design of Reinforced Concrete with FRP Bars. CRC Press.

[7] Hollaway, L., & Teng, J. G. (2008). Strengthening and Repairing Structures Using FRP Composites. CRC Press.

[8] Bank, L. C. (2006/2012). Composites for Construction: Structural Design with FRP Materials. Wiley.

[9] Teng, J.G., Chen, J.F., Smith, S.T., & Lam, L. (2002). FRP-Strengthened RC Structures. Wiley.

[10] Al-Saadi, N. T. K., et al. (2019). Mechanical properties of basalt-FRP bars for reinforced concrete. Structures, 20, 420–432.

[11] Koutas, L. N., Bournas, D. A., & Triantafillou, T. C. (2019). State-of-the-art review on strengthening masonry structures with TRM. Journal of Composites for Construction, 23(1), 03118001. https://doi.org/10.1061/(ASCE)CC.1943-5614.0000882

[12] De Felice, G., et al. (2020). Mechanical behavior of FRCM composites for strengthening masonry. Composites Part B: Engineering, 201, 108396. https://doi.org/10.1016/j.compositesb.2020.108396

[13] Papanicolaou, C. G., Triantafillou, T. C., et al. (2008). Comparison of textile-reinforced mortar (TRM) and FRP for strengthening masonry. Materials and Structures, 41(1), 143–157. https://doi.org/10.1617/s11527-007-9217-6

[14] Koronis, G., Silva, A., & Fontul, M. (2013). Green composites: A review of the mechanical performance of natural-fibre composites. Composites Part B: Engineering, 44(1), 120–127. https://doi.org/10.1016/j.compositesb.2012.07.002

[15] Faruk, O., Bledzki, A. K., Fink, H.-P., & Sain, M. (2012). Biocomposites reinforced with natural fibers: 2000–2010. Progress in Polymer Science, 37(11), 1552–1596. https://doi.org/10.1016/j.progpolymsci.2012.04.003

[16] Nurazzi, N. M., et al. (2021). A review of the mechanical performance of hybrid natural fibre polymer composites. Polymers, 13(13), 2170. https://doi.org/10.3390/polym13132170

[17] Suriani, M. J., et al. (2021). A review of natural fiber-reinforced hybrid composites: processing, properties, applications, and costs. Polymers, 13(20), 3517. https://doi.org/10.3390/polym13203517

[18] Valizadeh, A. et al. (2022). Life-cycle assessment of fiber-reinforced polymers for construction: A review. Sustainability, 14(19), 11887. https://doi.org/10.3390/su141911887

[19] Zeng, J. J., et al. (2024). Recyclable thermoplastic FRP bars for reinforced concrete: Status and opportunities. Composite Structures, 324, [118514].

[20] Das, S. (2011). Life cycle assessment of carbon fiber-reinforced polymer composites. The International Journal of Life Cycle Assessment, 16(3), 268–282. https://doi.org/10.1007/s11367-011-0264-z

[21] Boem, I., D'Ambrisi, A., Feo, L., et al. (2022). A review of TRM-strengthened masonry subjected to in-plane and out-of-plane loads. Buildings, 12(9), 1307. https://doi.org/10.3390/buildings12091307

[22] Girijappa, Y. G. T., et al. (2019). Natural fibers as sustainable reinforcements in polymer composites: A review. Frontiers in Materials, 6, 226. https://doi.org/10.3389/fmats.2019.00226

[23] Tatar, J., & Brenkus, N. R. (2021). Long-term performance of FRP-strengthened RC bridge girders in a coastal environment: A 12-year field study. (Report/article).

[24] Cadenazzi, T., Lee, H., Suraneni, P., Nolan, S., & Nanni, A. (2022). Life-cycle cost and life-cycle assessment of a GFRP-reinforced concrete flood control structure. Journal of Construction Engineering and Management, 148(11), 04022149.

[25] Ragni, D., Canestrari, F., Allou, F., Petit, C., & Millien, A. (2020). The fatigue behavior of geogrid-reinforced asphalt interlayers. Sustainability, 12(11), 4381.

[26] Bournas, D. A., Papalou, A., & Triantafillou, T. (2015). Seismic strengthening using textile-reinforced mortar (TRM): Experimental and analytical insights. Journal of Structural Engineering, 141(3), A4014013.

[27] (Heritage Masonry TRM) Refer to [13] for foundational applications of TRM in heritage structures.

[28] ACI Committee 440. (2015). 440.1R-15: Guide for the Design and Construction of Concrete Reinforced with FRP Bars. American Concrete Institute.

[29] Sbahieh, S., et al. (2023). A comparative life cycle assessment of fiber-reinforced polymers as a sustainable reinforcement option in concrete beams. Frontiers in Built Environment, 9, Article 1194121.

Keyword Index

About the Editors

Bolanle Deborah IKOTUN is an Associate Professor at the University of South Africa (UNISA) in the Department of Civil and Environmental Engineering and Building Science. With over 20 years of experience in academia, her research focuses on sustainable and eco-friendly construction materials. She has studied extensively on concrete material characterisation, concrete mechanical and durability properties testing and concrete quality optimisation techniques by exploring the use of industrial and agricultural wastes. Driven by the vision of incorporating sustainable waste materials into construction materials engineering, she has addressed critical issues like carbon emissions and resource depletion in the construction industry. She has investigated various waste materials, including fly ash, steel slag, ferrochrome slag, rubber crumbs, glass wastes, plastic wastes, wood ash, palm kernel shell, wastewater sludge and paper sludge. She has produced over 90 peer-reviewed publications. She has presented at national and international conferences and collaborated across multidisciplinary fields. She has served as a reviewer for various reputable journals in the field of construction materials.

Her research activities have earned her international and local recognition including the 2010 Developing Researcher Award and the 2016 Resilience in Research Award at UNISA. Others include the Best Paper Award (International Conference of Manufacturing Engineering and Engineering Management,4-6 July 2018, London, U.K) and Sustainable Concrete Ambassador Award (1st International Conference on Advances in Cement and Concrete Research, 15-18

November 2022, LAUTECH, Nigeria). Her research on sustainable materials for concrete production has also been funded and supported by Volkswagenstiftung, managed by Bundesanstalt für Materialforschung und -prüfung (BAM), Germany (2015, 2016), NRF through the Knowledge, Interchange and Collaboration Programme (KIC) (2016, 2017, 2019, 2023) and Women in Research support grant for project (2018-2021). With the aim of transferring skills and promoting sustainability, she has supervised/co-supervised over 40 postgraduate students to completion.

She is an editorial member for the Discovery Civil Engineering journal and has served as a Managing Guest Editor for "Materials Today: proceedings", Volume 86, in 2022. She has served as a Chair of the Research sub-committee at UJ, Civil Engineering Technology, Doornfontein Advisory board (2021-2023). Chair of the Departmental Research and Innovation Committee (DRIC) (2019-to date); Deputy Chair of the School of Engineering Research and Innovation Committee (2021-2023); the Chair of the School of Engineering Research and Graduate Studies Committee (2024-to date) and the head of the Sustainable Green Concrete Research group (2020-to date).

Walied A. ELSAIGH is an Associate Professor in the Department of Civil &Environmental Engineering and Building Science at the University of South Africa (UNISA). He obtained his PhD in Engineering from the University of Pretoria, South Africa. His three-decades career spans academia and industry across several countries, including South Africa, Sudan, Saudi Arabia, and the United States. His research mainly centers on concrete materials, pavement materials, pavement modelling and the beneficiation of waste materials in construction. He published research in reputable journals, mentored master's doctoral students, and served the academic community as a reviewer for journals and conferences.

Akeem Ayinde RAHEEM is a Professor of Building from Ladoke Akintola University of Technology (LAUTECH), Ogbomoso, Nigeria. He holds a Ph.D. in Building Structures from Obafemi Awolowo University Ile-Ife, Nigeria. He is presently the Head of Building Department (2021- till date). As an active researcher and project leader, he has led some funded research and has more than eighty (80) articles in reputable journals locally and internationally, to his credit. He has successfully supervised Seven (7) PhD theses, twenty (20) M. Tech. dissertations and over 500 B.Tech. Students' projects while many are on-going. His current research interest is in incorporation of agriculture residues (such as rice husk ash, wood ash, corn cob ash, etc,) in cement and concrete.

Prof. Raheem has served as a reviewer to notable local and international journals such as USEP, Journal of Research Information in Civil Engineering (RICE), University of Ilorin, Nigeria; Construction Research Journal, Department of Building, University of Lagos, Nigeria; Journal of Construction and Building Materials (Elsevier), United Kingdom; Sustainability – Open Access Journal, Switzerland, etc. He has served as Guest Editor for Materials Today Proceedings, Volume 86, 2022. He was the Pioneer Editor in Chief of LAUJOCES from 2017 to 2025 and produced 15 consecutive volumes of the journal within the period. He is the Team Leader, Cement and Concrete (CEMCON) Research Group; LAUTECH, Ogbomoso, Nigeria.

www.ingramcontent.com/pod-product-compliance
Ingram Content Group UK Ltd.
Pitfield, Milton Keynes, MK11 3LW, UK
UKHW020657041026
15590UKWH00001B/323

9 781644 904046